The Whipple Museum of the History of Science

The diverse objects of The Whipple Museum of the History of Science's internationally renowned collection are brought into sharp relief by a number of highly regarded historians of science in fourteen essays. Each chapter focuses on a specific instrument or group of objects, ranging from an English medieval astrolabe to a modern agricultural 'seed source indicator' to a curious collection of plaster chicken heads. The contributors employ a range of historiographical and methodological approaches to demonstrate the various ways in which the material culture of science can be researched and understood. The essays show how the study of scientific objects – including instruments and models – offers a window into cultures of scientific practice not afforded by textual sources alone. This title is also available as Open Access on Cambridge Core.

JOSHUA NALL is Curator of Modern Sciences in the Whipple Museum. His first book, published in 2019, is *News from Mars: Mass Media and the Forging of a New Astronomy, 1860–1910*.

LIBA TAUB is Curator and Director of the Whipple Museum, Professor of History and Philosophy of Science, and a Fellow of Newnham College. She received the Joseph H. Hazen Education Prize from the History of Science Society and a Pilkington Teaching Prize from the University of Cambridge, recognising her collections-based teaching.

FRANCES WILLMOTH† (1957–2017) was Archivist at Jesus College, Cambridge. She edited the three-volume edition of John Flamsteed's correspondence.

Robert Stewart Whipple (1871–1953)

The
Whipple Museum
of the History of Science

Objects and Investigations,
to Celebrate the 75th Anniversary of
R. S. Whipple's Gift to the
University of Cambridge

Edited by

JOSHUA NALL
University of Cambridge

LIBA TAUB
University of Cambridge

FRANCES WILLMOTH†
Jesus College, Cambridge

CAMBRIDGE
UNIVERSITY PRESS

CAMBRIDGE
UNIVERSITY PRESS

University Printing House, Cambridge CB2 8BS, United Kingdom

One Liberty Plaza, 20th Floor, New York, NY 10006, USA

477 Williamstown Road, Port Melbourne, VIC 3207, Australia

314-321, 3rd Floor, Plot 3, Splendor Forum, Jasola District Centre, New Delhi - 110025, India

79 Anson Road, #06-04/06, Singapore 079906

Cambridge University Press is part of the University of Cambridge.

It furthers the University's mission by disseminating knowledge in the pursuit of
education, learning and research at the highest international levels of excellence.

www.cambridge.org
Information on this title: www.cambridge.org/9781108498272
DOI: 10.1017/9781108633628

First published 2019

A catalogue record for this publication is available from the British Library

Library of Congress Cataloging in Publication data
Names: Nall, Joshua, 1982– | Taub, Liba Chaia, 1954– | Willmoth, Frances, 1957–2017
Title: TheWhipple Museum of the History of Science : objects and investigations, to celebrate the
    75th anniversary of R. S. Whipple's gift to the University of Cambridge / edited by Joshua Nall
    (University of Cambridge), Liba Taub (University of Cambridge), Frances Willmoth
    (Jesus College, Cambridge).
Description: Cambridge ; New York, NY : Cambridge University Press, 2019. | Includes
    bibliographical references and index.
Identifiers: LCCN 2019004820 | ISBN 9781108498272 (hardback : alk. paper)
Subjects: LCSH: Whipple Museum of the History of Science. | Science–Great Britain–History. |
    Science museums–Great Britain. | Whipple, Robert S., 1871-1953.
Classification: LCC Q180.G7 W445 2019 | DDC 502.8/4–dc23
LC record available at https://lccn.loc.gov/2019004820

ISBN 978-1-108-49827-2 Hardback

Contents

Illustrations

Contributors

JIM BENNETT was appointed Curator of the Whipple Museum of the History of Science in 1979. His connection goes back further since, on discovering history and philosophy of science as an undergraduate in 1967, he read books in the Whipple Library, then housed in a small section of the Main Gallery. Here the librarian, Rita van der Straeten, served tea to favoured readers in the distinctive 'Beryl Ware' crockery of the time, while the enigmatic Lady Rosemary Fitzgerald curated the Museum. When challenged to declare why on earth he was leaving in 1994, following a fond and emotional speech, he explained that this was the only way he could think of to stop the annual departmental magic lantern show with original slides, which, like serving tea in library reading rooms, was becoming disreputable.

CHARLOTTE CONNELLY has been the curator of the Polar Museum at the Scott Polar Research Institute since 2015, and works closely with colleagues at the Whipple Museum to deliver collections-based teaching to students across the University of Cambridge. Following her under-graduate degree in History and Philosophy of Science at University College London, Charlotte began her curatorial training by undertaking a BT Connected Earth research internship at the Science Museum in 2009 before completing a postgraduate degree in Museum Studies at the University of Leicester. In 2010 she returned to the Science Museum, where she was part of the curatorial team behind the Information Age gallery. Alongside her curatorial work, she has been based at the University of Cambridge since 2014 when she joined the Department of History and Philosophy of Science as a PhD student.

HASOK CHANG is the Hans Rausing Professor of History and Philoso-phy of Science at the University of Cambridge. He received his degrees from Caltech and Stanford, and has taught at University College London. He is the author of *Is Water H₂O? Evidence, Realism and Plur-alism* (2012) and *Inventing Temperature: Measurement and Scientific Progress* (2004). He is a co-founder of the Society for Philosophy of

Science in Practice (SPSP) and the Committee for Integrated History and Philosophy of Science.

HELEN ANNE CURRY is Peter Lipton Senior Lecturer at the Department of History and Philosophy of Science, University of Cambridge, where her teaching and research encompass the histories of modern science, agriculture, and environment. She is the author of *Evolution Made to Order* (2016) and co-editor of *Worlds of Natural History* (Cambridge, 2018). Given her interest in seed collections of all kinds, Joshua Nall enticed her into researching the history of a seed herbarium acquired by the Whipple Museum. The result is that the museum knows a little more about the herbarium, and she knows a lot (possibly too much) more about the history of forage crops.

CATHERINE EAGLETON is Associate Director of Curatorial Affairs at the National Museum of American History, Smithsonian Institution, where she leads the collections and curatorial team who are responsible for the museum's collection of around 1.8 million objects and three shelf-miles of archives, including important collections in the history of science, technology and medicine. She began her research into the history of scientific instruments when studying History and Philosophy of Science at Cambridge, leading first to an undergraduate dissertation focusing on one of the instruments in the Whipple Museum, then to an MPhil, and finally a PhD completed in 1995 (all supervised by the current Curator and Director of the Whipple Museum). Before taking up her current position, she worked at the Science Museum, British Museum, and British Library, and alongside those curatorial positions she has continued to research and teach on the material culture of science. Her chapter in this volume attempts to answer one of the questions her students have most frequently asked during her lectures on medieval astronomical instruments and practical classes with Whipple Museum instruments.

SEB FALK is a Research Fellow of Girton College, University of Cambridge. He came to the Department of History and Philosophy of Science for an MPhil in 2011, and wrote his first essay on an encyclo-paedic Spanish globe in the Whipple Museum collection. His article in this collection is based on his MPhil dissertation. He took leave from his PhD in the Department in order to complete a six-month internship in the Whipple Museum (funded by the Arts and Humanities Research Council). During this time, he researched the early history of the Museum; his investigation into a replica planetary model nicknamed 'King Arthur's Table' was published in the Royal Society journal *Notes and Records*, 68.2 (2014). When not researching the religious contexts of medieval sciences, he loves teaching members of the public how to use astrolabes.

MATTHEW GREEN completed his MPhil in the Department of History and Philosophy of Science in 2018; his chapter in this volume began as an MPhil essay supervised by Liba Taub. After graduating, he undertook a paid internship at the Whipple Museum as Collections Assistant, working on the Dillon Weston and Michael Clark collections.

BORIS JARDINE began his career as a historian of science at the Whipple Museum. Following employment there he has studied History and Philosophy of Science at the Universities of Leeds and Cambridge. His current research project (funded by the Leverhulme Trust and Isaac Newton Trust) deals with the history of the New Museums Site in Cambridge, and the role of different building types, specimens, and equipment in the period of the 'laboratory revolution' in science teaching and research. He has published on many Whipple objects, including Charles Darwin's achromatic microscope (Wh.3788) and a gauging rule by Henry Sutton (Wh.6239).

MICHAEL F. MCGOVERN is a PhD candidate in History of Science at Princeton University. While pursuing his MPhil in History and Philosophy of Science and Medicine at Cambridge in the 2013–14 academic year, he got side-tracked from his interest in big mainframe computers and started writing about the Whipple's remarkable Hookham collection of handheld calculators. He stayed on as a museum assistant for the summer following the completion of his degree, and produced the 'Calculating Devices' section of the Whipple's Explore website. Before settling down for further graduate studies in Princeton, he worked as a software developer for a large Midwest industrial supply company outside Chicago, Illinois, where he spent much of his time developing a history-of-computing exhibit for the company.

ADAM MOSLEY is Associate Professor in the History Department at Swansea University. Before moving to Wales, he spent many years working behind the scenes at the Whipple Museum as an MPhil and PhD student in the Department of History and Philosophy of Science, and then as a Research Fellow at Trinity College. Thanks in large part to that experience, he teaches the history of collecting and museums as well as incorporating study of the material culture of the sciences in his attempts to understand the early-modern mathematical disciplines.

JOSHUA NALL is Curator of Modern Sciences in the Whipple Museum. He first joined the Museum as a paid summer intern in 2007 following the completion of an MPhil in the Museum's parent department. He was soon beguiled by the Whipple's incredible collections, in particular a spectacular manuscript globe of Mars (Wh.6211). Researching this object led to a PhD on late-nineteenth-century debates over life on the red planet, following which he returned to the Museum as a curator. He has published on a range of Whipple objects, including some fake instruments (Wh.0365, Wh.0776, and Wh.1144), an imposing dividing engine (Wh.3270), the Museum's exceptional collection of teaching and research models, and its extensive collection of scientific trade literature. His work on Mars appears in his first book, *News from Mars: Mass Media and the Forging of a New Astronomy, 1860–1910* (2019).

SIMON SCHAFFER is Professor of History of Science in the Department of History and Philosophy of Science, University of Cambridge. He has long been fascinated by the Babbage engine in the Whipple collection. He co-edited *Material Culture of Enlightenment Arts and Sciences* (2017) and *Aesthetics of Universal Knowledge* (2017). In 2018 he was awarded a medal by the Scientific Instrument Society and the Dan David Prize.

HENRY SCHMIDT was an MPhil student in the Department of History and Philosophy of Science, University of Cambridge, from 2015 to 2016. He is now a doctoral student at the University of California, Berkeley. His interests primarily lie in the social and intellectual history of Earth-system sciences and ecology.

ANNE SECORD is an editor of *The Correspondence of Charles Darwin* and an Affiliated Research Scholar in the Department of History and Philosophy of Science, University of Cambridge. Her interest in working-class natural history in nineteenth-century Britain was greatly enhanced by the Whipple Museum's purchase of a copy of Edward Hobson's *Musci Britannici*. She is completing a book that explores social class, observation, and skill in nineteenth-century natural history. Her edition of Gilbert White's *Natural History of Selborne* was published in 2013.

LIBA TAUB is Curator and Director of the Whipple Museum, having taken up her post in 1995. She is also Professor of History and Philosophy of Science. Since the last Whipple volume, she has edited *On Scientific Instruments*, a special issue of *Studies in History and Philosophy of Science*, 40.4 (2009); and the Focus section on *The History of*

Scientific Instruments, Isis 102 (2011): 689–96. She has also written extensively on ancient Greek and Roman science.

TABITHA THOMAS (NÉE BURDEN) studied Natural Sciences at Newnham College, Cambridge (BA, MSci, 2018). She first entered the Department of History and Philosophy of Science in Part IB and continued there for Parts II and III of her degree. In her final year she engaged closely with the Whipple Museum, using a data-driven approach to study how much of its collection was acquired and formed, first, from the gift of Robert Stuart Whipple (on which the chapter in this volume is based), and secondly, from the Cavendish Laboratory in the 1950s and 1970s (which was the focus of her dissertation). She is now living in Oxford with her husband, Douglas.

FRANCES WILLMOTH† (1957–2017) completed her PhD in 1990 in the Department of History and Philosophy of Science, University of Cambridge, under the supervision of Jim Bennett. Her dissertation was a biography of Sir Jonas Moore, practical mathematician, surveyor, and founder of the Royal Observatory at Greenwich. She went on to work on an edition of John Flamsteed's correspondence (three volumes, 1995–2001). Together with Liba Taub, she edited *The Whipple Museum of the History of Science: Instruments and Interpretations, to Celebrate the 60th Anniversary of R. S. Whipple's Gift to the University of Cambridge*, published by the Museum in 2006. She served as Archivist at Jesus College, Cambridge, where she also directed studies in History and Philosophy of Science.

CAITLIN DONAHUE WYLIE is Assistant Professor of Science, Technology, and Society at the University of Virginia. Caitlin first visited the Whipple Museum as an undergraduate, and was so charmed by the glass fungi that she later applied to the Department of History and Philosophy of Science as a graduate student. As an MPhil student, she spent many happy hours gazing at the collection, including a cheerful monkey-shaped number plate displayed alongside more traditional calculating devices. She wrote an MPhil essay about this charming machine, which became a chapter in this Festschrift. She is grateful to Eleanor Robson, Liba Taub, Joshua Nall, and Ruth Horry, all of whom taught her how to handle and make sense of scientific instruments.

Acknowledgements

The production of this volume, celebrating the seventy-fifth anniversary of the founding of the Whipple Museum of the History of Science, has been, like everything else that takes place at the Whipple, very much a joint effort.

We are grateful to all of the authors for their contributions. They have enjoyed different sorts of relationships to the Museum, and their essays here are intended to provide a picture of the breadth and richness of the collections, as well as the very varied approaches to study that have been based on and inspired by Whipple holdings.

The collation of the volume itself has been helped substantially by Whipple Museum staff and by friends of the Museum, including Mary Brazelton, Toby Bryant, Rosanna Evans, Matthew Green, Tamara Hug, Boris Jardine, Steve Kruse, Simon Schaffer, Alison Smith, and Claire Wallace. Lucy Rhymer and Cassi Roberts at Cambridge University Press were wonderfully patient and helpful. We thank you all.

We are unable to personally thank several close friends of the Museum, whom we nevertheless mention with heartfelt gratitude for their generosity, shared knowledge, support, and friendship: Anita McConnell (1936–2016), Gerard L'Estrange Turner (1926–2012) and Don Unwin (1918–2015).

Finally, we mark the unwavering contribution of our close friend, colleague, and co-editor, Frances Willmoth (1957–2017). Encouraged by the success of the first Whipple Festschrift, marking the sixtieth anniversary, Frances conspired with us to organise and edit the present volume. Displaying her characteristically dry wit, she confessed her pleasure that she would not be in a position to be required to proofread this time. We, however, are very sad that she is not still with us, and not only to see the completed volume. Any typographical errors here are no fault of hers. We are exceedingly grateful for the inspiration and support she gave us in undertaking this project.

Liba Taub and Joshua Nall

Introduction

LIBA TAUB AND JOSHUA NALL

This volume celebrates the seventy-fifth anniversary of the founding of the Whipple Museum of the History of Science, through the gift of the collection of Robert Stewart Whipple to the University of Cambridge. This is the second Festschrift to celebrate a major anniversary of the Whipple Museum, following the first marking its sixtieth.[1] The founding of the Museum pre-dates the establishment of the Department of History and Philosophy of Science (HPS), one of the leading centres for science studies in the world. The Museum is now at the heart of the Department, and has a central role in teaching, training, research, and publication, as well as outreach. Together, the Whipple and HPS are internationally recognised as an exemplary centre for research on the material culture of science. The pre-eminence of the collection and the widely acknowledged leadership of the Whipple provide a unique environment for the study of the substance of science. The essays contained in this volume showcase recent research fuelled by the Museum's rich and varied holdings.

In 1944, Robert Whipple (1871–1953) presented his collection of more than 1,000 scientific instruments and related objects, and a similar number of rare books, to the University. In November of that year, an exhibition was held in the East Room of the Old Schools to mark the official presentation of Whipple's gift. Since the time of its founding, many historic scientific instruments in the possession of the University and the colleges have been generously loaned or transferred to the Museum. In addition, numerous items have been acquired through a special fund established by Whipple's bequest, along with the aid of other benefactors and

---

1 L. Taub and F. Willmoth (eds.), *The Whipple Museum of the History of Science: Instruments and Interpretations, to Celebrate the 60th Anniversary of R. S. Whipple's Gift to the University of Cambridge* (Cambridge: Whipple Museum of the History of Science, 2006).

various funding bodies.[2] Numerous objects have also entered the Museum through the generosity of donors; the reputation of the collection serves as a magnet in attracting those with relevant interests. With holdings dating from the medieval period to the present day, the Whipple's collections of instruments, models, books, and images illuminate the rich cultural significance of scientific pursuits, as well as the practice of science in Cambridge.

In 1995, the Secretary of State for Culture, Media and Sport recognised the international importance of the Whipple's collection by awarding it 'Designated' status. In 2013, a survey conducted of the University of Cambridge Museums discovered that visitors to the Whipple find it the 'most intellectually stimulating' experience of all of the University museums. Our visitors enjoy learning about history of science in its many dimensions, investigated through objects and inside a vibrant academic department. This volume explores some of the ways in which Robert Whipple's vision for his collection have been realised, demonstrating its use in integrated teaching, research, and outreach.

Robert Stewart Whipple and History of Science

Robert Whipple (frontispiece) had a life-long connection with the world of scientific instruments. His father, George Mathews Whipple, was a scientist and instrument specialist, serving for much of his life as Superintendent of Kew Observatory. Whipple himself started his working life as an assistant at Kew, later leaving to become assistant manager for the major London instrument-maker L. P. Casella. He came to Cambridge in 1898 to serve as personal assistant to Horace Darwin (youngest surviving son of Charles Darwin), the co-founder of the Cambridge Scientific Instrument Company. Whipple would have a stellar career at the firm, rising to become Managing Director and eventually its Chairman.

Whipple was involved in numerous learned societies and institutions, being a Founder-Fellow of the Institute of Physics, a Fellow of the Physical Society – where he served as Vice-President and Honorary Treasurer – and President of the British Optical Instrument Manufacturers' Association, amongst others. His interest in the practice of science and its relationship to the development of its

---

2 See, for example, J. A. Bennett, *A Decade of Accessions: Selected Instruments Acquired by the Whipple Museum between 1980 and 1990* (Cambridge: Whipple Museum of the History of Science, 1992).

instrumentation lead him to amass an outstanding collection of antique scientific instruments.

Whipple was not alone in his enthusiasm for the history of science; in the first half of the twentieth century, its importance was being increasingly acknowledged by the academic world. At Cambridge in 1936 there was an exhibition of the historical scientific apparatus owned by various colleges and departments within the University, organised by R. T. Gunther, who in 1937 published his *Early Science in Cambridge*, the result of his survey of surviving scientific instruments there. Soon after this initiative, a History of Science Lectures Committee was established. This committee, and the Cambridge Philosophical Society (which is celebrating its own 200th anniversary in 2019), were involved in negotiations concerning Whipple's wish to donate his collection to form the basis of a museum within the University.[3] The desire for the development of history of science as a subject of study and research was emphasised throughout, as is exemplified by the memorandum submitted to the University concerning the founding of the Whipple Museum:

> Since Cambridge is pre-eminent in her tradition of associating teaching with the active prosecution of research ... it is important that the museum should be much more than a well-arranged repository of historic scientific apparatus. It should be designed and maintained as a valuable teaching instrument and a cultural accessory to modern research.[4]

In this way, Whipple's gift of his collection of antique instruments and rare books in 1944 was part of a larger effort to establish history of science as a subject within the University.

3 J. A. Bennett, 'Museums and the Establishment of the History of Science at Oxford and Cambridge', *British Journal for the History of Science*, 30.1 (1997), pp. 29–46; and A.-K. Mayer, 'Setting Up a Discipline: Conflicting Agendas of the Cambridge History of Science Committee 1936–50', *Studies in History and Philosophy of Science Part A*, 31.4 (2000), pp. 665–89.
4 'Memorandum on the Future of the History of Science as a Subject of Study and Research in the University with Proposals for the Creation of a University Museum and a University Department of the History of Science', 9 February 1944. A full transcript of this memorandum is given (on pp. 12–17) in F. Willmoth, 'Documents from the Founding and early history of the Whipple Museum', Part I of L. Taub and F. Willmoth (eds.), *The Whipple Museum of the History of Science: Instruments and Interpretations, to Celebrate the 60th Anniversary of R. S. Whipple's Gift to the University of Cambridge* (Cambridge: Whipple Museum of the History of Science, 2006), pp. 11–55.

The Whipple Collection, Museum, and Library

Because of lack of space within the post-war University, the Whipple collection was initially stored in various buildings, including the basement of the Fitzwilliam Museum, Girton College, and two rooms in Corn Exchange Street. Once the Whipple Museum had been established, in addition to housing Whipple's own collection it soon became the home of many of the scientific artefacts used and preserved in University departments and laboratories, as well as in Cambridge colleges.

In 1959, the growing collection moved to its permanent home, the old Perse School Hall, now the Main Gallery of the Whipple Museum. In 1973–5, extensive work restored the Perse Hall to its original form and created a separate library for the Department of History and Philosophy of Science. (In 2008, the Whipple Library moved into a new home, the renovated Heycock Lecture Theatre.) At the centre of this Library are the rare books donated by Robert Whipple himself. In keeping with his intentions, the Whipple Museum and the Whipple Library together play an active role in teaching and research. The rare book collection includes famous works such as Isaac Newton's *Principia Mathematica* and Christiaan Huygens's *Horologium oscillatorium*, detailing the invention of the pendulum clock. Most importantly, the collection includes many rare publications on scientific instruments, ranging from texts on medieval instruments for astronomical observations to trade literature for early-twentieth-century industrial technology.[5]

The Whipple Research Model

One of the hallmarks of research undertaken at the Whipple is the active study of instruments alongside related textual and visual material, including those books describing the design and use of instruments. The physical proximity of the Whipple Museum and the Whipple Library within HPS makes such study possible, a point emphasised in several of the chapters in this volume.

Speaking in 2012 at the launch of the Science Museum Group's Research and Public History Department, Ludmilla Jordanova offered insights as a Trustee of the Science Museum that resonate

---

5 Silvia De Renzi, *Instruments in Print: Books from the Whipple Collection* (Cambridge: Whipple Museum of the History of Science, 2000).

with many of us trained in history of science, especially those concerned with collections of historical scientific material:

> I was mainly taught about abstract ideas; now it is taken as read that embodied knowledge, material and visual culture, and the close analysis of social practices are central to our field. But we must confess that the *full* potential of integrating museum collections and the expertise of museum professionals into academic understanding is yet to be realised.[6]

At a time when academic historians of science are increasingly acknowledging the importance of material and visual culture, few have had opportunities to work on such material first-hand. The result, as a number of scholars have noted, has been a 'material turn' that is often not very materialist.[7] At the same time, in many museums object-based research is something of a luxury; curators are often too busy with other important tasks to study the objects in their care. Collections-based research is often done to deadlines, such as the opening of an exhibition, and is often not made public beyond the lifetime of a show.

The Whipple Museum is very proud, therefore, that it has over many years been the base of a remarkably rich and varied research output, communicated to a wide range of audiences via a number of formats, including exhibitions, talks, and podcasts, as well as print and digital media. Much of this research has been conducted by undergraduate, MPhil, and PhD students, supervised by Whipple curators and HPS academic staff. In particular, the MPhil essay – 5,000 words, produced over six to eight weeks – has provided a wonderful template for enabling original object-based research to be undertaken in a form that can be shared readily with others. Usually, students have no prior experience of working with material culture, yet produce excellent work of broad benefit. A key feature of the Whipple method is to encourage researchers to take a 'deep dive' into our holdings and engage directly with the object(s). Research

6 Speech by Ludmilla Jordanova on the occasion of the launch of the Research and Public History Department, Science Museum Group, London, 18 September 2012; quoted with permission.
7 See, for example, J. J. Corn, 'Object Lessons/Object Myths? What Historians of Technology Learn from Things', in D. W. Kingery (ed.), *Learning from Things: Method and Theory of Material Culture Studies* (Washington: Smithsonian Institution, 1996), pp. 35–54; Bruno Latour, 'Can We Get Our Materialism Back, Please?', *Isis*, 98.1 (2007), pp. 138–42; and K. Anderson, M. Frappier, E. Neswald, and H. Trim, 'Reading Instruments: Objects, Texts and Museums', *Science and Education*, 22.5 (2013), pp. 1167–89.

students meet with Whipple curators to view, choose, study, and discuss objects together. While this involves a sort of match-making process – requiring time to understand students' other interests and backgrounds – the results are often wonderfully unpredictable. The first few objects considered as candidates for research may not ultimately be chosen, and what some may regard as rather mundane and ordinary objects become fascinating subjects of research in the hands of the right researcher. Other members of the Department and museum staff are often also involved, and museum staff play a key role in providing all researchers with access to the collection and relevant documentation. There is a commitment of staff at all levels to work closely with researchers, and everyone gains and learns through the process. Importantly, in return knowledge of the Museum's holdings is greatly enhanced.

The physical location of the Museum at the centre of the HPS Department and our work to make as much of the collection as possible visually accessible (and not only virtually) encourages engagement with otherwise 'unknown' objects. Researchers and students also benefit from having access to the past work done on the collection, providing exemplars of what it is possible to do, even in a relatively short span of time. In some cases, past work serves as the springboard for a new study. The richness of our holdings allows a variety of resources to be available to researchers, including other related objects, ephemera, photographs, and written material such as instruction manuals, makers' trade catalogues, and published papers.

We are pleased, as a University of Cambridge museum, to make this research accessible in many ways, including through student-produced displays and through the placing of student work in our galleries next to the objects that have been investigated. A wealth of student research is also accessible on the Museum's Explore website.[8] All these presentations are 'signed' by their creators, highlighting that the information provided is an interpretation, and not simply information. An appendix to this volume gives a comprehensive list of undergraduate, MPhil, and doctoral work undertaken on the collection over the past two decades. Since the appearance of the Museum's first Festschrift, we have also been gratified to see a wealth of scholarship based on the Whipple collections published in

8 www.whipplemuseum.cam.ac.uk/explore-whipple-collections.

a wide range of journals and books.[9] Furthermore, we are very proud that an impressive number of those who have studied and worked in the Whipple have gone on to professional careers in museums and libraries around the world, working with material culture.

Objects and Investigations

The following chapters – which are ordered broadly chronologically in terms of the objects and books they study – focus on diverse objects in the Whipple Museum's collection, ranging from an English medieval astrolabe to a modern agricultural 'seed source indicator' to a curious collection of plaster chicken heads. The chapters' authors employ a range of historiographical and methodological approaches in their studies, enabling this volume to display not only the extraordinary range of the Whipple's collection, but also the

9 Though no doubt not a comprehensive list, such works include M. Keene, '"Every Boy & Girl a Scientist": Instruments for Children in Interwar Britain', *Isis*, 98.2 (2007), pp. 266–89; L. Taub (ed.), 'On Scientific Instruments', special issue of *Studies in History and Philosophy of Science Part A*, 40.4 (2009) (the articles by K. Taylor, S. Al-Gailani, B. Jardine, R. W. Scheffler, and K. de Soysa in this special issue all study Whipple Museum objects); M. J. Barany, 'Great Pyramid Metrology and the Material Politics of Basalt', *Spontaneous Generations*, 4.1 (2010), pp. 45–60; C. Eagleton, *Monks, Manuscripts and Sundials: The Navicula in Medieval England* (Leiden: Brill, 2010); L. Taub (ed.), 'Focus: The History of Scientific Instruments', special section of *Isis*, 120.4 (2011), pp. 689–729; S. Falk, 'A Spanish Globe: Origins and Interpretations', *Globe Studies*, 59/60 (2014), pp. 142–59; S. Falk, 'The Scholar As Craftsman: Derek de Solla Price and the Reconstruction of a Medieval Instrument', *Notes and Records of the Royal Society*, 68.2 (2014), pp. 111–34; J. Davis and M. Lowne, 'An Early English Astrolabe at Gonville & Caius College, Cambridge, and Walter of Elveden's *Kalendarium*', *Journal for the History of Astronomy*, 46 (2015), pp. 257–90; D. E. Dunning, 'What Are Models for? Alexander Crum Brown's Knitted Mathematical Surfaces', *Mathematical Intelligencer*, 37.2 (2015), pp. 62–70; J. Poskett, 'Sounding in Silence: Men, Machines and the Changing Environment of Naval Discipline, 1796–1815', *British Journal for the History of Science*, 48.2 (2015), pp. 213–32; B. Jardine, 'Henry Sutton's Collaboration with John Reynolds (Gauger, Assayer and Clerk at the Royal Mint)', *Bulletin of the Scientific Instrument Society*, 130 (2016), pp. 4–6; J. Nall and L. Taub, 'Three-Dimensional Models', in Bernard Lightman (ed.), *A Companion to the History of Science* (Chichester: Wiley Blackwell, 2016), pp. 572–86; J. Nall and L. Taub, 'Selling by the Book: British Scientific Trade Literature after 1800', in A. D. Morrison-Low, S. J. Schechner, and P. Brenni (eds.), *How Scientific Instruments Have Changed Hands* (Leiden: Brill, 2016), pp. 21–42; B. Jardine, J. Nall, and J. Hyslop, 'More Than Mensing? Revisiting the Question of Fake Scientific Instruments', *Bulletin of the Scientific Instrument Society*, 132 (2017), pp. 22–9; B. Jardine, 'State of the Field: Paper Tools', *Studies in History and Philosophy of Science*, 64 (2017), pp. 53–63; and J. Nall, '"Certainly Made by Ramsden": The Long History of the Whipple Museum's Dividing Engine', *Bulletin of the Scientific Instrument Society*, 137 (2018), pp. 40–3.

various ways in which the material culture of science can be researched and understood. Just like manuscript and published works, scientific objects can be studied closely as individual entities, scrutinised and 'read' to reveal crucial traces of past scientific work. Yet, as the chapters by Seb Falk, Anne Secord, and Jim Bennett demonstrate, such tight focus on individual things and their makers (in their cases an English medieval astrolabe, a single bound set of dried moss specimens, and Henry Sutton, respectively) is most effective when conducted in comparison with complementary sources, including the wealth of books that describe instruments and explain their uses. Indeed, many of the studies in this volume analyse a broad collection of sources, considering *en masse* a type of instrument and its associated print culture. Even though the objects studied by Catherine Eagleton (medieval portable astronomical instruments), Adam Mosley (early-modern mathematical and cosmographical instruments), Charlotte Connelly and Hasok Chang (Victorian and Edwardian galvanometers), and Michael McGovern (1970s programmable pocket calculators) are very different, the authors demonstrate that starting with a few objects and working outwards to consider a broader group or class offers a window onto cultures of scientific practice that is not afforded by textual sources alone.

Whether considering objects individually or as a group, what unites these investigations is not only the ability of material culture to reveal new information about past science, but also its ability to act as a signpost to wider stories. The chapters in this volume remind us that museum objects save in material form traces of the past that are often missing from conventional textual records. Scientific practitioners are, after all, unreliable chroniclers of their own work, and the material culture they leave behind very often preserves aspects of their practices and broader social milieu that were never recorded on paper. Though such objects may not be straightforwardly legible – a simple key to be read and understood – they almost invariably offer up hints and clues that point the historian in new and heretofore unexpected directions, often extending well beyond the thing itself and towards that thing's place in wider social and cultural contexts. Such objects can be as previously obscure as Helen Curry's 'seed source indicator', Caitlin Wylie's 'educated monkey' calculating toy, Matthew Green's plaster chicken heads, or Henry Schmidt's cloud camera. Or they can be as monumental as Simon Schaffer's fragment of Babbage's famous difference engine. All, in this sense, are equally worthy of preservation and study, in that their very survival points the scholar towards events in past science that might have otherwise remained overlooked. In every case, scientific instruments prove both malleable enough to have many lives, yet robust enough to preserve those lives' dependence on materiality, design, and labour.

The practice that underpinned the saving of so many objects in the Whipple Museum is, of course, collecting. Whether motivated by curiosity, scholarship, the urge to preserve, or simply the thrill of the chase, collectors like Robert Whipple gathered objects that then formed the basis for many of the world's major history of science museums. Such practices of collecting are, therefore, themselves worthy of study.[10] As the chapters by Tabitha Thomas and Boris Jardine demonstrate, what did and did not make it into collections, and how the emerging marketplace for collectable scientific antiques shaped the historical record we now have, are important questions for scholars of scientific material culture. As Thomas and Jardine both make clear, instruments change hands as commodities, and both Whipple and his contemporary Lewis Evans (whose collection formed the basis for the Museum of the History of Science at the University of Oxford) were major players in a growing marketplace for such collectibles. The choices they and others like them made had a significant impact on the scholarship exemplified by the works in this volume.

We are very grateful to all of the contributors to this volume. As in the first Festschrift, it is intended to demonstrate both the richness of our holdings and also the very special intellectual opportunities afforded by having an actively working museum open to the public at the centre of a university department focused on history and philosophy of science. Throughout its existence, the Whipple Museum has striven to develop its capabilities in ways that fulfil the intentions and ambition of its farsighted founder, and of the University of Cambridge when it had the foresight to accept Whipple's generous gift. We hope that the work presented here will serve as exemplars and stimulation for future generations of students and scholars, inspired by the Whipple collection and by the active synergies at work within the Museum, the Whipple Library and the Department of History and Philosophy of Science.

10 See, for example, P. de Clerq and A. J. Turner (eds.), 'Origins and Evolution of Collecting Scientific Instruments', special issue of *Journal of the History of Collections*, 7.2 (1995); and S. J. M. M. Alberti and C. Berkowitz (eds.), 'Shaping Scientific Instrument Collections', special issue of *Journal of the History of Collections*, 31.3 (2019).

1 ❧ Sacred Astronomy? Beyond the Stars on a Whipple Astrolabe*

SEB FALK

It has occasionally been my privilege to act as a stand-in gallery attendant in the Whipple Museum. This has afforded precious opportunities to observe visitors, who seem not to feel my scrutiny as they explore the atmospheric main gallery. Almost invariably they wander clockwise. They may pause first at the horses' teeth or glass fungi. But they are guaranteed to stop, and to stare, at the astrolabes case.

Astrolabes seem to hold a fascination for museum visitors, even – perhaps especially – if they have no understanding of their workings. A mathematical instrument that is as beautiful as it is precise, a medieval astrolabe can be appreciated on multiple levels, scientific or artistic. This is not as anachronistic as it might appear: when they were made, too, astrolabes – at least the ones that survive in museum collections – were ornate status symbols as well as functional tools. Even so, it is often hard to imagine the contexts in which these devices were first designed and used. Behind glass, their three-dimensionality and mutability obscured by the fixed presentation of one face to the observer, they may epitomise the 'decontextualised commodities' deplored by Ludmilla Jordanova.[1] Even for those of us who study them, they seem to recede into mystery even as new methods of analysis allow us to get closer to them than ever before: as the newly delineated complexities of their long lives blur simple

* For her support and guidance of my research into scientific instruments, I am grateful to Liba Taub. I would also like to thank Steve Kruse, Josh Nall, and Claire Wallace at the Whipple Museum, Oliver Cooke (British Museum) and Mark Statham (Gonville & Caius College) for facilitating access to astrolabes, and Nigel Morgan and Katie Eagleton for their advice. I have drawn extensively on the (published and unpublished) work of John Davis, and I am immensely grateful for his generous assistance.

1 L. Jordanova, 'Objects of Knowledge: A Historical Perspective on Museums', in Peter Vergo (ed.), *The New Museology* (London: Reaktion, 1989), pp. 22–40, on p. 25.

ascription, or as once-prized historic objects turn out to be modern fakes.[2] It is thus perhaps not surprising that, at least until recently, approaches to astrolabes have been narrowly antiquarian.[3] Understanding the conditions and motivations of their use was seen as less important than seeking ever greater precision about the time and place of their production. Needless to say, in order to use an object to illuminate its context we first need to know where and when that context was. Yet, even when we lack certainty about their provenance, there remain ways that astrolabes can be understood and can help us to better understand the Middle Ages more generally.

This chapter focuses on one astrolabe in the Whipple Museum's collection, Wh.1264 (Figure 1.1), as a way of highlighting these issues. It is an object that has not been extensively studied: it is not clear when or how it came to be in the Whipple collection, and it was not included in the foundational catalogues of astrolabes.[4] Some studies have considered it, but mainly as a way of elucidating other instruments.[5] However, it has recently played a supporting role in a detailed treatment of another instrument in Cambridge, and it has been included in an extensive programme of metallurgical analysis carried out by John Davis.[6] Such new methods as X-ray fluorescence

2 B. Jardine, J. Nall, and J. Hyslop, 'More Than Mensing? Revisiting the Question of Fake Scientific Instruments', *Bulletin of the Scientific Instrument Society*, 132 (2017), pp. 22–9.

3 These were epitomised by R. T. Gunther in his *Astrolabes of the World* (Oxford: Oxford University Press, 1932); *Early Science in Oxford* (Oxford: Oxford University Press, 1923); and *Early Science in Cambridge* (Oxford: Oxford University Press, 1937). For the influence of such approaches on the early development of the Whipple Museum, see S. Falk, 'The Scholar as Craftsman: Derek de Solla Price and the Reconstruction of a Medieval Instrument', *Notes and Records of the Royal Society*, 68 (2014), pp. 111–34.

4 Gunther, *Astrolabes of the World*; D. J. Price, 'An International Checklist of Astrolabes', *Archives internationales d'histoire des sciences*, 32 (1955), pp. 243–63; and S. L. Gibbs, J. A. Henderson, and D. J. de Solla Price, *Computerized Checklist of Astrolabes* (New Haven: Yale University Press, 1973). It is included, with the briefest description, in David Bryden's catalogue of sundials at the Whipple Museum: D. J. Bryden, *The Whipple Museum of the History of Science, Catalogue 6: Sundials and Related Instruments* (Cambridge: Whipple Museum of the History of Science, 1988), no. 342.

5 O. Gingerich, 'Zoomorphic Astrolabes and the Introduction of Arabic Star Names into Europe', *Annals of the New York Academy of Sciences*, 500 (1987), pp. 89–104; and C. Eagleton, '"Chaucer's Own Astrolabe": Text, Image and Object', *Studies in History and Philosophy of Science Part A*, 38 (2007), pp. 303–26.

6 J. Davis and M. Lowne, 'An Early English Astrolabe at Gonville & Caius College, Cambridge, and Walter of Elveden's *Kalendarium*', *Journal for the History of Astronomy*, 46 (2015), pp. 257–90. I am grateful to J. Davis for sharing the results of his endeavours with me.

Figure 1.1 Wh.1264, an English astrolabe, *c.* 1350. Image © Whipple Museum.

(XRF) analysis, diffraction analysis, and scanning radiography have the potential to revolutionise our understanding of instruments. Hard data about their chemical composition or metallic microstructure can, in combination with more traditional comparative techniques, support theories about their age, geographical origins, and methods of production, as well as testing old broad-brush dating tools such as precession data.[7]

Yet pinpointing the age and geographical origins of an astrolabe is problematic, for two contrasting reasons. First, these were never static objects. They moved freely across the national boundaries

---

7 In principle, the astrolabe rete and calendars should reflect the state of the skies at the time the astrolabe was made, and the position of the first point of Aries has often been used as an indication of this, but this approach is unreliable. See Gingerich, 'Zoomorphic Astrolabes and the Introduction of Arabic Star Names into Europe', p. 89; and G. L'Estrange Turner, 'A Critique of the Use of the First Point of Aries in Dating Astrolabes', in G. L'Estrange Turner, *Renaissance Astrolabes and Their Makers* (Aldershot: Ashgate, 2003), Part III, pp. 548–54.

marked on modern maps – and as they moved, they changed. Parts of these instruments – always intended to be dismantled and reconfigured – were lost; new parts were added; new engravings were made, altering the purposes or appearance of the instruments. Some may almost be regarded as compilations, or as having been composed and later re-edited. When we talk of astrolabes having replacement parts we may picture insensitive Victorian curators, and indeed astrolabes in British museums contain their fair share of nineteenth-century brass. Yet we must reflect that parts were most likely to be lost or broken when the instruments were in most active use. XRF analysis would seem to support this, as we find different parts of instruments containing quite different – but still medieval – alloys. Secondly, a precise guess of a date and place of origin, or even ascribing an instrument to a named individual, may overlook the continuity of artistic and particularly scientific trends across time and context. Contemporary scholars were remarkably uninterested in the geographical or even religious origins of scientific instruments or ideas.[8]

Nevertheless, even within such broader trends we find local specificities. One example of this is the religious motivation for scientific inquiry. Links between Christianity and astronomy were long underestimated, and although no serious historian now subscribes to the idea of a 'warfare of science with theology', historians may still disagree about how far Christian faith inspired an understanding of nature, or was simply set aside by natural philosophers.[9] Astrolabes have a part to play in exploring such questions. Just as an image of an instrument might symbolise learning in an illuminated bible (Figure 1.2), so the inclusion of religious information on an astrolabe could allow its patron or maker to express his devotional preferences.[10] This need not have been in an explicitly religious setting like a monastery; it seems to have occurred as much on instruments

---

8   O. Pederson has shown how unconcerned commentators were with the nationality of Johannes de Sacrobosco. See O. Pederson, 'In Quest of Sacrobosco', *Journal for the History of Astronomy*, 16 (1985), pp. 175–220.

9   See A. D. White, *A History of the Warfare of Science with Theology in Christendom* (New York: Appleton, 1896). See also the debate between E. Grant and A. Cunningham in the pages of *Early Science and Medicine*, 5 (2000), pp. 258–300.

10   On devotional motivations for practising astronomy, see S. Falk, 'Improving Instruments: Equatoria, Astrolabes, and the Practices of Monastic Astronomy in Late Medieval England', unpublished PhD thesis, University of Cambridge (2016), pp. 13–41.

Figure 1.2 Solomon observing the stars, from a Franciscan Bible. The message here is ambiguous: the historiated initial adorns the opening to the Book of Ecclesiastes, in which the wise Solomon admonishes that 'in much wisdom is much grief: and he that increaseth knowledge increaseth sorrow' (1:18). Reproduced courtesy of the Bibliothèque Nationale de France (MS Latin 16745 (*c.* 1170–80), fol. 108).

made for lay patrons, and in any case the links between the larger monasteries and the universities and royal court were strong across the late medieval period.

The Whipple's English Astrolabe

Wh.1264 is an ideal object to show how such devotional preferences might be expressed. It has usually been dated to the late fourteenth century, and is among the larger Western astrolabes known from this period: its mater is 295 mm in diameter, and 40 mm thick; the entire instrument including its suspension ring and throne measures 348 mm in length. The mater was constructed by riveting a cast rim (with a depth of 5 mm) onto the backplate, with twenty-three regularly spaced pins that have been driven through the front. The throne is set into the rim and fixed in place with two rivets, though this joint has become a little loose. The throne is very small and plain: a round boss that is almost completely covered by the shackle; the bail is in the T–H form common to astrolabes of this period. The astrolabe is held together with a plain pin and horse, including three modern washers (one metal, two plastic) – it is not known when these were added. It has a double graduated rule atop the rete, and an alidade with pinhole sights at the back.

It was manufactured from a fairly typical medieval latten, an alloy of copper and zinc with smaller quantities of tin and lead. XRF analysis of the instrument by Davis shows that it contains an unusually low level of zinc (7.7 per cent) compared with other astrolabes of the period which are more likely to have 10–15 per cent.[11] The rete has slightly higher levels of zinc, showing how variable the smelting process could be, and the alidade, rule, and pin have significantly higher zinc levels (*c.* 20 per cent) which suggest these may be later replacement parts. The horse is certainly made of a modern brass.

There are no separate tympans for specific latitudes; the only stereographic projection is engraved within the womb of the mater. It is not labelled for a specific place or latitude, but the distance between the zenith and the celestial pole indicates that it was produced for use at latitude 52°. This corresponds to locations in central England where astronomy was extensively practised, such as the university of Oxford and monastery of St Albans; however, Davis and Lowne, connecting it with an astrolabe at Gonville & Caius College, have suggested that it may have been made for use at Norwich.[12] The almucantars, which mark celestial altitude, are drawn and labelled every two degrees: as closely spaced as, and more frequently labelled than, on any catalogued astrolabe. This would have made it exceptionally user-friendly when it came to finding the locations of stars. Yet this 600-year-old instrument was surely used in different ways at different times. Engraved and labelled among the almucantars with a finer tool and later script are the Great Houses, useful for astrology; much more crudely, hammered points just inside the rim were used to add the first few letters of the name of each month, as well as four dots in the shape of a diamond, twice between each month name and the next (Figure 1.3).

The absence of interchangeable tympans (plates) for different latitudes makes Wh.1264's origins and purpose harder to identify. The presence of modern washers to prevent the rete, rule, and alidade from rotating too loosely suggests that the astrolabe previously had tympans which have been lost. However, tympans must be held in place within the womb of the mater; this was usually accomplished by making the tympans with tangs that fit into a slot in the rim, though some later astrolabes instead had lugs in the rim and notches in the tympans. This astrolabe has neither system, and a stereographic projection is, somewhat unusually, engraved in the

---

11 Davis and Lowne, 'An Early English Astrolabe at Gonville & Caius College, Cambridge, and Walter of Elveden's *Kalendarium*', p. 280; and J. Davis, private correspondence, 6 April 2018.
12 Davis and Lowne, 'An Early English Astrolabe at Gonville & Caius College, Cambridge, and Walter of Elveden's *Kalendarium*', p. 257.

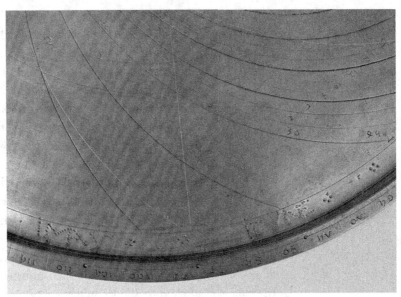

Figure 1.3 Detail from the womb of Wh.1264, showing the equator, almucantars, and unequal hours, and a finer Great House line (with corrected '6'). Note also the hammered-in month names and diamonds. Image © Whipple Museum.

womb, so it may be questioned whether it ever had separate tympans. In addition, the single stereographic projection has, unusually, neither a named location nor a latitude. These details would have been omissible if there was no need to distinguish between different projections; if, perhaps, its user had no plans to travel with it. On the other hand, if an astrolabe was intended for use at a single latitude, the mater could be reduced to a single plate, as we find on Wh.4552, a near neighbour in the Whipple's current display. The fact that Wh.1264 has a recessed womb surrounded by a rim suggests that it was at least intended to be equipped with tympans. In any case, astrolabes without tympans are rare, whereas it is relatively common for tympans to have been lost from astrolabes now on display in museums. Lacking any other evidence, we must assume that this is the case with this instrument. How the tympans would have been secured in place is not clear, though since the throne is a little loose it is possible that it was originally fitted differently, and that the refitted throne has filled a slot that was previously located just beneath, as is customary. Alternatively, perhaps the astrolabe is incomplete: its maker may have failed to fit the womb with lugs, just as he failed to mark the latitude; or, conceivably, he chose to add a rim for aesthetic reasons.

Tympans are not the only notable absence from this astrolabe. It is also missing any engraving within the top inner semicircle on the back (apart from a roughly scratched 'Hd'). In Western astrolabes from this period it is fairly common to see an unequal-hour scale there. John North has called the inclusion of these lines an 'empty ritual', noting how rarely the scales are accurately engraved or supplied with a counterpart giving solar positions; it might be added that such scales are usually unnecessary, since they are commonly also on the front of the astrolabe.[13] Their appearance on the back may indeed be ritualistic, reminding users of the astrolabe's time-keeping function and perhaps privileging that over its parallel astro-nomical uses. In this context, it is also notable that the rim of Wh.1264 is labelled with 360 degrees, rather than the twenty-four hours that were a common feature of Western astrolabes in this period.[14] One may, then, suggest that its maker was relatively uninterested in timekeeping functions. Needless to say, it can still be used to tell the time with some precision, during the day or night, at any season of the year. It has unequal-hour lines on the front, and the rule is graduated to allow conversion between equal and unequal hours, according to the midday solar altitude, at the latitude for which the astrolabe was made. The lack of an equal-hour scale on the rim certainly makes Wh.1264 less user-friendly for timekeeping, but even if the maker of this astrolabe was more interested in astronomical uses, or wanted to use the 360-degree scale on the rim to represent a conceptualisation of the cosmos as a geometrical entity, such intentions might not be reflected in the way it was used. Certainly, the 360-degree scale by no means precludes its use as a time-telling device.

Stars and Almucantars

It is possible to characterise the back of the astrolabe, with its calendar of feast days and surveyor's shadow square, as representing terrestrial things; the front, in contrast, carries the net of stars and so looks more directly towards the heavens. The rete has been con-sidered by a few scholars who have sought to develop typologies of

---

13 J. North, 'Astrolabes and the Hour-Line Ritual', in J. North, *Stars, Minds and Fate: Essays in Ancient and Medieval Cosmology* (London: Hambledon, 1989), pp. 221–2, on p. 221. First published in *Journal for the History of Arabic Science*, 5 (1981), pp. 113–14.
14 The astrolabe illustrated in Chaucer's *Treatise* has the latter arrangement. See G. Chaucer, *A Treatise on the Astrolabe* (c. 1391), ed. S. Eisner (Norman: University of Oklahoma Press, 2002), pp. 142–3.

astrolabes according to their shapes, symbolism, and the stars they contain.[15] Wh.1264 fits into a group of astrolabes with quatrefoil and demi-quatrefoil motifs on their retes, which have been distinguished from other instruments whose retes are dominated by a Y-shape within the ecliptic circle. The latter group are sometimes characterised as 'Chaucerian' because the same Y-shape appears in illustrations within some early copies of the *Treatise on the Astrolabe*, but it is not clear whether the illustrations imitate the astrolabes, vice versa, or both in different cases.[16] Those astrolabes adorned with architectural decoration such as quatrefoils have been persuasively linked with similar examples of church architecture as a way of localising their production (or adaptation); such comparisons by themselves may be unconvincing, but can add important support to origins hypotheses based on other parts of the instruments.

The stars marked on astrolabe retes do not necessarily correlate closely with the decoration of their supporting framework. They have been analysed in terms of the selection of stars included, the positions given, and the names used. Gingerich has called the fourteenth century 'a key period in the transmission of Arabic star names into common English usage', and we certainly find these Arabic star names on Wh.1264.[17] (Many of these Arabic names, such as Altair and Vega, are still in common use today.) The lists of stars chosen were first systematically analysed as a series of 'types' by Paul Kunitzsch, and his Type VIII corresponds most closely to the Whipple rete.[18] This list, Kunitzsch demonstrates, combines one that appeared in Spain in the late tenth century and another compiled by John of London in 1246, in Paris. It contains forty-nine stars, forty-one of which appear on the rete of Wh.1264 (see Table 1.1).[19]

15 Gingerich, 'Zoomorphic Astrolabes and the Introduction of Arabic Star Names into Europe'; D. A. King, 'An Ordered List of European Astrolabes to *ca.* 1500', in D. A. King, *Astrolabes from Medieval Europe* (Farnham: Ashgate, 2011), p. xii; and J. Davis, 'Fit for a King: Decoding the Great Sloane Astrolabe and Other English Astrolabes with "Quatrefoil" Retes', *Medieval Encounters*, 23 (2017), pp. 311–54.

16 Eagleton, 'Chaucer's Own Astrolabe'; J. Bennett and G. Strano, 'The So-Called "Chaucer Astrolabe" from the Koelliker Collection, Milan', *Nuncius*, 29 (2014), 179–229.

17 Gingerich, 'Zoomorphic Astrolabes and the Introduction of Arabic Star Names into Europe', 96.

18 P. Kunitzsch, *Typen von Sternverzeichnissen in astronomischen Handschriften des zehnten bis vierzehnten Jahrhunderts* (Wiesbaden: Otto Harrassowitz, 1966).

19 An almost identical list of stars 'to be placed on the astrolabe' survives in an early-fourteenth-century collection of astronomical and astrological texts from the monastery of Bury St Edmunds: Cambridge University Library MS Add.6860, ff. 70v–71r.

TABLE 1.1 List of stars marked on rete of Wh.1264

Star Name on Wh.1264	Modern Name	Kunitzsch Type VIII Number
Mirak	β Andromedae	1
Batuchaythos	ζ Ceti	2
Cenok	α Arietis	4
Menkar	α Ceti	6
Algeneb	α Persei	7
Augetenar	τ Eridani	8
Aldeboram	α Tauri	9
Alhaok	α Aurigae	10
Rigil	β Orionis	11
Elgeuze	α Orionis	12
Alhabor	α Canis Majoris	13
[unlabelled pointer]	α Geminorum	14
Algomeiza	α Canis Minoris	15
Markeb	κ Velorum	16
[unlabelled pointer]	μ Ursae Majoris	17
Alfard	α Hydrae	19
Cor	α Leonis	20
[unnamed bird]	Corvus	22
Edub	α Ursae Majoris	23
Cauda	β Leonis	24
Algorab	γ Corvi	25
Alehimek	α Virginis	26
Benenaz	η Ursae Majoris	27
[unlabelled pointer]	? μ ≈ Lib 20, δ ≈ −18	–
Alramek	α Bootis	28
Elfeca [broken off][a]	α Coronae Borealis	29
Yed	δ Ophiuchi	31
Alacrab	α Scorpii	32
Alhawe	α Ophiuchi	33
Thaben	γ Draconis	34
Wega	α Lyrae	35
Althayr	α Aquilae	36
Delfin	ε Delphini	37
Aldigege	α Cygni	39
Aldera	α Cephei	42
Musida Equi	ε Pegasi	43
Denebalgedi	δ Capricorni	44
Cenok	δ Aquarii	45
Humerus Equi	β Pegasi	46
Alferas	α Andromedae	47
Denebchaytos	β Ceti	48
Skeder	α Cassiopeiae	49

[a] The pointer is broken, leaving only 'El'. Gingerich ('Zoomorphic Astrolabes and the Introduction of Arabic Star Names into Europe') noted this as Elfeca without further comment; perhaps the rete was broken after he studied it.

Two stars are labelled Cenok. Four are unnamed, including one in the shape of a pretty bird resembling a song thrush, possibly intended to represent the constellation Corvus (the star Algorab, γ Corvi, is also shown and labelled).[20]

The zoomorphic pointer for Corvus made it unnecessary to engrave a name: the shape was its own label. Here zoomorphism served a mnemonic function, but elsewhere on astrolabes in this period it served an aesthetic one. Compared with some others of the period, most notably the Sloane astrolabe in the British Museum, Wh.1264's decoration is sparse, but a few other pointers do suggest zoomorphs. The Paris workshop of Jean Fusoris was later to popularise a sparser style, but the fourteenth century in England was clearly a period when astrolabe-makers were keen to display their aesthetic, as well as geometrical, skills.

Saints and Calendars

If we turn the astrolabe over, we encounter what might be termed the 'terrestrial' side of the astrolabe (Figure 1.4). This is mundane in two senses: features such as the shadow square highlight uses such as surveying, while the dual reference calendars make the astrolabe as much almanac as instrument. In decorative terms there is little to remark here, though it is notable that care has been taken over the names of the months and zodiac signs; the Gothic-style lettering here is considerably more elaborate than the simpler, more archaic capitals used for the star names on the rete. More worthy of comment are the circles of saints' names, feast dates, and dominical letters that form the inner rings of the Julian calendar. These are a relatively common feature of astrolabes produced in the fourteenth century; it seems calendrical functions became less important later.[21] Until very recently, no historian has given more than cursory consideration to the feast days featured on astrolabes.[22] But they are far from space-fillers: even small astrolabes from this period, such as Gonville &

20 See also Table 1 in Gingerich, 'Zoomorphic Astrolabes and the Introduction of Arabic Star Names into Europe'. Gingerich noted only two unlabelled stars, including the bird. One of the ones he omitted is identifiable as Kunitzsch 14 (α Geminorum, known as Razalgeuze). The other cannot be identified with any star in Kunitzsch's lists.
21 K. de Soysa, 'The Decline and Fall of the Astrolabe', unpublished M.Phil. essay, University of Cambridge (2000), pp. 7–8.
22 J. Davis has recently begun to rectify this. See, for example, J. Davis, 'Dating an Astrolabe from its Calendar Scales', *Bulletin of the Scientific Instrument Society*, 135 (2017), pp. 2–7.

Figure 1.4 Detail from the back of Wh.1264, showing the calendar of feasts for March. Note the uncial-Gothic capital 'M'. Image © Whipple Museum.

Caius Astrolabe B, squeeze in a calendar with as many feasts as possible. It has been more than a century since the antiquarian and chancellor of the diocese of Carlisle Richard Saul Ferguson wrote that 'it is to be wished that some expert in hagiology would examine the . . . calendars [on English astrolabes] and endeavour to ascertain the principle of selection'.[23] The rest of this chapter represents a tentative initial answer to his plea. Feast-day calendars certainly have the capacity to impart valuable understanding not only of the instruments themselves, but of the society that produced them.

Examination of the calendars and analysis of their sources must start from an awareness that astrolabe-makers chose well-known feasts to fill almost all the twenty-four to forty-eight spaces on their instruments. This reflects the fact that in fourteenth-century England the date was more often reckoned with reference to saints' days than using the Roman system. Thus the majority of dates marked on the astrolabes are feasts celebrated throughout Christendom, and we should see their function on the astrolabe as a method of measuring the course of the year parallel to the twelve months and zodiac signs, quite apart from their devotional significance. Nevertheless, the variety that remains can tell us much about the influences and interests of astrolabe-makers.

Table 1.2 lists the feast days marked on Wh.1264, and compares them with ten other astrolabes attributed to fourteenth-century England. It is noteworthy that, of the eight dates that appear on all

23 R. S. Ferguson, 'On an Astrolabe Planisphere of English Make', *Archaeologia*, 52 (1890), pp. 75–89.

TABLE 1.2 Comparative calendar of feast days on Wh.1264 and ten other astrolabes

Date	Dominical Letter	Feast Day[a]	English Translation and Notes	Others[b]
Ianuarius				
6	f	Ephia dñi	Epiphany	10
22	a	vincentus	Vincent of Saragossa (d. c. 304)	5
25	d	pauli	Paul	8
Februarius				
*	*	puriff[c]	Purification of Mary (Candlemas)	9
5	a	agathe	Agatha of Sicily (d. c. 251)	1
22	d	petri	Chair of St Peter	10
24	f	math	Matthias	7
Marcius				
2	e	cedde	Chad, bishop of the Mercians (d. 672)	3
12	a	gregor'	Gregory the Great	9
25	g	anñciacō ma	Feast of the Annunciation	10
Aprilis				
4	g[d]	ambrosii	Ambrose (d. 397)	9
23	a	georgii	George	4
25	c	marcii	Mark	10
Mayius				
1	b	phelip ia	Philip and James	6
3	d	crucis	Holy Cross (Finding)	7
19	f	dunst'	Dunstan (d. 988)	7
26	f	aug'	Augustine of Canterbury	7
Iunius				
11	a	barnab'	Barnabas	10
17	g	botulf'	Botulph (d. c. 680)	2
24	g	Johis bap	John the Baptist (Nativity)	8
29	e	petri	Peter (and Paul)	8
Iulius				
7	f	thome	Translation of St Thomas of Canterbury	6
20	e	marga	Margaret of Antioch (d. 304)	6
22	g	magdal[e]	Mary Magdalene	4
25	c	Iacob	James	8
Augustus				
1	c	petri	St Peter in Chains	4
10	e	laur'	Lawrence of Rome (d. 258)	10
15	c	marie	Assumption of Mary	8
24	e	barth	Bartholomew	9
29	c	joh	John the Baptist (Beheading)	3
Septemb-				
8	f	marie	Mary (Nativity)	9
14	e	crucis	Feast of the Cross	7
21	e	mathei	Matthew	9
29	f	mich	Michael(mas)	8

TABLE 1.2 (*cont.*)

Date	Dominical Letter	Feast Day[a]	English Translation and Notes	Others[b]
October				
9	b	dionisii	Denis (d. *c.* 250)	9
18	d	luce	Luke	10
28	g	simonis iude	Simon and Jude	10
Novemb-				
1	d	omni scōrum	All Saints	8
11	g	mart'	Martinmas (Martin of Tours, d. 397)	9
23	e	clem'	Clement of Rome (d. *c.* 98)	1
30	e	andi'	Andrew	4
December				
6	d	Nichol'	Nicholas	7
8	f	mar'	Immaculate Conception of Mary	5
13	d	lucie	Lucy	4
21	e	thom'	Thomas	9
25	b	Nat' d'	Feast of the Nativity	9

[a] The transcription of feast names is as close as possible to what we see on the astrolabe. However, I have expanded some common superscript abbreviations (-ri and -ru-).

[b] The number of other (possibly) English astrolabes on which this appears (out of ten). The others are (1) Oxford, Museum of the History of Science 47869, 'the Painswick astrolabe' (#299 in the 'International Checklist of Astrolabes' first compiled by Price in 1955); (2) Chicago, Adler Planetarium M-26 (#200 = #295); (3) Cambridge, Gonville & Caius College Astrolabe B (#301); (4) London, British Museum 1914, 0219.1 (#298); (5) British Museum SLMathInstr.54, 'The Sloane astrolabe' (#290); (6) British Museum 1909, 0617.1 (dated 1326) (#291); (7) Liège, Musée de la vie Wallonne (#457); (8) London, Science Museum, inv. no. 1880–26 (#293); (9) Innsbruck, Tiroler Landesmuseen Ferdinandeum (#2579); and (10) Astrolabe formerly in a Belgian private collection, present location unknown but included in Georges Baptiste (ed.), *La mesure du temps dans les collections belges: Catalogue et sélection des pièces* (Brussels: Generale Bankmaatschappij, 1984), p. 37 (#4518). I am grateful to John Davis for sharing images of nos. 7, 8, and 9, and providing invaluable information about no. 10. See also Davis, 'Fit for a King', pp. 337–9.

[c] No date or dominical letter is given (the date of this feast is 2 February). The day appears to have been written 'pufiff', and the first 'f' subsequently changed to an 'r'.

[d] This seems to be a mistake: the correct dominical letter would be 'c'.

[e] Written 'magdat', and subsequently corrected.

eleven instruments, seven commemorate people or events named in the Bible; the last, St Lawrence, was a third-century holy man venerated across Europe. How were the remaining feast days selected? As a fairly large astrolabe, Wh.1264 had room for forty-six, more than most other instruments. Even so, none of its feasts is unique to it. This is probably because astrolabe-makers adhered closely to a small number of calendars that circulated in fourteenth-century England.

The source of saints' days in this period was the Sarum calendar, part of the liturgy instituted by the bishops of Salisbury in the eleventh century. This was adopted as the calendar of daily use across England in the late thirteenth century, and astronomers who compiled their own calendars were generally faithful to it.[24] In the fourteenth century the most notable of these were Walter of Elveden and especially, later in the century, John Somer and Nicholas of Lynn. The last two were both cited by Chaucer in his *Treatise on the Astrolabe*. Somer's calendar survives in at least thirty-three complete and nine partial copies, while there are twenty-one of Nicholas's.[25] These calendars name feasts for almost every day of the year, so it is not surprising that almost all the feasts featured on the astrolabes examined for this study were listed by both Somer and Nicholas.

It is likely that such calendars provided important source material for these astrolabes, though the calendars of Somer and Nicholas themselves, which both begin in 1387, were probably produced after Wh.1264. An astronomical calendar such as this would have been useful not only to provide the basic data of feast days, but also to draw out the precisely aligned Roman and zodiacal calendars which together produce a solar equatorium. Indeed, Davis has shown that it may be possible to identify the calendrical source of an astrolabe by comparing their values of solar longitude.[26] However, when it comes to the saints' days it is hard to propose a single source. Autograph versions of the calendars of Walter, Somer, and Nicholas do not survive, and the extant copies vary somewhat in their calendrical coverage. Yet there were almost certainly some saints, such as Botulph or Chad, which appear on Wh.1264 and other astrolabes but were not included in the manuscript calendars.[27] Thus we can observe the makers of these astrolabes exercising a degree of personal choice.

24 N. Morgan, 'The Introduction of the Sarum Calendar into the Dioceses of England in the Thirteenth Century', in M. Prestwich, R. Britnell, and R. Frame (eds.), *Thirteenth Century England VIII: Proceedings of the Durham Conference 1999* (Woodbridge: Boydell, 2001), pp. 179–206, on p. 184.

25 L. Mooney (ed.), *The Kalendarium of John Somer* (composed 1380) (Athens: University of Georgia Press, 1998), p. 18; and S. Eisner (ed.), *The Kalendarium of Nicholas of Lynn* (completed 1386) (London: Scolar Press, 1980).

26 Davis, 'Dating an Astrolabe from its Calendar Scales'.

27 They do not appear in the base manuscripts chosen by their modern editors, only in later copies.

Botulph and Chad were also originally absent from the Sarum calendar. Botulph's feast of 17 June was added to the calendar in many dioceses, but Chad, who had been bishop of the Mercians in the seventh century (and was a patron saint of astronomers), was unlikely to appear in calendars outside the dioceses of Coventry, Lichfield, and Lincoln until around 1400. The see of Lincoln included Oxford, where both Somer and Nicholas were active at the end of the fourteenth century, but it appears that neither astronomer chose to honour this local saint. It seems Somer was working fairly uncritically from the Sarum calendar: despite dedicating his *Kalendarium* to Sts Francis of Assisi, Anthony of Padua, and Louis of Toulouse, he did not include any of their feast days in the calendar itself.

So it is clear that some calendars, and perhaps some parts of every calendar, were populated indiscriminately with saints chosen from a standard list. Even so, the question of choice is crucial – and is often ignored by historians intent on proving the sources of astrolabe data. Beyond the almost ubiquitous fixed feasts, from Epiphany to Christmas, astrolabe-makers had a free choice of what saints to include; most astrolabes include at least one somewhat obscure feast, such as Scholastica or Perpetua, that is unique to that instrument.[28] On what basis did makers exercise this choice? The basic calendrical function, marking out the regular passing of days across the year for reference purposes, was certainly a consideration; it was evidently critical to the maker of Gonville & Caius College Astrolabe B (who may have been Walter of Elveden himself); that astrolabe's calendar marks a consistent two feasts per month.[29] The author of the British Museum's Sloane Astrolabe appears to have made a special effort to include feasts on the first day of the month, achieving this in six months by marking such obscure celebrations as the Translation of St Remigius, and St Egidius's (Giles's) day.[30] Beyond this, though, it must simply have come down to personal devotional preference. If the maker of the Caius astrolabe had only been concerned to mark

28 St Scholastica's day (10 February) appears only on the Painswick astrolabe; St Perpetua's day (7 March) appears only on Science Museum inv. no. 1880–26.
29 This is the persuasive identification of Davis and Lowne, 'An Early English Astrolabe at Gonville & Caius College, Cambridge, and Walter of Elveden's *Kalendarium*'.
30 These two feasts also appear on astrolabes closely associated with the Sloane: Liège MVW and Science Museum 1880–26 (St Remigius is only on the former). Davis has made a close study of these three astrolabes, and suggests that Giles was a saint of particular interest to Richard de Bury, who may well have commissioned the Sloane astrolabe (Davis, 'Fit for a King', p. 343).

passing time, he would surely not have omitted Christmas; doing so left space in December for Nicholas and Thomas the Apostle. Thus the regime of two feasts each month did not prevent this astrolabist from making individual choices. Very few astrolabes (and none of our eleven) have a maker's name attached to them, so we cannot know whether they were more often made by the person who wished to use them, or commissioned from a skilled craftsman. However, since demand was, until the fifteenth century, insufficient to create a livelihood for professional makers, it is most probable that these astrolabes were the product of the personal choices of the first astronomers to use them.[31]

In all but one case, they exercised their choice with the inclusion of some English saints. Ten of the eleven astrolabes used in this study include at least three saints who were English, or had a particular following in this country.[32] Sts Dunstan, Augustine of Canterbury, Thomas of Canterbury, Margaret of Antioch (whose unusually strong cult led to the dedication of fifty-eight churches to her in Norfolk alone), and George all appear on at least five out of those ten.[33] Among the English saints represented less often are Alban, Aldhelm, Alphege, Botulph, Cuthbert, King Edmund, King Edward the Confessor, Frideswide, Guthlac, Hugh of Lincoln, and Swithin.

To draw some conclusions about how the astrolabe-makers chose their saints, we should consider what information, beyond lists in calendars, they would have had about them. Whether based in the monastery or the university, the makers of these astrolabes would

31 A. J. Turner, *Early Scientific Instruments: Europe 1400–1800* (London: Sotheby's, 1987), p. 27.
32 The exception is the Adler Planetarium's M-26. It was identified as English (*c.* 1250) by the Websters in their 1998 catalogue (R. Webster and M. Webster, *Western Astrolabes* (Chicago: Adler Planetarium, 1998), p. 40), but its first cataloguer, M. Engelmann, suggested it was 'probably French' and from around 1550 (M. Engelmann, *Sammlung Mensing: Altwissenschaftliche Instrumente: Katalog* (Amsterdam: Muller, 1924), p. 26). Its split personality is such that R. T. Gunther included it twice in his *Astrolabes of the World* (pp. 348, 472): once as a French 'astrological astrolabe' and once as a fourteenth-century English instrument. In a recent reassessment (J. Davis, 'A Royal English Medieval Astrolabe Made for Use in Northern Italy', *Journal for the History of Astronomy*, 48 (2017), pp. 3–32), J. Davis concludes that it is English, but made (and later modified) for use in continental Europe.
33 On Margaret of Antioch, see D. Farmer, *The Oxford Dictionary of Saints* (Oxford: Oxford University Press, 2003) and Davis and Lowne, 'An Early English Astrolabe at Gonville & Caius College, Cambridge, and Walter of Elveden's *Kalendarium*', p. 273. St Thomas Becket is generally represented by his translation on 7 July; only one astrolabe also gives his martyrdom on 29 December.

almost certainly have read saints' lives as part of their education. Indeed, it is likely that they would have written some hagiography themselves, as an exercise in grammar and rhetoric which helped keep hagiographical practices alive. English saints would have been frequent subjects of these hagiographies, one of whose purposes was to promote local saints and their shrines. Through hagiography the faithful were encouraged to learn from the lives of the saints, and to imitate their exemplary actions. If hagiographies were, as Heffernan explains, 'a catechetical tool much like the stained glass which surrounded and instructed the faithful in their participation at the liturgy', we can see these astrolabes fulfilling the same function: an aide-memoire that encouraged the devout astronomer to meditate on the saints in heaven as he looked to the sky.[34] If we accept a link between hagiographies and the astrolabes, it is hardly surprising to see Dunstan on so many of the latter: four hagiographies were written in barely more than 100 years after his death (d. 988); the last of these, by Eadmer, was much copied and rewritten.[35] Likewise, the popularity of the hagiographies of Martin of Tours, by Sulpicius Severus and later Gregory of Tours, made it almost inevitable that he would appear on ten of the eleven astrolabes (indeed three feature him twice, including his translation on 4 July as well as the more familiar November Martinmas festival).[36] It is also tempting to suggest that the special associations of certain saints would encourage astrolabe-makers to choose them. For example, an astronomer making an astrolabe for didactic purposes might include St Catherine, the patron of education and scholarship; indeed, she features on eight of the eleven astrolabes in our group. And a maker with sore eyes after painstakingly engraving azimuths and almucantars with perfect precision might be tempted to choose St Lucy, who appears on five of the eleven, because of her power to intercede on behalf of those suffering from visual problems.

The Whipple astrolabe, and four others in this group, provide evidence for the cult of a saint whose popularity grew exponentially in the fourteenth century: St George. Increasingly revered in the

34 T. J. Heffernan, *Sacred Biography: Saints and Their Biographers in the Middle Ages* (Oxford: Oxford University Press, 1988), p. 6.
35 E. van Houts, 'Review of A. J. Turner and B. J. Muir (eds.), *Eadmer of Canterbury: Lives and Miracles of Saints Oda, Dunstan, and Oswald*', *English Historical Review*, 123 (2008), pp. 1515–16.
36 On the cult of Martin, see T. F. X. Noble and T. Head, *Soldiers of Christ: Saints and Saints' Lives from Late Antiquity and the Early Middle Ages* (London: Sheed & Ward, 1995), p. xxvi.

West after he was associated with the First Crusaders' victory at Antioch in 1098, George became a symbol of chivalry across Christendom.[37] Edward I of England, who had himself taken the cross, was a devotee, and he invoked George's support for what he saw as his crusade against the Welsh dragon by having George's red cross stitched onto the bracers and pennoncels of his soldiers.[38] The success of his campaigns against the Welsh led him to honour St George in peacetime too: in 1285 he gave gold figures of George and Edward the Confessor to the shrine of George at Canterbury. Still, it was not until the reign of his grandson Edward III (1327–77) that George took on a role as protector of the whole English nation; it was confirmed by his installation as the patron of the new Order of the Garter, founded in the late 1340s. It is an open question how far people identified with the kingdom or nation in the High Middle Ages, but insofar as there was a sense of belonging to a nation, George was strongly associated with that national sense.[39]

Of course, loyalties could be much more local. Margaret of Antioch, featured on Wh.1264 and six other astrolabes of our group, has already been mentioned as a saint with a particular local following in Norfolk. Another is St Clement, whose feast on 23 November appears only on the Whipple instrument and Gonville & Caius Astrolabe B. In a recent article, Davis and Lowne note that twenty-two of the twenty-four feast days on Caius B also appear in Wh.1264's longer list. Suggesting a close relationship between the two instruments, they point out that the great church at Terrington St Clement in Norfolk was built and refounded in 1348 by Edmund Gonville, the founder of what would become Gonville & Caius College and an associate of Walter of Elveden.[40] Likewise, they posit a link between the slightly unusual appearance of the beheading of John the Baptist (29 August) on both instruments and the depiction of that event in the painted stone bosses of Norwich Cathedral cloisters. If such precise localisation is not a coincidence, we should hardly be surprised to find that we land on consecrated ground.

37 D. S. Fox, *Saint George: The Saint with Three Faces* (Windsor Forest: Kensal, 1983), p. 63.
38 J. Good, *The Cult of St George in Medieval England* (Woodbridge: Boydell, 2009), pp. 52–8.
39 Good, *The Cult of St George*, pp. 4–5.
40 Davis and Lowne, 'An Early English Astrolabe at Gonville & Caius College, Cambridge, and Walter of Elveden's *Kalendarium*', p. 274.

Conclusions

It is clear that astrolabes played a complex role in medieval culture: not only did they perform a staggering variety of functions; they could also be objects of prestige, aids to spiritual contemplation, and even, in the case of Peter Abelard and Héloïse, function as a name for a child.[41] In the case of the Whipple astrolabe, it is hard to imagine a purpose other than practical. Its medium size, large enough for precision and stability, but not too large to hold at arm's length, makes it ideal for observations. Its iconography is relatively spare: instead of lavishing attention there, its maker has saved his energy for the engraving of closely spaced almucantars and precisely marked eccentric calendars. Its astrological functions appear to have been expanded by the later addition of Great Houses. All in all, it is an extremely user-friendly device. That does not preclude its employment as a teaching tool, either by its first or subsequent users, and, of course, it is likely that its user had some kind of devotional motivation for his astronomical observations, just as its maker would have exercised personal devotional preferences in his choice of saints.[42] For what it is worth, it is reasonable to call it English: at least, it was certainly made for an Englishman. And a date in the middle of the fourteenth century fits with the metallurgical analysis, quatrefoil tracing, and eccentric calendar, as well as the choices of saints.[43]

Recent studies have shown that a combination of contrasting analyses can yield new insights even into much-studied instruments. Even though astrolabes have always been popular with researchers, there is certainly scope for new approaches that combine techno-logically advanced methods such as XRF, or the use of computer-aided design tools to analyse the geometry of their engraving, with more traditional tools of assessment such as palaeography, iconog-raphy, and textual analysis. There is also plenty more to be said about the backs of astrolabes. These remain rather ignored in studies,

---

41 Astralabe [*sic*], or (Petrus) Astralabius, was born to the famous lovers *c.* 1117–18. See P. Abelard, *Historia Calamitatum*, ed. J. Monfrin (Paris: Vrin, 1962), p. 74; and *The Letters of Abelard and Heloise*, trans. B. Radice (Har-mondsworth: Penguin, 1974), pp. 285–7.

42 J. Bennett has noted how, at least in the early-modern period, instruments that started as observational might become didactic. See J. Bennett, 'Early Modern Mathematical Instruments', *Isis*, 102 (2011), pp. 697–705, on p. 698.

43 Davis has suggested that eccentric calendars fell out of fashion in England in the mid-fourteenth century (Davis, 'Dating an Astrolabe from its Calendar Scales', p. 3).

and illustrated catalogues all too often picture only the front of instruments, making it harder to survey astrolabe backs without examining instruments in person or commissioning expensive new photography.[44]

Museum visitors sometimes suffer in the same way. While some museums are able to display astrolabes in freestanding cabinets, or make creative use of mirrors so that both sides are accessible, in most cases only one side can be seen, and curators perhaps understandably opt to display the more visually arresting front. If astrolabes like Wh.1264 are to be reassessed by researchers, this should feed into museum practice; but it should also take account of recent work in museology and instrument studies, starting with what Ken Arnold and Thomas Söderqvist have called 'a more forthright concern with [instruments'] immediate presence'.[45] This would not be to succumb to what Jordanova scorned as 'childish awe',[46] but, as W. David Kingery has argued, 'the warm emotional and aesthetic content of objects should share the spotlight with their cold practical and cognitive aspects in a holistic approach to material culture'.[47] A museum of the history of science is always likely to privilege the technology over its context, but it is vital to show that both are changed by their interaction. If museums were to permit a more active engagement with astrolabes in their context, it might encourage scholars to pay less attention to the astronomical theories they demonstrated, and more to the varied ways, and reasons why, they were commissioned, made, and used in the Middle Ages.

44 Falk, 'Improving Instruments', pp. 46–76. Some recent articles by Davis, cited above, have paid new attention to the neglected backs.
45 K. Arnold and T. Söderqvist, 'Medical Instruments in Museums: Immediate Impressions and Historical Meanings', *Isis*, 102 (2011), pp. 718–29, on p. 718.
46 Jordanova, 'Objects of Knowledge', p. 40.
47 W. D. Kingery (ed.), 'Introduction' to *Learning from Things: Method and Theory of Material Culture Studies* (Washington: Smithsonian Institution, 1996), pp. 1–15, on pp. 4–5.

2 ✌ What Were Portable Astronomical Instruments Used for in Late-Medieval England, and How Much Were They Actually Carried Around?

CATHERINE EAGLETON

Telling the time in medieval England was not always straightforward. There were growing numbers of mechanical clocks, which had a significant impact on medieval society,[1] as well as a number of astronomical instruments that used the motion of the sun to tell the time: astrolabes, quadrants, and sundials. Writing in the late fourteenth century, Jean Froissart praised the clock in his poem *L'orloge amoureus*, describing it as beautiful and remarkable, pleasing and profitable, because it shows the hours night and day, even when there is no sun.[2] But is this really a statement of frustration at the challenges of telling the time on a cloudy day, or something more literary, more ambiguous, or more abstract? Looking specifically at the astronomical and timekeeping instruments that survive from the medieval period, many are portable and could have been carried around to tell the time. But were they? Or, were they made and used for other purposes? These questions run through much of the scholarship on such instruments. Examples of these instruments –

---

1 G. Dohrn-van Rossum, *History of the Hour: Clocks and Modern Temporal Orders*, trans. T. Dunlap (Chicago: Chicago University Press, 1996). See also L. R. Mooney, 'The Cock and the Clock: Telling the Time in Chaucer's Day', *Studies in the Age of Chaucer*, 15 (1993), pp. 91–109, on the methods and instruments used for telling the time in the late Middle Ages.

2 P. F. Dembowski (ed.), *Le paradis d'amour: L'orloge amoureus* (Geneva: Droz, 1986), p. 83 (lines 1–6):

> ... l'orloge est, au vrai considerer,
> Un instrument tres bel et tres notable
> Et s'est aussi plaisant et pourfitable,
> Car nuit et jour les heures nous aprent
> Par la soubtilleté qu'elle comprent,
> En l'absence meïsme dou soleil.
> Dont on doit mieuls priser son appareil,
> Ce que les aultres instrumens ne font pas,
> Tant soient fait par art et par compass.

and mechanical clocks – survive in museum collections, but it is difficult to determine how they were used, and for what purpose.

Derek J. de Solla Price, the Whipple Museum's first serious instrument scholar, studied a range of different instrument types in his career.[3] In 1980 Price published an influential article that outlines his view of the uses of medieval astronomical instruments, suggesting that their practical value may have been that they were 'ideas made brass' rather than instruments for observation. If someone understood the structure of the heavens and the motion of the heavenly bodies, then they could understand the construction of an astrolabe and how the various lines on the instrument related to the heavens, and the instrument could function as an embodiment of that knowledge:

> These devices ... were tangible models that served the same purpose as geometric diagrams or mathematical or other symbolism in later theories. They were embodied explanation of the way that things worked ... I suggest that tangible modelling as a species of comprehension comes nearer to the 'purpose' of armillary spheres or star and earth globes than to imagine they had a prime utility as devices for teaching or for reference.[4]

Francis Maddison, who was Curator of the Museum of the History of Science in Oxford, took a similar view, and in an article based around a typology of the most important medieval instruments explained that 'none of the ... instruments discussed above was of much use to any practical profession, except that of teaching astronomy', but that they were instead (with the exception of the magnetic compass and mechanical clock) an application of theoretical astronomical knowledge.[5]

Over a number of years, I used these two quotes as part of lectures and teaching sessions in the Department of History and Philosophy

---

3 Aspects of this work are described in detail in Chapter 9 by Boris Jardine. See also Seb Falk, 'The Scholar as Craftsman: Derek de Solla Price and the Reconstruction of a Medieval Instrument', *Notes and Records of the Royal Society*, 68 (2014), pp. 111–34.

4 D. J. de Solla Price, 'Philosophical Mechanism and Mechanical Philosophy: Some Notes towards a Philosophy of Scientific Instruments', *Annali dell'Istituto e Museo di Storia della Scienza di Firenze*, 5 (1980), p. 76.

5 F. R. Maddison, *Medieval Scientific Instruments and the Development of Navigational Instruments in the Fifteenth and Sixteenth Centuries* (Coimbra: Agrupamento de Estudos de Cartografia Antiga, 1969), p. 20. The instruments in the typology are the armillary sphere, the equatorium, the torquetum, the planispheric astrolabe, the quadrant, other types of sundial (under which heading he lists the navicula, horizontal dial, and universal equinoctial dial), the magnetic compass, and the mechanical clock.

of Science about medieval instruments. Some of those, thanks to the Whipple Museum collection and the support of museum staff, included hands-on engagement by the students with the instruments,[6] and those sessions helped to engage undergraduate students in a subject that can be difficult and alien. However, the students also challenged me to explain what a medieval astrolabe like the one they were looking at (see Figure 1.1) was actually used for, and they looked at an astrolabe alongside a cylinder dial and asked about the difference between the two. Prompted by my students' questions, I here return to questions I considered earlier, now based on new evidence that has become available in the last two decades.

This chapter is the result of that reconsideration of fundamental questions about the uses and users of astronomical instruments in medieval England. I argue that we should look again at the practical uses of some types of instruments, and consider whether some were carried around to tell the time, to be used for practical purposes alongside the symbolic, teaching, and other functions that Price and Maddison articulate. Instruments could be 'ideas made brass' but they could also be of practical use, and I suggest that it was precisely this combination that may have made some instruments more important, or more common, in the period. Looking carefully at the provenance of objects, and at the reasons why they are or are not preserved in museum collections, and considering the full range of types of evidence available for the study of instruments paints a fuller picture.

Evidence from Texts

The challenges around answering questions about what instruments were used for relate in part to the available evidence. There are many hundreds of manuscripts that include technical works about astronomical and timekeeping instruments, describing how to make and use them.[7] Many of these manuscript volumes are compilations of

6 These included Wh.1264, the medieval astrolabe analysed in Chapter 1 by Seb Falk.
7 On scientific and medical books in later-medieval England see, for example, P. M. Jones, 'Medicine and Science', in L. Hellinga and J. B. Trapp (eds.), *The Cambridge History of the Book in Britain* (Cambridge: Cambridge University Press, 1999), pp. 433–48; and S. Livesey, 'Transitions 1: Scientific Writing in the Latin Middle Ages', in A. Hunter (ed.), *Thornton and Tully's Scientific Books, Libraries and Collectors: A Study of Bibliography and the Book Trade in Relation to the History of Science*, 4th edn (Aldershot: Ashgate, 2000), pp. 72–98. The US National Library of Medicine's *IndexCat* provides keyword search across the electronic versions of the classic catalogues compiled by Lynn Thorndike and

astronomical texts along with texts about instruments, mostly in Latin, but, from the later fourteenth century, with some in English. Some manuscripts are rubricated and some are illustrated with diagrams, but hardly any have pictorial images, indicating a scholarly or practical purpose more than a decorative one. One of the most well-known of these technical works was the *Treatise on the Astrolabe* written in English by Geoffrey Chaucer in the late fourteenth century. That text sets out the parts of an astrolabe, and then details more than forty observations and calculations that the instrument could perform.[8] Other astrolabe treatises, as well as texts about other astronomical instruments, often also include detailed instructions for their construction, linking the lines on the instrument to the lines of the model of the heavens on which they were based. In many cases, although the instructions in these manuscripts are clear and can be followed, they are not straightforward practical manuals, but oriented towards an understanding of the theory of astronomy as much as to the use of an instrument to tell the time, or to survey a field.

One possibility for assessing what instruments were used for in the medieval period is to look at who owned them, and for this there are scattered and fragmentary references in wills and probate inventories. For example, the 1434 will of John de Manthorp, vicar of Hayton, East Yorkshire, in northern England, includes an astrolabe and a calendar, and the will of John Hurt in 1476 bequeathed to the Cambridge University Gotham Loan Chest a book about astronomical instruments that was already in the chest as security against a loan.[9] Surviving probate inventories for Oxford University in the fifteenth century do not include any references to objects identifiable as astrolabes or sundials, nor do they list any books identifiable as

Pearl Kibre (for Latin scientific manuscripts) and by Linda Ehrsam Voigts and Patricia Deery Kurtz (for Middle English), but, although this is a useful starting point, it is still incomplete and must be supplemented by individual libraries' catalogues: www.nlm.nih.gov/hmd/indexcat/index.html (accessed 2 February 2018).

8 S. Eisner, *A Treatise on the Astrolabe, Volume VI: The Prose Treatises, Variorum Editions of the Works of Geoffrey Chaucer* (Norman: University of Oklahoma Press, 2002).

9 S. Cavanaugh, 'Books Privately Owned in England, 1300–1450', unpublished PhD dissertation, University of Pennsylvania, 1980, pp. 561–2; and P. D. Clarke, *The University and College Libraries of Cambridge* (London: British Library, 2002), p. 704, no. 14.

manuscripts about instruments.[10] However, there is evidence from
the fourteenth century that Oxford scholars did own instruments –
Simon Bredon (d. 1372) and William Rede (d. 1385) both
bequeathed astronomical books and instruments to individuals and
to Oxford colleges.[11]

None of these sources relating to the ownership of instruments
gives clear information about what those people did with the instruments and manuscripts. Some non-technical works connect Oxford
students and scholars to the use of astrolabes, and seem at first read
to provide some evidence for what they did with them, including a
story of the drunken behaviour of an Oxford student named Robert
Dobbys, describing his use of an astrolabe to find his way home,
sometime in the 1420s:

> It is related of him that one night after a deep carouse, when on
> his way from Carfax to Merton, he found it advisable to take his
> bearings. Whipping out his astrolabe he observed the altitude
> of the stars, but, on getting the view of the firmament through
> the sights, he fancied that sky and stars were rushing down
> upon him. Stepping quickly aside he quietly fell into a large
> pond. 'Ah, ah', says he, 'now I'm in a nice soft bed I will rest in
> the Lord.' Recalled to his senses when the cold struck through,
> he rose from the watery couch and proceeded to his room
> where he retired to bed fully clothed. On the morrow, in answer

10 H. Anstey (ed.), *Epistolae Academicae Oxoniensis*, 2 vols. (Oxford: Her Majesty's
    Stationery Office, 1868), vol. 2: pp. 514–15, 525, 531–2, 543–6, 557–62, 565–67,
    579–85, 592–7, 599–600, 604–15, 622–5, 627–31, 638–67, 671–3, 698, 704–8,
    711–13. One entry is unclear: the 'bursa cum uno "diall" de ligno' owned by
    John Lashowe might have been a small wooden sundial kept in a purse (p. 663).
    Some of these records give evidence for mathematical interests, for example,
    Thomas Cooper owned a geometrical book (p. 516). Several of the inventories
    record the ownership of musical instruments, for example, Sir John Lydbery
    owned a lute, valued at 6d (p. 698), and John Hosear owned a harp valued at 4d
    (p. 705).
11 In Simon Bredon's will (dated 1368) he leaves his large astrolabe to Merton
    College and the small astrolabe to William Rede. See M. Powicke, *The Medieval
    Books of Merton College* (Oxford: Clarendon Press, 1931); and Cavanaugh,
    'Books Privately Owned in England', p. 129. William Rede's will of 1 August
    1382 includes a large number of astronomical and mathematical books, left to
    his relatives and to various Oxford Colleges. At the end is an indenture listing a
    number of instruments to be given to Merton College: 'Preterea dictus vener-
    abilis pater dedit et assignauit eiusdem calicem deauratum, et decem alia
    instrumenta, videlicet albionem, equatorium planetarum, quadrantem, chilin-
    drum, speram materialem, speram solidam, tabulam ymaginum celestium,
    cartam maris, lapidem calculatorium, et tabulas dealbatas pro tabulatione
    librorum.' See J. D. North, *Richard of Wallingford*, 3 vols. (Oxford: Oxford
    University Press, 1976), vol. 3, appendix 15, pp. 132–5.

to kind inquiries, he denied all knowledge of the pond. Thus were his feckless drunken ways amply proved.[12]

A clue that this may not be a straightforward account of the practical use of an astrolabe is in the last few words – which mirror the wording of academic argument in the medieval period. Indeed, this story about Dobbys was told as part of the disputations that qualified the medieval student to be granted a degree, by the 'Father of the Act', who introduced the inceptors by poking fun at them.[13] It also harks back to the ancient story told about Thales of Miletus, who was so busy looking at the stars that he fell into a ditch.[14] The story is clearly supposed to be funny, at least to the learned Oxford audience who heard it, so perhaps the joke is that this is precisely *not* how an astrolabe was used? Oxford scholars would know that it is only a short distance from Carfax to Merton, and so the only reason you might use an astrolabe is if you (like Dobbys, as painted by this story) were a fool.

Written a few decades earlier, Geoffrey Chaucer's *Canterbury Tales* is a more well-known literary work that includes a number of references to pilgrims using astronomical and other instruments. Chaucer's characters might use a cylinder dial to determine that it is time to eat, or be described as owning an astrolabe, but these

---

12 Oxford, Magdalen College, MS 38, ff. 41v–42r. This translation is from S. Gibson, 'The Order of Disputations', *Bodleian Quarterly Record*, 6, no. 65 (1930), pp. 107–12, on p. 108. See also C. Eagleton, J. Rampling, and D. Banham, 'Masters of Incompetence: Learned Humour in 15th-Century Oxford', in J. Rampling, D Banham, and N. Jardine (eds.), *Recipes for Disaster* (Cambridge: Whipple Museum of the History of Science, 2010).

13 For biographical details of Robert Dobbys, see A. B. Emden, *Biographical Register of the University of Oxford*, 3 vols. (Oxford: Clarendon Press, 1957), vol. 1, pp. 579–80. Gibson 'The Order of Disputations', pp. 107–12; and S. Gibson, 'Appendix C: The Order of Disputations', in S. Gibson (ed.), *Statuta antiqua Universitatis Oxoniensis* (Oxford: Clarendon Press, 1931), pp. 643–7, describe the procedure of inception, the people involved, and the records in Magdalen MS 38.

14 R. H. Hicks, translation of Diogenes Laertius, *Lives of Eminent Philosophers*, 2 vols. (Cambridge: Harvard University Press, 1972), p. 35. The old woman who answered his cries said 'How can you expect to know all about the heavens, Thales, when you cannot even see what is just before your feet?' The stories told by Diogenes Laertius were widely available in fifteenth-century England, both directly, and in works based on them, such as Walter Burley's fourteenth-century work *De vita et moribus philosophorum*: see Walter Burley (attr.), *Liber de vita et moribus philosophorum*, ed. H. Knust (Tübingen: Bibliothek des Litterarischen Vereins in Stuttgart, 1886). A similar story, of an astronomer who is so busy looking at the stars that he falls into a pit, is told in Chaucer's 'Miller's Tale': see L. D. Benson (ed.), *The Riverside Chaucer* (Oxford: Oxford University Press, 1987), p. 71, lines 3457–61.

references are not straightforward to interpret.[15] Other medieval writers with courtly audiences included references to astronomy and its instruments in their works, too: for example, Gower included an astronomical section in the *Confessio Amantis*, and Robert Henryson's Middle Scots version of Aesop's Fables includes the astrolabe, quadrant, and almanac as things that can teach someone about the heavens.[16] In all these cases, perhaps the clearest thing they indicate is that courtly audiences knew enough about astronomy and its instruments to understand their use in this literary context.

Some clues as to what instruments were seen as useful for can be found by considering other objects that they are described as, or seem to be, equivalent to. Henryson's poem lists an astrolabe, quadrant, or almanac together when talking about learning astronomy, and Chaucer's student Nicholas (in 'The Miller's Tale' in the *Canterbury Tales*) has not only an astrolabe, but also a manuscript of Ptolemy's Almagest, augrim (calculating) stones, among various other objects kept at the head of his bed.[17] In a letter written to William Worcester (clerk to the Norfolk knight Sir John Fastolf) in May 1449, his correspondent John Crop asks about an augrim table, or a book of augrim, or an astrolabe, again indicating that there were connections made between these types of instruments in the period that we perhaps do not as strongly make in museums, where books and manuscripts may no longer be kept with astronomical instruments like astrolabes, or objects for calculation and accounting.[18]

15 A recent special issue of *The Chaucer Review*, 43.4 (2009), is dedicated to Chaucer's references to, and interest in, time and its measurement. The classic study is J. D. North, *Chaucer's Universe* (Oxford: Clarendon Press, 1988). In a more recent analysis, Sara Schechner concluded that none of Chaucer's pilgrims was likely to have both the means to own, and the skill to use, an instrument like an astrolabe: S. J. Schechner, 'Astrolabes and Medieval Travel', in R. Bork and A. Kann (eds.), *The Art, Science, and Technology of Medieval Travel* (Aldershot: Ashgate, 2008), pp. 181–210, on p. 204.

16 On Gower, see J. Simpson, *Sciences and the Self in Medieval Poetry: Alan of Lille's 'Anticlaudianus' and John Gower's 'Confessio Amantis'* (Cambridge: Cambridge University Press, 1995). On Henryson, see Robert Henryson, *The Morall Fabillis of Esope the Phrygian*, ed. G. G. Smith (Edinburgh and London: Scottish Text Society, 1906), p. 49; and A. Hanham and J. C. Eade, 'Foxy Astrology in Henryson', *Parergon*, 24 (1979), pp. 25–9.

17 Benson, *The Riverside Chaucer*, pp. 68–78, lines 3208–11.

18 See R. Beadle and C. Richmond (eds.), *Paston Letters and Papers of the Fifteenth Century*, part III (London: Early English Text Society, 2005). Worcester is known to have been interested in astronomical and calendrical subjects, and owned a number of manuscripts on these subjects, as well as on historical and antiquarian matters. See K. B. McFarlane, 'William Worcester: A Preliminary Survey', in K. B. McFarlane (ed.), *England in the Fifteenth Century* (London:

Finally, there is strong evidence from medieval England that manuscripts were kept together with the instruments they describe in libraries – astronomical instruments are the only objects other than books that appear in late-medieval English booklists. Library lists include details of the instruments held, which include astrolabes, quadrants, and in one case a navicula sundial. I have argued in a previous publication that this is the case because both are regarded as sources of information about astronomy and the achievements of great astronomers, with the instruments complementing the books. This was not only a connection made in abstract, but in practice, since for example Merton College library lent out astronomical instruments in the same way as it lent out astronomical books to fellows of the college. And, interestingly, in the few listings that give this detail, instruments and books were estimated as being of similar value to each other.[19]

Instruments and Archaeology

Medieval astronomical and timekeeping instruments are today not only rarely found in libraries, but are more often found in museums and collections around the world.[20] Hundreds of astrolabes, quadrants, sundials, and related instruments survive from the Middle East and from Europe, and detailed research on them has revealed much about how they were made, and where. Most instruments from medieval Europe are neither signed nor dated, and their identification relies upon close study of the objects themselves, looking at the letter and number forms marked on them, as well as construction marks, lines, and clues that sometimes help to identify the maker or workshop.[21]

Comparing the astronomical instruments preserved in museums with those described in the manuscript texts indicates that, as might be expected, metal instruments are much more likely to have

Hambledon Press, 1981), pp. 199–224, on p. 199; and J. Harvey, *William Worcester: Itineraries* (Oxford: Clarendon Press, 1969), p. 240.

19 C. Eagleton, 'John Whethamstede, Abbot of St Albans, on the Discovery of the Liberal Arts and Their Tools. Or, Why Were Astronomical Instruments in Late-Medieval Libraries?', *Mediaevalia*, 29.1 (2009), pp. 109–36.

20 D. A. King, 'Medieval Astronomical Instruments: A Catalogue in Preparation', *Bulletin of the Scientific Instrument Society*, 31 (1991), pp. 3–7.

21 For example, the close study of three English quadrants that enabled them to be dated, and linked to the court of Richard II: S. Ackermann and J. Cherry, 'Richard II, John Holland and Three Medieval Quadrants', *Annals of Science*, 56 (1999), pp. 3–23.

survived than those made from other materials. It is likely, for example, that paper and wooden instruments were more widely made and used in the medieval period, and a commonly copied text on the cylinder dial specifies that it should be made from a piece of boxwood.[22] Other, often larger or more complex, instruments may only rarely have been made;[23] on the other hand, there are instruments that were used but not widely written about, like sandglasses and simple compass dials.[24]

This bias in the survival of instruments comes in part from the fact that it is more likely that wooden or paper instruments were disposed of, or became damaged, since they were less durable than instruments made of brass. Occasionally, it is possible to identify medieval instruments that were altered in the sixteenth century, indicating perhaps that they were still then being used.[25] By the seventeenth century, scholars and antiquaries had begun to collect astronomical manuscripts, and sometimes instruments too. For example, there were three scholars and collectors in the eighteenth century who were interested in the navicula sundial, one of whom made his own instrument following the designs preserved in medieval manuscript texts (Figure 2.1). In the eighteenth and nineteenth centuries an increasing interest in the medieval period also had an impact both on the collecting and on the study of astronomical instruments from the period.[26] Across all periods, however, collectors of astronomical instruments seem to have preferred more

22 C. Kren, 'The Traveller's Dial in the Late Middle Ages', *Technology and Culture*, 18 (1977), pp. 419–35, discusses the construction of the cylinder dial, questioning whether they were made in large numbers but do not survive, or whether the cylinder existed primarily as a written text.

23 For example, the equatorium, on which see S. Falk, 'A Merton College Equatorium: Text, Translation, Commentary', *SCIAMVS*, 17 (2016), pp. 121–59; S. Falk, 'Learning Medieval Astronomy through Tables: The Case of the Equatorie of the Planetis', *Centaurus*, 58 (2016), pp. 6–25; and Falk, 'The Scholar as Craftsman'.

24 On the iconography of sandglasses, see D. G. Boullin, 'An Iconographic Study of Sandglasses', *Nuncius*, 4 (1989), pp. 67–85. On references to them in accounts of navies and merchants, see P. F. Naish, 'The Dyoll and the Bearing Dial', *Journal of the Institute of Navigation*, 7 (1954), pp. 205–8. On compass dials, which may have been imported from German-speaking areas of Europe, see E. G. R. Taylor, *The Haven-Finding Art* (London: Hollis and Carter, 1956), p. 173.

25 G. L'Estrange Turner, 'Charles Whitwell's Addition, *c.* 1595, to a Fourteenth Century Quadrant', *Antiquaries Journal*, 85 (1995), pp. 454–5.

26 J. H. Leopold, 'Collecting Instruments in Protestant Europe before 1800', *Journal of the History of Collections*, 7 (1995), pp. 151–7, describes the collecting of astronomical instruments in *Kunstkammer* and outlines the early English collections of the seventeenth and eighteenth centuries.

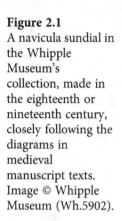

Figure 2.1
A navicula sundial in the Whipple Museum's collection, made in the eighteenth or nineteenth century, closely following the diagrams in medieval manuscript texts. Image © Whipple Museum (Wh.5902).

complex and decorative instruments over those that were more functional, which almost certainly skewed what survives.

However, while traces of ownership of manuscripts can sometimes be found in annotations made in them by later owners, few instruments have detailed provenance recorded on them or about them. Few medieval astrolabes, quadrants, or sundials in museum collections are associated with specific places or people beyond the eighteenth or nineteenth centuries, hampering efforts to consider what these instruments in particular might (or might not) have been used for and how they might have come to be preserved in museum collections. Complementing museum collections, however, there is some archaeological evidence relating to medieval astronomical instruments, which has so far not been analysed in order to consider the uses of medieval instruments.

The archaeological evidence for England and Wales is particularly useful thanks to the Portable Antiquities Scheme (PAS). This voluntary scheme records hundreds of thousands of small finds, including many objects that are not required to be processed as Treasure, often made by metal detectorists, but which can nonetheless create a valuable picture of the material culture of a place or period. The PAS is supported by a network of Finds Liaison Officers who work with finders and local communities to ensure the quality of the data and images in the database, as well as to promote good practice among metal-detector communities, including the need to gain permission from the landowner, and the importance of recording context and precise locations for finds.[27]

Some common types of objects dominate those recorded by the PAS: on 14 July 2017, there were more than 380,000 coins in the database, along with more than 41,000 buckles and 34,000 brooches. These are the kinds of things that people probably lost while moving around their local area, or travelling further afield (often known as 'stray finds'), rather than objects that were deliberately concealed or hoarded. They are also dominated by base-metal objects that are relatively easily located with a metal detector. Finds in the PAS database are categorised by time period, with more than 298,000 Roman-period objects, more than 171,000 medieval objects, and more than 141,000 post-medieval objects recorded at the time of writing. Looking to some less common types of objects, and thinking back to Chaucer's pilgrims travelling to or from Canterbury, there are 461 pilgrim badges recorded,[28] and sixty-one book-related objects (including book clasps), which suggests that some people carried books around with them.[29]

Turning to astronomical and timekeeping instruments, the numbers are smaller: 110 sundials are recorded from medieval and post-medieval periods, the majority of which are ring dials from the seventeenth and eighteenth centuries. In many cases, the PAS database cross-references major museum collections, including the British Museum and the Museum of the History of Science, Oxford, giving some (albeit limited) consistency of dating and identification.[30] Crucially, finds in the PAS database have locations associated with them, which enables an assessment of where these types of objects were used, or where they were lost (Figure 2.2). Taking these data, and removing those objects that are dated, or are likely to date, from after *c.* 1500, gives a small dataset of finds that can be combined

27 www.finds.org.uk and R. Bland, M. Lewis, D. Pett, I. Richardson, K. Robbins, and R. Webley, 'The Treasure Act and Portable Antiquities Scheme in England and Wales', in Gabriel Moshenska (ed.), *Key Concepts in Public Archaeology* (London: UCL Press, 2017), pp. 107–21.

28 https://finds.org.uk/database/search/results/q/pilgrim+badge/show/100 (accessed 14 July 2017).

29 https://finds.org.uk/database/search/results/q/book+clasp/show/100 (accessed 14 July 2017).

30 John Davis has been active in helping to identify things that are or might be sundials or other mathematical and astronomical instruments, which means those records are often better than average. He has also published a number of articles about particular instruments recorded by the Portable Antiquities Scheme: see the publication list at www.flowton-dials.co.uk/publications/ (accessed 12 May 2018).

Figure 2.2 Map of all finds whose database record includes the keyword 'sundial' in the Portable Antiquities Scheme database, on 14 July 2017. Some finds have their find spot protected and are not mapped, but they have less precise information on the location at which they were found in the database record text. Map of search results from Portable Antiquities Scheme database www.finds.org.uk (CC-BY 4.0) using Google Maps, with data from Open Street Map www .openstreetmap.org (CC-BY-SA).

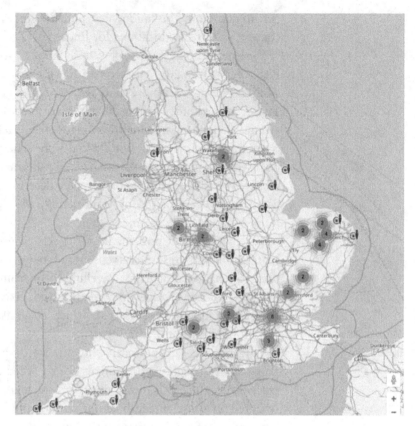

with evidence from museum collections, and information from texts, to begin to assess whether portable astronomical instruments were carried around, or whether they more often stayed inside the libraries, studies, and other places in which they were kept.

Astrolabes and Compass Dials

The astrolabe, in some ways the quintessential medieval astronomical instrument, has been the focus of much substantial research. In a recently published listing drawing on many years of detailed work, David King lists the European astrolabes from before *c.* 1500 that are known to survive, grouping them by type.[31] Some English instruments can be grouped together, and it has been argued that some

---

31 D. A. King, 'European Astrolabes to *ca.* 1500: An Ordered List', *Medieval Encounters*, 23 (2017), pp. 355–64.

should be associated with the royal court,[32] whereas others have long connections to Oxford or Cambridge colleges that suggest learned contexts for their ownership.[33]

Reviewing a range of evidence, including the latitudes for which astrolabe plates were made, unusual markings or parts on surviving instruments, and literary and other references to the instrument, Sara Schechner concluded that there is little evidence of medieval travellers taking an astrolabe on the road, or to sea, until the late fifteenth century.[34] Astrolabes and texts about them did move around Europe, she explains, but once they were owned by someone it seems that they were not taken outside their study for more practical purposes like timekeeping, despite the instrument being well-adapted for that purpose as well as for astrological medicine or for surveying. The evidence from the PAS backs this up, with no astrolabes or fragments of astrolabes recorded in the database. (One item was listed in the database with the keyword 'astrolabe' (the object top left in Figure 2.3, found at a site where more than 226 medieval objects have been recorded), but John Davis has more recently reidentified it as part of a nocturnal.[35])

If astrolabes were complex, expensive, and often too large to be carried around easily, the same was not true for simple compass dials for telling the time from the sun at a particular (fixed latitude). These instruments, or parts of them, are the majority of medieval astronomical and timekeeping instruments recorded by the PAS. Some of these examples, when complete, were probably compass dials with nocturnals on their lids, which could be used at night as well as during the day.[36] These instruments (those in the all but the top row

---

32 J. Davis, 'A Royal English Medieval Astrolabe Made for Use in Northern Italy', *Journal for the History of Astronomy*, 48.1 (2017), pp. 3–32; and J. Davis, 'Fit for a King: Decoding the Great Sloane Astrolabe and Other English Astrolabes with "Quatrefoil" Retes', *Medieval Encounters*, 23 (2017), pp. 311–54.

33 For example, Merton College, Oxford, which still has instruments in its old library: A. Chapman, 'Merton College and Its Astrolabes', *Postmaster and The Merton Record* (October 1992), pp. 88–100. A fourteenth-century English astrolabe at Gonville & Caius College, Cambridge, has been associated with Walter Elveden, who wrote a *Kalendarium* in 1327 and was associated with the founders of the College: see J. Davis and M. Lowne, 'An Early English astrolabe at Gonville & Caius College, Cambridge, and Walter of Elveden's Kalendarium', *Journal for the History of Astronomy*, 46.3 (2015), pp. 257–90.

34 Schechner, 'Astrolabes and Medieval Travel', p. 204.

35 J. Davis, comment at https://finds.org.uk/database/artefacts/record/id/192080 (accessed 2 February 2018).

36 An early-sixteenth-century drawing by Urs Graf shows a man holding a compass dial which, in common with several of the surviving instruments, has a nocturnal on the lid so that the owner could tell the time by day or by night:

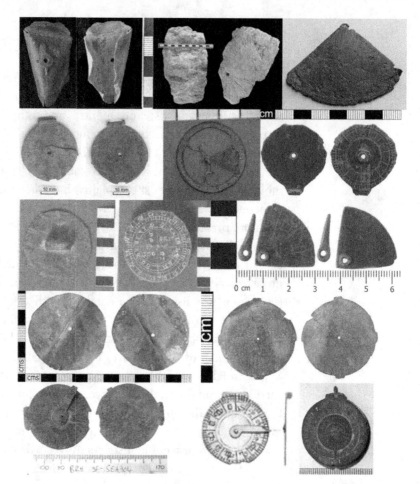

Figure 2.3 Thirteen astronomical or time-telling instruments (or parts of them) that are likely to be late-medieval in date, recorded by the Portable Antiquities Scheme. Composite image from www.finds.org.uk (CC-BY 4.0).

of Figure 2.3) are similar in their size, form, and decorative details to the few examples of medieval English compass dials that are preserved in museum collections. Indeed, some of the examples in museum collections display clear evidence of corrosion consistent with being buried in the ground, so perhaps they, too, are objects that were found, even though their archaeological context is no longer known or was at the time of finding and/or acquisition not recorded. There are finds of similar simple compass dials in other parts of Europe, dating to the fifteenth century, indicating that this pattern is also repeated outside England.[37]

F. A. B. Ward, 'An Early Pocket Sundial Illustrated in Art', *Antiquarian Horology*, 11 (1979), pp. 484–7.

37 For example, R. Salzer, 'Mobility Ahead of Its Time: A Fifteenth-Century Austrian Pocket Sundial as a Trailblazing Instrument for Time Measurement

For these two very different instruments, the archaeological evidence suggests two very different kinds of usage: astrolabes probably were not carried around much, but compass dials were. Astrolabes may have been more associated with studying the heavens, or for timing religious observances,[38] than with more practical needs, while compass dials, although they still worked by observation of the motion of the sun through the heavens, were more practical in their function. Compass dials were probably more common, and carried around (and lost) more frequently.

Quadrants and Naviculae

I turn now to the first of two types of English astronomical instrument that, in the medieval period, seem to have existed between the extremes of the astrolabe and the compass dial. Medieval-period quadrants of various kinds are preserved in museum collections, and there are also archaeological finds. An exceptional find was made in 2005, when, during an excavation of a sealed soil deposit at the back of a medieval inn building in Canterbury, a quadrant was found and dated to 1388 by close examination of the tables engraved on it.[39] This instrument was made for latitude 52°, appropriate for Southern England; it has no stars marked on it, simplifying the instrument but meaning that it could not have been used for finding the time at night.

Following widespread news reporting of this quadrant (and its subsequent sale price), another quadrant was found by a metal detectorist in 2014, near Chetwode, Buckinghamshire (Figure 2.4). This quadrant was found in a field through which there is a footpath running to a medieval priory, indicating perhaps that the person who lost it was travelling to or from the priory.[40] John Davis has

on Travels', in M. C. Beaudry and T. G. Parno (eds.), *Archaeologies of Mobility and Movement* (New York: Springer, 2013), pp. 65–79.

38 On religious uses for astrolabes in this period, see Chapter 1 by Seb Falk.

39 The quadrant found at the 'House of Agnes' in Canterbury, in 2005, is British Museum 2008,8017.1, and images of it can be seen in their online collections database at www.britishmuseum.org/research/collection_online/search.aspx. On this quadrant, including its dating, see E. Dekker, 'With His Sharp Look Perseth the Sonne: A New Quadrant from Canterbury', *Annals of Science*, 65.2 (2008), pp. 201–20. For more detailed archaeological and site details see A. Linklater and E. Dekker, 'The Discovery of a Quadrant Novus at the House of Agnes, St Dunstan's Street, Canterbury', *Archaeologia Cantiana*, 130 (2010), pp. 65–82.

40 www.christies.com/features/The-Thrill-of-Discovery-Finding-the-Chetwode-Quadrant-6620-1.aspx (accessed 2 February 2018).

Figure 2.4 The Chetwode quadrant, BERK-C673DD, image from www.finds.org.uk database (CC-BY 4.0).

argued that it is similar to another metal-detectorist find of part of a medieval quadrant in Norfolk, and suggests that the two instruments may have been produced to the same design, or even in the same workshop.[41]

Looking at an instrument with similar size and functionality to the quadrants, the navicula sundial, there is a similar pattern. Four medieval naviculae are preserved in museum collections,[42] and these instruments share – both with each other and with the manuscript texts on each navicula – a remarkably consistent design.[43] Some recorded provenance information, even if only partial, exists for all of them. The Oxford navicula was given to Lewis Evans in 1898 by the Curator of the Norwich Museum, and then given by Evans to the Museum of the History of Science.[44] The Greenwich navicula, in the collections of the National Maritime Museum, was found by metal detectorists near Sibton Abbey, Suffolk. The Geneva navicula was owned in the eighteenth century by John Wilson (1719–83), an antiquarian and collector who lived in the North of England, near Sheffield. This instrument remained in his family until it was sold at auction to the Musée d'Histoire des Sciences in Geneva. Its excellent

41 J. Davis, 'The Chetwode Quadrant: A Medieval Unequal-Hour Instrument', *Bulletin of the British Sundial Society*, 27.2 (2015), pp. 2–6.
42 In Oxford (History of Science Museum) www.mhs.ox.ac.uk/object/inv/54358; Greenwich (National Maritime Museum) http://collections.rmg.co.uk/collec tions/objects/211073.html; Geneva (Museum of the History of Science) http:// institutions.ville-geneve.ch/fileadmin/user_upload/mhn/images/votre_visite/site_ mhs/aide_cadrans_e.pdf; and Florence (Galileo Museum) https://catalogue.museo galileo.it/object/NaviculaDeVenetiis.html (all links accessed 12 May 2018).
43 C. Eagleton, *Monks, Manuscripts, and Sundials: The Navicula in Medieval England* (Leiden: Brill, 2010), especially Chapter 5 on the design of the instrument and the relationship between instrument, text, and image.
44 Provenance notes in the record of a meeting at the National Maritime Museum, Greenwich, 1992, which are in the instrument's accession file at the History of Science Museum, Oxford. The meeting was called to assess the Geneva navicula, then being researched before it was sold, since its larger size had created some doubts about whether it should be dated to the fifteenth or the eighteenth century.

condition makes it most likely that it passed from owner to owner until it came into the possession of John Wilson in the eighteenth century, although there are no records of where or how Wilson might have acquired such an instrument, other than a reference to his having collected 'some things which were scarcely worth preservation' by a near-contemporary commentator.[45]

Another navicula, of a similar size to the Geneva navicula, and with very similar decorative features to Greenwich's example, was found in 2017.[46] It is about the same size as the larger Geneva instrument, and has similar letter and number shapes, as well as similar positioning of those markings in relation to the hour lines on the front of the instrument. Despite their different sizes, this newly found instrument also has much in common with the Greenwich navicula, details that are mostly not specified in the manuscript texts about the navicula. This privately owned navicula was most probably found by a metal detectorist in Yorkshire, although details are somewhat uncertain; nevertheless, its archaeological context is clear from the corrosion on the instrument.[47] Indeed, only the Florence and Geneva naviculae, those that have provenance linking them to collections in the eighteenth century and earlier, do not seem to have surface corrosion consistent with having been buried in the ground.

Despite the information being incomplete and to some degree uncertain, the places where the naviculae have been found can be compared with lists of towns and their latitudes that appear on some of the instruments, as well as in manuscripts describing how to make and use them. The same places appear in manuscripts and on the instruments, albeit with fewer places listed on the instruments than in the written sources.[48] And the places where naviculae were found, or have provenance linking them to, are within the same range of latitudes as these lists of places (Figure 2.5). The places themselves are also striking, with Sibton Abbey (the find spot for the Greenwich navicula) adding to the evidence from manuscripts and library booklists that points to the study and use of astronomical and

---

45 J. Hunter, *Hunter's Hallamshire* (London: privately printed, 1819), p. 276.

46 This instrument is described and illustrated in J. Davis, 'The Navicula: Made in Medieval East Anglia?', *Bulletin of the British Sundial Society*, 29.2 (2017), pp. 15–23.

47 Information on provenance and likely find spot provided by the instrument's owner, personal correspondence with the author.

48 Eagleton, *Monks, Manuscripts, and Sundials*, Chapter 4 on latitude lists.

Figure 2.5 Places marked on the back of surviving naviculae (York, Northampton, Oxford, London, Winchester, and Exeter) and places where naviculae were found or have provenance linking them to (Vale of York, Norfolk, and Sibton Abbey in Suffolk). Map of search results from Portable Antiquities Scheme database (www.finds.org.uk) using Google Maps, with map data from Open Street Map (www .openstreetmap.org, CC-BY-SA), and GeoBasis-DE/BKG ((c) 2009).

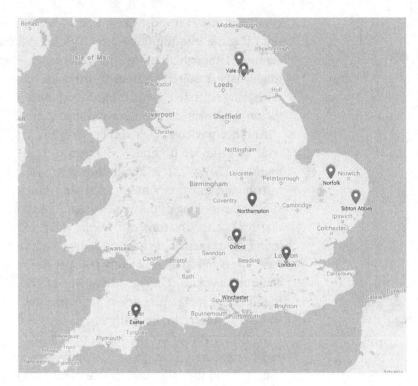

timekeeping instruments by people in monasteries and other religious institutions.[49] The picture is far from complete, although the manuscript sources and surviving instruments together suggest that the navicula was a fairly common type of portable sundial in medieval England. The fact that two of them have been found by metal detectorists suggests that they were not only made and studied for their connection to astronomical theory, but also carried around.

Considering texts and instruments together – and, for the latter, including archaeological finds along with objects in museum collections – points to a more complex pattern of uses and usage for astronomical instruments in medieval England than any of those sources alone can give. There is little evidence that astrolabes, that quintessential astronomical instrument of the period, were carried around, and this is perhaps unsurprising given the size and weight,

---

49 John Davis's argument that the navicula was a specifically East Anglian instrument seems rather thin, however, when looking at all the available evidence.

and likely value, of the brass astrolabes that are preserved in museums. It is possible that paper or wooden astrolabes were more affordable and more portable, but these are unlikely to have survived. These instruments, then, perhaps stayed in libraries, monasteries, colleges, and people's houses, and were just the kind of 'tangible model' that Price described.[50] On the other hand, there are finds indicating that simple compass dials may have been fairly commonly used, and carried around, but not written about in manuscript texts. Between these two extremes – of astrolabes and compass dials – archaeological context combines with other evidence to suggest that quadrants and naviculae might have been two types of instruments that were portable, and actually were carried around. At the same time, they still embodied the kind of astronomical learning that astrolabes do, albeit with less complexity. So, they were written about and studied in learned contexts like monasteries and universities, and kept in libraries along with astronomical books, but they were also carried around (and occasionally lost).

Conclusion

If we know that an instrument is carried around, this does not necessarily mean that we can determine what practical purpose it had for its owner. If someone carried his quadrant, perhaps in a leather case, or his navicula, was it decorative, acting as a symbol or representation of his astronomical learning? Or, were there practical reasons why people travelling around medieval England might have needed to use timekeeping instruments of these kinds? It is tempting to imagine, as in Chaucer's *Canterbury Tales*, pilgrims determining the time using instruments, and there is some evidence for astronomical instruments being used in religious contexts,[51] but it is important to consider astrological medicine as perhaps the primary context within which small and portable instruments like quadrants and naviculae found practical use. Time, whether calculated or observed using an instrument, or derived using tables, was important for the efficacy of herbal remedies, and medical treatises were often

---

50 This is not to say that, in those locations, there couldn't also have been practical use made of astrolabes. They were useful for calculations, among other things, as noted by one Middle English writer of a *compotus* text: L. Braswell-Means, 'Ffor As Moche As Yche Man May Not Haue the Astrolabe: Popular Middle English Variations on the Compotus', *Speculum*, 67.3 (1992), pp. 595–623.

51 The Chetwode quadrant was found near a priory, but it is not clear whether this links it to religious uses or contexts. See also Chapter 1 by Seb Falk.

combined with astronomical works in manuscript volumes, attesting to the importance of this connection. Hilary Carey has argued that in medieval England astrology was the preserve of three groups: people at court, those in academic institutions, and medical men.[52] It is striking, then, that these are three groups for whom there is good evidence of interest in astronomical instruments at the time.[53] Indeed, one of the manuscripts containing a text on how to make and use the navicula sundial includes this work on an instrument along with astrological and medical texts. A bifolium from this manuscript seems to have been carried around folded, indicating that these parts in particular (including text on whether a patient will die, and whether an expected baby is a boy or a girl) were needed for reference separately from the rest of the volume.[54] Folded almanacs of various kinds survive from late-medieval England, containing calendrical and astrological information aimed at medical practitioners.[55] Alongside these portable manuscript texts, it is easy to imagine a small instrument like a quadrant or a navicula being useful, either to tell the time or to simplify calculations relating to the hours of the day.

It is hard to assess how widespread the use of astronomical instruments was in medieval astrological medicine, and how many practitioners might have carried around timekeeping instruments. This example is an outline of a group of people who could – and perhaps did – use instruments, more than a definitive study of the many functions and purposes of these objects. However, it is clear from the limited archaeological evidence available that some kinds of instruments were carried around, and piecing together the provenance recorded for medieval instruments held in museum collections, along with the locations of finds of instruments (and parts of instruments) in the last twenty years, suggests that Price and Maddison may have been too categorical in their assessment of the practical utility of instruments. Indeed, the possibility that some types of astronomical instrument could both demonstrate and symbolise

52 H. M. Carey, *Courting Disaster: Astrology at the English Court and University in the Later Middle Ages* (London: Macmillan, 1992), p. 56.

53 Courtly and academic interest in astronomical instruments is clear in the literary and other references to them, outlined above. For examples of physicians interested in astronomical instruments, see also L. White, 'Medical Astrologers and Late-Medieval Technology', *Viator*, 6 (1975), pp. 295–308.

54 Trinity College, Cambridge, MS O.5.26, ff. 92–3.

55 H. M. Carey, 'What Is the Folded Almanac? The Form and Function of a Key Manuscript Source for Astro-medical Practice in Later Medieval England', *Social History of Medicine*, 16.3 (2003), pp. 481–509.

learning about the heavens, and also be of practical use, might explain why quadrants and naviculae seem to have been made in larger numbers, and to similar or standardised designs, than has in the past been recognised. These instruments were, I would argue, both 'ideas made brass' and portable, practical, objects for the people who owned them. As more finds are made, and as provenance research and detailed cataloguing of museum collections continues, it seems likely that this contextual information – relating to the history of the object from when it was made to when it was collected by a museum – will further enrich the understanding that can be gained from close study of the surviving instruments and texts.

3 ᵱᴁ 'Sundials and Other Cosmographical Instruments': Historical Categories and Historians' Categories in the Study of Mathematical Instruments and Disciplines

ADAM MOSLEY

In 1635, the Scots-born Jesuit Hugh Sempill published a twelve-book text on the mathematical disciplines.[1] Sempill devoted book seven of this work to the subject of cosmography; subsequent books consider what he described as the constituent elemental and celestial parts of that discipline, namely geography (book eight), hydrography and meteorology (book nine), astronomy (book ten), and astrology and calendrics (books eleven and twelve). Chapter eleven of book ten is entitled 'Of Sundials and Other Cosmographical Instruments'.[2]

This one chapter, easily overlooked amid the wealth of material regarding the mathematical disciplines in the early modern period, is of considerable interest to historians of science and curators of scientific instruments. At first sight, it constitutes an extraordinary vindication of the claim, advanced by former Whipple Museum Curator Jim Bennett, that sundials were cosmographical devices in the long sixteenth century.[3] Bennett presents the Renaissance discipline of cosmography as a key to unlocking the true meaning of these objects, all too frequently understood merely as time-telling devices.

1 H. Sempilius, *De Mathematicis Disciplinis Libri Duodecim* (Antwerp: Ex Officiana Plantiniana, 1635).
2 H. Sempilius, 'De Horologiis sciotericis & aliis instrumentis Cosmographicis', in *De Mathematicis Disciplinis* (Antwerp: Ex Officiana Plantiniana, 1635), p. 226. On Sempill (or Semple), see E. L. Ortiz, 'Sempill, Hugh (1596–1654), Mathematician', in H. G. C. Matthew and B. Harrison (eds.), *Oxford Dictionary of National Biography* (Oxford: Oxford University Press, 2004), http://doi.org/10.1093/ref:odnb/25072. Some aspects of his extraordinary career are dealt with in passing in D. Worthington, *Scots in the Habsburg Service, 1618–1649* (Leiden: Brill, 2004).
3 J. Bennett, 'Sundials and the Rise and Decline of Cosmography in the Long Sixteenth Century', *Bulletin of the Scientific Instrument Society*, 101 (2009), pp. 4–9; and J. Bennett, 'Cosmography and the Meaning of Sundials', in M. Biagioli and J. Riskin (eds.), *Nature Engaged: Science in Practice from the Renaissance to the Present* (New York: Palgrave Macmillan, 2012), pp. 249–62.

By associating sundials with cosmography, he seeks to demonstrate how they were not only part of a strong mathematical tradition that was intellectually stimulating for its many enthusiastic participants, but also intimately connected to broader cultural changes such as the European overseas expansion and the competitive pursuit of power, territory, and commercial advantage entailed by that enterprise. Bennett's arguments challenge scholars to produce much richer histories of sundials and dialling than have so far been generated. They also imply that curators at institutions with rich collections of dials – institutions like the Whipple Museum – might profitably rethink how best to display and interpret them for a visiting public.[4]

This chapter will revisit Bennett's arguments regarding sundials not only in the light of Sempill's text and the burgeoning literature on cosmography, but also by drawing on the collections of the Whipple Museum and the Whipple Library and the scholarship they have inspired. Like Bennett's inquiries into the connection between dialling and cosmography, my own studies of this subject have their origin in our ongoing attempts to make sense of the mathematical culture of the early modern period.[5] That culture was generative not only of instruments and texts, but also of texts about instruments, instruments reproduced from texts and their accompanying images, images that functioned as instruments, and instrument–book hybrids.[6] Efforts to understand it, therefore, typically cross

4 For dials acquired by R. S. Whipple and the Whipple Museum up until the mid 1980s, see D. J. Bryden, *The Whipple Museum of the History of Science, Catalogue 6: Sundials and Related Instruments* (Cambridge: Whipple Museum of the History of Science, 1988).

5 See A. Mosley, 'Spheres and Texts on Spheres: The Book–Instrument Relationship and an Armillary Sphere in the Whipple Museum of History of Science', in L. Taub and F. Willmoth (eds.), *The Whipple Museum of the History of Science: Instruments and Interpretations to Celebrate the 60th Anniversary of R. S. Whipple's Gift to the University of Cambridge* (Cambridge: Whipple Museum of the History of Science, 2006), pp. 301–18; A. Mosley, 'Objects of Knowledge: Mathematics and Models in Sixteenth-Century Cosmology and Astronomy', in I. Maclean and S. Kusukawa (eds.), *Transmitting Knowledge: Words, Images, and Instruments in Early Modern Europe* (Oxford: Oxford University Press, 2006), pp. 193–216; and A. Mosley, 'Objects, Texts and Images in the History of Science', *Studies in History & Philosophy of Science*, 28 (2007), pp. 289–302.

6 The literature exploring the intersections of book, instrument, and image is now substantial. See, for example, O. Gingerich, 'Astronomical Paper Instruments with Moving Parts', in R. G. W. Anderson, J. A. Bennett, and W. F. Ryan (eds.), *Making Instruments Count: Essays on Historical Scientific Instruments Presented to Gerard L'Estrange Turner* (Aldershot: Variorum, 1993), pp. 63–74; D. J. Bryden, 'The Instrument-Maker and the Printer: Paper Instruments Made in Seventeenth Century London', *Bulletin of the Scientific Instrument Society*, 55 (1997), 3–15; S. De Renzi, *Instruments in Print: Books from the Whipple*

backwards and forwards across the adjacent realms of text, image, and instrument, and are frequently drawn to the points of closest overlap. This approach is hardly unique to graduates of the Whipple school of instrument studies, but it is one that the environment of the Whipple Museum and Library – like other institutions whose collections encompass both instruments and books – is especially conducive to developing. Our common past association with the Whipple can help to explain why Jim Bennett and I are exercised by similar issues in the history of Renaissance mathematics and mathematical instruments, and employ a similar technique, attentive to multiple *kinds* of sources, in attempting to resolve them.

But while our preoccupations and our methods our similar, our conclusions sometimes differ. Here, by retracing Bennett's steps through the cosmographical literature of sixteenth-century Europe, I shall suggest some problems with his argument that sundials, in particular, can be associated with cosmography. I shall then use Sempill's account to explore, more generally, the advantages and disadvantages of labelling certain objects as 'cosmographical instruments', suggesting that such designations, when used at all in the period, were idiosyncratically applied. In addition, I shall re-examine the question of cosmography's supposed decline after 1600. For Bennett, the disappearance of cosmography is another way in which cosmography and dialling might be associated. He suggests that as cosmography faded its astronomical component found a new home in the vibrant dialling tradition of subsequent centuries.[7] Because, I shall argue, it actually persisted as a category of knowledge and a set of activities even into the twentieth century, attempts to employ

---

Collection (Cambridge: Whipple Museum of the History of Science, 2005); C. Eagleton and B. Jardine, 'Collections and Projections: Henry Sutton's Paper Instruments', *Journal of the History of Collections*, 17 (2005), pp. 1–13; A. Marr, 'The Production and Distribution of Mutio Oddi's *Dello squadro* (1625)', in I. Maclean and S. Kusukawa (eds.), *Transmitting Knowledge: Words, Images, and Instruments in Early Modern Europe* (Oxford: Oxford University Press, 2006), pp. 165–92; Katie Taylor, 'A "Practique Discipline"? Mathematical Arts in John Blagrave's *The Mathematical Jewel* (1585)', *Journal for the History of Astronomy*, 41.3 (2010), pp. 329–53; S. K. Schmidt, *Altered and Adorned: Using Renaissance Prints in Daily Life* (Chicago: Art Institute of Chicago, 2011), pp. 73–82; S. Gessner, 'The Use of Printed Images for Instrument-Making at the Arsenius Workshop', in N. Jardine and I. Fay (eds.), *Observing the World through Images: Diagrams and Figures in the Early Modern Arts and Sciences* (Leiden: Brill, 2014), pp. 124–52; and B. Jardine, 'State of the Field: Paper Tools', *Studies in History & Philosophy of Science*, 64 (2017), pp. 53–63.

7 Bennett, 'Sundials and the Rise and Decline of Cosmography in the Long Sixteenth Century', p. 9.

'cosmographical' as a term of the historian's art risk confusing, rather than clarifying, our accounts of past scientific practice. We need to be particularly attentive to the *variety* of ways in which such categories were deployed at different times and places, and avoid overwriting them with our own, even – perhaps especially – when we seek to recruit our terms from those used in the past.

Sundials and Cosmography

That sundials should have been considered cosmographical instruments during the sixteenth and seventeenth centuries is entirely plausible on *prima facie* grounds. Renaissance cosmography's foundational text was Ptolemy's *Geography* – a guide to mapping the Earth by longitude and latitude, with an accompanying gazetteer of places, from which both world and regional maps could be drawn. On completing the first translation of the text from Greek into Latin in the early fifteenth century, the Florentine humanist Jacopo Angeli elected to rename this work the *Cosmography*, after the word *cosmos*, arguing that the text concerned both the heavens and the Earth.[8] As depicted in one of the best-known and most frequently republished cosmographic works of the sixteenth century, Peter Apian's *Cosmographicus liber* (1524), the truth of Angeli's claim rested on the fact that coordinate mapping depends upon the projection onto the surface of the Earth of fundamental divisions of the celestial sphere (Figure 3.1).[9] These lines define the equator and the tropics, and allow meridians passing through the poles to be drawn. Texts like Apian's were also concerned with the apparent annual motion of the Sun along the sphere, known as the ecliptic, and with the sphere's daily rotation. They related these phenomena to the surface of the Earth by, for example, discussing the ancient division of the globe into *klimata* – latitudinal bands defined by the maximum length of the day. Thus cosmography was fundamentally concerned with using projective geometry to connect the heavens and the Earth and, frequently, to relate solar motion, terrestrial location, and time. Sundials are devices constructed using projective geometry to relate

---

8 See C. Burnett (trans.), 'Jacopo Angeli's Introduction to his Translation into Latin of the *Geography*', in C. Burnett and Z. Shalev (eds.), *Ptolemy's Geography in the Renaissance* (London: The Warburg Institute, 2011), pp. 225–29.

9 For convenience, I cite the first edition of the text revised by Gemma Frisius. See P. Apian and G. Frisius, *Cosmographicus liber mathematici, studiosi correctus, ac erroribus vindicatus per Gemmam Phrysiam* (Antwerp: in aedibus Rolandi Bollaert, 1529), fol. IIv.

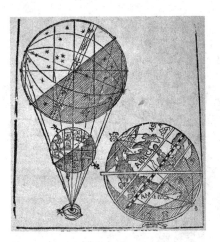

Figure 3.1 Peter Apian's visual representation of the discipline of cosmography, from Peter Apian and Gemma Frisius, *Cosmographia* (Antwerp, 1584), p. 2. Image © Whipple Library (95:50).

solar position and time for one or more terrestrial locations.[10] It is therefore easy to see cosmography and dialling as closely related, both conceptually and technically.

Bennett strengthens this *prima facie* case for considering sundials cosmographical devices by adducing various other kinds of historical evidence. He notes, for example, that Apian's *Cosmographicus liber* includes a sundial among the several paper instruments that it contains: a universal rectilinear altitude dial, which he identifies as belonging to a class of instruments sometimes referred to by the name 'organum Ptolomei', or 'instrument of Ptolemy' (Figure 3.2).[11] Within the pages of the book, this device clearly served a didactic rather than an immediate time-telling function. Apian instructed his reader how to use it to perform a range of operations connecting time, solar motion, and terrestrial location.[12] It was, therefore, an application rather than an explication of the art of dialling, employed to demonstrate some of the fundamental relationships at the heart of cosmography. Nevertheless, Bennett suggests, its presence in Apian's influential text connected that work with the subject's ancient authority, Claudius Ptolemy, via a sundial. Indeed, Bennett posits that the name 'organum Ptolomei' was rooted in Ptolemy's

10 See A. Turner, 'Sundials: History and Classification', *History of Science*, 27 (1989), pp. 303–18.

11 Apian and Frisius, *Cosmographicus liber*, fol. XIIv. On the use of such devices in this text, see S. van den Broecke, 'The Use of Visual Media in Renaissance Cosmography: The Cosmography of Peter Apian and Gemma Frisius', *Paedagogica Historica*, 56 (2000), pp. 130–50; and M. Gaida, 'Reading *Cosmographia*: Peter Apian's Book–Instrument Hybrid and the Rise of the Mathematical Amateur in the Sixteenth Century', *Early Science and Medicine*, 21 (2016), pp. 277–302.

12 Apian and Frisius, *Cosmographicus liber*, fols. Xr–XIIr.

Figure 3.2 The
paper universal
altitude dial
constructed in
Apian's textbook
cosmography,
from Peter Apian
and Gemma Frisius,
Cosmographia
(Antwerp, 1584),
p. 25. Image
© Whipple Library
(95:50).

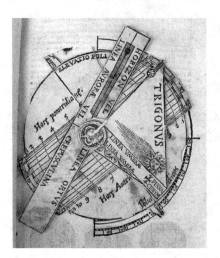

contemporary status as a cosmographical author, rather than a writer on astronomy.[13] Dialling, on this evidence, was fundamentally intertwined with cosmography.

Bennett strengthens the case for considering sundials cosmographical by demonstrating that a large number of individuals he identifies as cosmographers also wrote on dials, or were involved in their production. Besides Apian, the examples include Gemma Frisius, who revised and augmented Apian's work, Egnazio Danti, Sebastian Münster, Oronce Finé, and Gerard Mercator.[14] The full list of cosmographers Bennett provides does indeed seem too extensive for the overlap between cosmography and dialling to be attributed merely to coincidence. Once again, therefore, a strong connection between sundials and cosmography seems to have been established.

Arguments such as these are cumulatively powerful, but the evidence deployed in them is somewhat circumstantial. And, even though the claim that sundials were cosmographical instruments can now be supported by reference to Sempill's text, analyses are not necessarily justified by the conclusion to which they lead. Closer scrutiny reveals problems with both sets of evidence. While it is clearly true that Ptolemy was the ghost at Apian's cosmographic feast, several elements of the *Cosmographicus liber* would have suggested his presence to a contemporary reader more clearly than Apian's inclusion of a paper dial in the book. Notably, the opening

---

13 Bennett, 'Sundials and the Rise and Decline of Cosmography in the Long Sixteenth Century', p. 7.
14 Bennett, 'Sundials and the Rise and Decline of Cosmography in the Long Sixteenth Century', p. 7.

chapter of his text, 'What Is Cosmography, and How Does It Differ from Geography and Chorography', was an adaptation of the first chapter of book one of the *Geography*, 'How Geography Differs from Chorography'.[15] And while that discussion did not identify Ptolemy by name, it did refer to a work by Johannes Werner, the *Paraphrases*, which was a summary of book one of the *Geography* and had been printed alongside Werner's translation of that book in 1514.[16] A later chapter of the *Cosmographicus liber* explicitly discussed the use and the form of Ptolemy's maps, while the gazetteer of places which appeared in the second part of the book strongly echoed the style of the *Geography*, listing the coordinates of the principal places of the world by region, as if to facilitate their use in the production of maps, rather than organising them alphabetically.[17]

To some extent, these elements worked to distance the *Cosmographicus liber* from Ptolemy's legacy, rather than to evoke it. Partly, no doubt, this was because of the shortcomings in the geographical knowledge of the ancients exposed by the New World discoveries; these 'new' parts of the world were briefly discussed in the text and incorporated into its coordinate lists.[18] But it may also have been because, by this point in time, 'cosmography' had already fallen out of fashion as the preferred title for Ptolemy's geographical work. From 1490 onwards, editors had begun to reinstate the title that Jacopo Angeli had changed, renaming Ptolemy's text the *Geography*.[19] Thus, as Apian's first chapter suggested, the subject of cosmography could no longer simply be considered co-extensive with the material covered by Ptolemy.

15 Apian and Frisius, *Cosmographicus liber*, fol. IIr: 'Quid sit Cosmographia et quo differat a Geographia & Chorographia'; c.f. C. Ptolemy, *Geographia* (Rome: Petrus de Turre, 1490), sig. a r: 'In quo differt Geographia a Chorographia'. Earlier editions of Ptolemy's text use *cosmographia* instead of *geographia* both in the title of the work and in this chapter.

16 Apian and Frisius, *Cosmographicus liber*, fol. IIIr: 'Geographia (ut Vernerus in paraphrasi ait) ...' The reference is to C. Ptolemy *et al.*, *In hoc opere haec continentur: Nova translatio primi libri geographiae Cl'. Ptolomaei ... In eundem primum librum geographiae Cl. Ptolomaei: argumenta paraphrases ...* (Nuremberg: Johann Stuchs, 1514).

17 Apian and Frisius, *Cosmographicus liber*, fol. XXXr: 'De usu tabularum Ptho. et qualiter uniuscuiusque regionis, loci aut oppidi situs in illis sit inveniendus'. For the gazetteer, see fols. XXXv–LIIr.

18 Apian and Frisius, *Cosmographicus liber*, fols. XXXIIIIr–v, LIv–LIIr.

19 Ptolemy, *Geographia* identifies 'geography' as the subject of the work in the opening chapter of the text, but uses 'cosmography' and 'geography' interchangeably for the title of the text in the incipits and explicits of the individual books; this indecisiveness is also evident in the editions of the text published in Rome in 1508 and 1509, and Venice in 1511.

Figure 3.3
A navicula dial,
1620. The geometry
underlying these
ship-shaped dials is
similar to that of
Apian's dial, as
shown in Figure 3.2,
the Regiomontanus
dial, and the *organa
ptolomei*. Image
© Whipple Museum
(Wh.0731).

Nevertheless, the references and allusions in the *Cosmographicus liber* were clear enough to hint at Ptolemy's continuing importance to cosmographical theory and practice.

In contrast, it would have taken a particularly astute reader, already familiar with the geometry and nomenclature of dials, to infer a strong connection between the paper dial in Apian's book and this ancient authority, especially since the text itself did not refer to the instrument as an *organum ptolomei*. An expert reader might have recognised the underlying relationship between the instrument in Apian's book, the ship-shaped dials known as navicula (Figure 3.3), the so-called Regiomontanus dial, and the instruments explicitly named as *organa ptolomei* in late-medieval manuscripts. As Catherine Eagleton has noted, the identity of the latter category of instruments is somewhat confused in the manuscript tradition: the description was applied in certain texts to a range of instruments with varying physical forms and omitted entirely from other manu-scripts describing devices with identical geometry.[20] The origins of the name are also not clear, although some manuscripts suggest that the designation 'organum' derives from the resemblance of the device, in some forms or at some stage of its production, to a *musical* instrument.[21] But the underlying geometry of these dials, as well as that of naviculae and the Regiomontanus dial, was treated in

20 C. Eagleton, *Monks, Manuscripts and Sundials: The Navicula in Medieval England* (Leiden: Brill, 2010), pp. 93–100.
21 Austrian National Library, Vienna, MS 5418, fol. 181r, as transcribed in Eagle-ton, *Monks, Manuscripts, and Sundials*, p. 273: 'Et in residuo forma cuiusdam organi musici relinquatur. Verum et ipsum nomine ut credo accepit'. Eagleton dates this manuscript, on p. 266 of her work, to the first half of the fifteenth century.

Ptolemy's work *On the Analemma*, known in the Latin West since the middle of the thirteenth century.[22] It is not impossible, therefore, that the *organum ptolomei* was named in recognition of Ptolemy's authority in *dialling*, rather than astronomy or cosmography. Apian himself might have understood as much; even before his appointment to the University of Ingolstadt in 1526, his studies at Vienna had introduced him to the remnants, and hence the legacy, of the cluster of mathematicians associated with these two institutions that had formed around the humanist scholar Conrad Celtis in the late fifteenth century.[23] Among them was Andreas Stiborius, whose writings included *Canons* for the use of the *organum ptolomei* – a text sometimes transmitted, naturally enough, alongside one describing the instrument.[24] Apian would have been well placed therefore, to conceive of a relationship between the devices that were described as *organa ptolomei* in Viennese manuscripts and the paper instrument with which he supplied his *Cosmographicus liber*, as well as to recognise the multiple associations with the ancient author of the *Geography* that this relationship suggested. But this connection was surely not perceptible, without explicit commentary, to the neophyte mathematician and geographer at whom Apian's cosmographical text appears to have been aimed. As we shall see, there are other grounds for supposing that the *Cosmographicus liber* strengthened the association of cosmography with dialling. But those reasons emerge most clearly when the status of instruments in the work is considered more generally.

The argument that a great many sixteenth-century cosmographers were active in the manufacture of sundials and/or the production of associated texts also seems problematic when inspected more closely. This argument holds most force precisely *if* we elect to employ the category of 'cosmographer' to describe these individuals active in

---

22 Eagleton, *Monks, Manuscripts, and Sundials*, p. 3. See, on Ptolemy's work, O. Neugebauer, 'Mathematical Methods in Ancient Astronomy', *Bulletin of the American Mathematical Society*, 54 (1948), 1013–41, on pp. 1030–4. As Neugebauer notes, the application of an analemma to the construction of sundials had previously been described in the ninth book of Vitruvius's *De Architectura*.

23 C. Schöner, *Mathematik und Astronomie an der Universität Ingolstadt im 15. und 16. Jahrhundert* (Berlin: Dunker & Humblot, 1994), pp. 233–84, 358–64.

24 D. Hayton, *The Crown and the Cosmos: Astrology and the Politics of Maximilian I* (Pittsburgh: University of Pittsburgh Press, 2015), pp. 71–7, especially p. 76. There are several copies of Stiborius's *Canons* in the Munich Bayerische Staatsbibliothek, including MSS. Clm 19689, Clm 24103, and Clm 24105. The latter, which also includes a copy of the text describing the *organum ptolomei*, is listed in Eagleton, *Monks, Manuscripts, and Sundials*, pp. 267–8.

dialling. But a more parsimonious explanation is suggested by the fact that both cosmography and gnomonics were mathematical pursuits. If these 'cosmographers' are instead identified as mathematical authors, professors of mathematics, mathematical instrument-makers, practical mathematicians – or, if we wish to employ the term, mathematical practitioners – then it hardly seems surprising that their interest in mathematics should have led them to undertake work in both of these fields. Indeed, the fact that many authors of cosmographic texts *also* wrote texts on sundials itself seems to suggest that cosmography and dialling were considered separate (or at least separable) forms of mathematical expertise. Oronce Finé's *De solaribus horologiis et quadrantibus* first appeared in his *Protomathesis* of 1532 *alongside*, not *within*, his treatment of cosmography.[25] And the dialling texts of the polymathic Sebastian Münster, including *Erklerung des newen Instruments der Sunnen* (1528) and *Compositio horologiorum* (1531), were likewise published separately from his encyclopaedic *Cosmographia* of 1544.[26] The genres were distinct.

That is not to say, however, that the category of 'cosmographer' is anachronistic or redundant. On the contrary, some individuals were indeed identified as 'cosmographers' in the long sixteenth century, and some of them either consciously embraced that label as a professional identity or had it thrust upon them. However, the role of the cosmographer, and the tasks that it entailed, varied from place to place in the period.[27] One such individual known to have constructed dials, Egnazio Danti, illustrates one form of cosmographical practice. Danti served first as 'ducal cosmographer' to Cosimo I de' Medici and later as 'papal cosmographer' to Gregory XIII – although whether these designations represent *descriptions* of his service to these princes, or *offices* that he occupied, is not entirely clear. In either case, whilst occupying these roles Danti was principally engaged in the design and production of lavishly painted maps and globes, created to adorn spaces within the Palazzo Vecchio in

25 O. Finaeus, *Protomathesis* (Paris: impensis Gerardi Morrhii, & Ioannis Petri, 1532), fols. 101r–156v, 157r–207r.

26 S. Münster, *Erklerung des newen Instruments der Sunnen, nach allen seinen Scheyben und Circkeln* (Oppenheim: Jakob Köbel, 1528); S. Münster, *Compositio horologiorum* (Basel: Heinrich Petri, 1531); and S. Münster, *Cosmographia: Bschreibung aller Lender* (Basel: Heinrich Petri, 1544).

27 See A. Mosley, 'The Cosmographer's Role in the Sixteenth Century: A Preliminary Study', *Archives internationales d'histoire des sciences*, 59 (2009), pp. 423–39.

Florence and in the Vatican.[28] He also designed and constructed other instruments, including astrolabes, sundials, and the anemoscope in the Vatican's Tower of the Winds.[29] For this reason, the production of such non-cartographic devices has sometimes been characterised as part of the task of the Renaissance cosmographer.[30] Yet given the opportunity to define cosmography in his *Le scienze matematiche ridotte in tavole* of 1577, produced when he was a professor of mathematics at Bologna, Danti did so quite narrowly: in Table XXIII, on the science of geography, he indicated that cosmography was the description of the Earth made with reference to the heavens (which is to say, using longitude and latitude), and of the heavens as well.[31] A different table entirely was devoted to 'Gnomonic Science', while the very first table in the text, 'Of the Division of the Mathematical Sciences' showed both gnomonics and geography as subjects subalternated to geometry.[32] In Danti's analysis, therefore, cosmography-geography and the theoretical understanding of sundials were related to one another as branches of mathematics, but were nevertheless distinct. Moreover, the practical activity of *making* dials, astronomical instruments, and other kinds of device was classified separately again, as a mechanical art, rather than a science.[33] Overall, the text strongly suggests that Danti himself did not consider everything that he did *whilst* a cosmographer as cosmographical, in the proper sense of that term, including the construction of dials.

28 F. Fiorani, *The Marvel of Maps: Art, Cartography and Politics in Renaissance Italy* (New Haven and London: Yale University Press, 2005); and M. Rosen, *The Mapping of Power in Renaissance Italy: Painted Cartographic Cycles in Social and Intellectual Context* (New York: Cambridge University Press, 2015).

29 F. Camerota, 'Egnazio Danti as a Builder of Gnomons: An Introduction', in Marco Beretta, Paolo Galluzi, and C. Triarco (eds.), *Musa Musaei: Studies on Scientific Instruments and Collections in Honour of Mara Miniati* (Florence: Olschki, 2005), pp. 93–115; and N. Courtwright, *The Papacy and the Art of Reform in Sixteenth-Century Rome: Gregory XIII's Tower of the Winds in the Vatican* (Cambridge: Cambridge University Press, 2003), especially pp. 28–32, 219–41. Given the longstanding association of wind-roses with cartography and navigation, and their presence in cosmographic textbooks, a strong *prima facie* case for considering the anemoscope a cosmographical instrument can also be made.

30 See, for example, Fiorani, *The Marvel of Maps*, pp. 41–51.

31 E. Danti, *Le scienze matematiche riddote in tavole* (Bologna: Appresso la Compagnia della Stampa, 1577), p. 44.

32 Danti, *Le scienze matematiche riddote in tavole*, p. 43: 'Tavola XXXI. Della Scienza Gnomonica'; pp. 2–3: 'Tavola Prima della Divisione delle Scient. Matematiche'.

33 Danti, *Le scienze matematiche riddote in tavole*, p. 3.

Other Cosmographical Instruments

If cosmography was not *always* identified with dialling, how then should we understand Sempill's claim that dials were cosmographical devices? It is helpful to recall that the text with which we began referred to 'Sundials and *other* cosmographical instruments', and to consider what that category meant to Sempill more generally. Sempill's chapter on the topic first identified waterclocks, sandglasses, and clocks driven by weights and cogs as mechanical devices, outside the realm of his discussion.[34] Amongst the 'scioteric' or 'gnomonic' devices which he considered cosmographical, however, and which told the time by shadows of sunlight or moonlight, or the observation of a star, he included astrolabes, pillar dials, astronomical rings, and quadrants. He abbreviated his discussion of other instruments 'by cosmographers, geographers, hydrographers, and astronomers', in recognition of the fact that there were too many to discuss.[35] But he identified amongst the principal ones celestial globes, armillary spheres, planispheres, and terrestrial globes.[36] Finally, having named a number of scholars who had 'left such devices to posterity', he listed 'various kinds of quadrants, radii, annula, cosmolabes, trigons, torqueta, mesolabes, mariners' compasses, azimuthal semicircles, parallactic rulers, armillaries, bipartite arcs, and sextants' as also belonging to this category.[37] He ended the chapter by promising to write of these instruments more fully in a forthcoming *Mathematical Dictionary* that he never actually produced.[38]

Just as with sundials, a *prima facie* case can be established for considering most of the instruments listed by Sempill as cosmographical devices. Nevertheless, it is conceivable that Sempill's understanding of the different sub-disciplines of mathematics, the relationship between them, and the status of their instruments, was idiosyncratic. In other words, Sempill's classification of sundials and other instruments as cosmographical may well have reflected his *own* understanding of this category, developed from first principles,

---

34 Sempilius, *De Mathematicis Disciplinis*, p. 226.
35 Sempilius, *De Mathematicis Disciplinis*, p. 227: 'a Cosmographis, Geographis, Hydrographis & Astronomis'.
36 Sempilius, *De Mathematicis Disciplinis*, pp. 227–8.
37 Sempilius, *De Mathematicis Disciplinis*, p. 228: 'posteris reliquerunt'; 'varia genera quadrantum, radiorum, annulorum, cosmolabiorum, trientium, torquetorum, mesolabiorum, nauticarum pyxidem, semicirculorum azimuthalium, regularum parallacticarum, armillarum, arcuum bipartitorum, sextantum'.
38 Sempilius, *De Mathematicis Disciplinis*, p. 228.

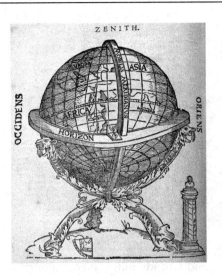

Figure 3.4 The 'cosmographical globe', with pillar dial, horary quadrant, and diptych compass dial, from Peter Apian and Gemma Frisius, *Cosmographia* (Antwerp, 1584), p. 46. Image © Whipple Library (95:50).

rather than a view that was generally held. However, as a Scottish Jesuit, based in Madrid, and publishing in Antwerp, the print capital of the Spanish Netherlands, Sempill had access to multiple forms of cosmographical tradition. And that he did indeed draw on a wide range of cosmographical texts and works by cosmographers is revealed by the indices of authors, ancient and modern, with which he furnished his book.[39] Consideration of just a few of his sources suggests why it would have been easy for him to identify such a wide range of instruments as cosmographical in kind.

Peter Apian was one of those cited by Sempill; and, of course, his *Cosmographicus liber* was frequently published in Antwerp from 1529 onwards, in multiple languages.[40] Even before it was augmented by Frisius, this work referenced numerous instruments, in both images and text. Two of these devices were explicitly labelled cosmographical: the 'cosmographical globe' (Figure 3.4) and the *speculum cosmographicum,* or 'cosmographical mirror' (Figure 3.5).[41] The former was the name given by Apian to the terrestrial globe divided by longitude and latitude; in most editions of the work, it was depicted on the title page, as well as in the chapter in which it was named and discussed. The *speculum cosmographicum* was a paper volvelle, pre-assembled in the book. Taking the form of a terrestrial planisphere surmounted with a rotatable ecliptic ring,

39 Sempilius, *De Mathematicis Disciplinis*, pp. 262–310.
40 F. van Ortroy, *Bibliographie de l'œuvre de Pierre Apian* (Amsterdam: Meridian Publishing, 1963), pp. 29–68.
41 Apian and Frisius, *Cosmographicus liber*, fols. XXIIIr–XXIIIIr, XXXIr–XXXIIr.

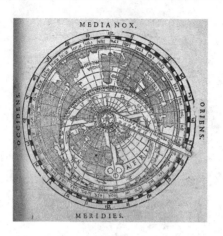

Figure 3.5 Peter Apian's *speculum cosmographicum* (or cosmographic mirror), from Peter Apian and Gemma Frisius, *Cosmographia* (Antwerp, 1584), p. 65. Image © Whipple Library (95:50).

hour circle, and latitude index, it was superficially very like a standard planispheric astrolabe, but furnished with a geographical rather than a celestial latitude plate. Any reader of the *Cosmographicus liber* attentive to these names might conclude that the text referenced precisely two kinds of cosmographical instrument, the globe and the *speculum*. These were indeed instruments particularly distinguished by the celestial and terrestrial divisions that they physically embodied, in quite a straightforward way.

Equally, however, the presence in Apian's text of a wider range of devices might lead a reader to suppose that other instruments could also be considered cosmographical in certain contexts of use. The work also contained an illustration of an armillary sphere, with a terrestrial globe divided by lines of longitude and latitude clearly visible at its centre.[42] It presented to the reader other paper devices, including the *instrumentum theoricae solis* – essentially a printed rendering of the reverse of a planispheric astrolabe, with eccentric calendar scales that could, in theory, be used to locate the place of the sun in the zodiac on any day of the year.[43] It discussed the use of the astronomical staff.[44] And it depicted a magnetic compass, and discussed its use, in more than one place in the text. Importantly, for the supposition that the work encouraged the association between cosmography and sundials, the cosmographical globe was typically depicted with three other devices: a pillar dial, an horary quadrant, and – resting on the horizon ring of the globe – a diptych compass

---

42 Apian and Frisius, *Cosmographicus liber*, fol. Vv.
43 Apian and Frisius, *Cosmographicus liber*, fol. Xv.
44 Apian and Frisius, *Cosmographicus liber*, fols. XVv–XVIv.

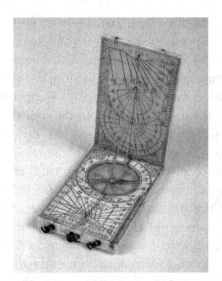

Figure 3.6 Ivory diptych sundial by the Nuremberg maker Johann Gebhert, 1556. Image © Whipple Museum (Wh.1681).

dial of the type for which Nuremberg was particularly well-known (Figure 3.6).[45] The compass dial was explicitly discussed in connection with the globe, as Apian explained how to use it to establish the meridian line and set the globe for a particular location.[46] Further devices were discussed in appendices and additions to the text; these varied as the book went through its numerous editions, but among them were the nocturnal and the ring-dial, or *annulus astronomicus*.[47] Any of these instruments could, surely, have been considered cosmographical by a reader of this text.

In the wake of Apian, and in the context of the increasing dissociation of cosmography from Ptolemy's *Geography*, variant forms of cosmographical authorship emerged. A textbook tradition developed that increasingly identified cosmography with the mathematical intersection of astronomy and geography. Since that intersection was largely co-extensive with the contents of treatises on the celestial sphere, these textbooks were commonly titled in such a way as to identify themselves as works in cosmography *or* spherical astronomy. Thus Finé's 1532 work was the *De cosmographia sive mundi sphaera*; Antoine Mizauld published *De mundi sphaera sive cosmographia* in 1552, also in Paris; Thomas Blundeville's *Exercises*

45 Apian and Frisius, *Cosmographicus liber*, fol. XXIIIIr. On Nuremberg diptych dials, see P. Gouk, *The Ivory Sundials of Nuremberg 1500–1700* (Cambridge: Whipple Museum of the History of Science, 1988).
46 Apian and Frisius, *Cosmographicus liber*, fol. XXVIr.
47 Apian and Frisius, *Cosmographicus liber*, fol. LIIIIv; for the *annulus astronomicus*, see P. Apian and G. Frisius, *Cosmographia* (Antwerp: Aegidius Coppenius, 1539), fol. LIIIr.

of 1594 included *A Plaine Treatise of the First Principles of Cosmo-graphie, and Specially of the Spheare*; and Rudolf Goclenius produced *Cosmographiae seu sphaera mundi descriptionis, hoc est astronomiae et geographiae rudimenta* in Marburg in 1599. All four of these authors were cited by Sempill.[48] Naturally their texts, and others like them, gave warrant to the idea that armillary spheres were cosmographical instruments; like their medieval exemplar, Sacrobosco's *De sphaera*, they commonly invoked the physical instrument (or an image of the instrument) as a pedagogical tool.[49]

More ambitious mathematical authors were able to enlarge their cosmographical textbooks by extending their coverage to include planetary astronomy as well. Several authors of such works, including Francesco Maurolico, Francesco Barozzi, and the Jesuit Giuseppe Biancani (Blancanus), were likewise cited by Sempill.[50] Of course, many of the textbook cosmographies included geographical as well as astronomical material. But rather different in kind were the encyclopaedic cosmographies prepared by Sebastian Münster and his French imitator, André Thevet, who also featured in Sempill's lists.[51] Their texts were descriptive geographies, incorporating both human and natural history, in the style of Strabo and Pliny the Elder, rather than Ptolemy.[52] The mathematics of the sphere that underpinned coordinate-based geography was present in these texts, but served as the very shallow foundation to a much more elaborate superstructure that prioritised detailed accounts of places, peoples, events, and resources. A third kind of cosmographical work was the *Atlas*, a genre inaugurated in 1595 by Gerard Mercator, another author acknowledged by Sempill.[53] Mercator's posthumously published text was only a partial realisation of his vision of cosmography, which encompassed a causal understanding

48 Sempilius, *De Mathematicis Disciplinis*, pp. 278–9.

49 Mosley, 'Spheres and Texts on Spheres', 313–17; Mosley, 'Objects of Knowledge', 211–14.

50 F. Maurolico, *Cosmographia* (Venice: apud haeredes Luca Antonio Iunta, 1543); F. Barozzi, *Cosmographia* (Venice: Gratiosus Perchachinus, 1585); and J. Blancanus, *Sphaera mundi, seu Cosmographia* (Bologna: Sebastianus Bonomius, 1620). For the citations, see Sempilius, *De Mathematicis Disciplinis*, pp. 278–79, 295.

51 Sempilius, *De Mathematicis Disciplinis*, pp. 279, 286.

52 Münster, *Cosmographia*, is discussed in M. McLean, *The Cosmographia of Sebastian Münster: Describing the World in the Reformation* (Aldershot: Ashgate, 2007). On A. Thevet, *La Cosmographie Universelle* (Paris: Guillaume Chaudiere, 1575), see F. Lestringant, *André Thevet: Cosmographe des derniers Valois* (Geneva: Droz, 1991), pp. 231–5.

53 Sempilius, *De Mathematicis Disciplinis*, p. 286.

of Creation, as well as mathematically founded descriptions of heaven and earth, and accounts of human history.[54] Beginning with a natural philosophy based on an exegesis of Genesis, this work suggested that cosmography was much more than the intersection of astronomy and geography required to produce the coordinate maps with which it was lavishly endowed. Indeed, this was the common characteristic of the extended textbooks, encyclopaedic cosmographies, and the *Atlas*: they pushed the boundaries of cosmography towards something more akin to the *sum* of astronomy and geography, although the precise dimensions of those components varied from case to case. Presumably, in the process, they could have authorised readers such as Sempill to consider astronomical *and* cartographic and surveying instruments as, in some sense, cosmographical.

A handful of authors cited by the Scottish Jesuit published works that explicitly referenced cosmographical instruments in their titles. Johannes Stoeffler's posthumous *Cosmographicae aliquot descriptiones* of 1537 referred to a *sphaera cosmographica*, meaning the terrestrial globe.[55] And Giovanni Paolo Gallucci published his *Della fabrica et uso di diverso stromenti di Astronomia et Cosmografia* in 1597 and 1598. This quarto text of 293 pages was itself a work of compilation, one that discussed the construction and use of twenty-seven distinct devices, including the planispheric astrolabe, the Rojas universal astrolabe, the astronomical staff of Peter Apian, the *specchio geografico* of Apian (i.e., his *speculum cosmographicum*), the horary quadrant of Johannes Stoeffler, the *annulus astronomicus* or ring-dial of Gemma Frisius, and several of the observing instruments described by Ptolemy in the *Almagest*.[56] This work too, therefore, offered contemporary readers considerable latitude in the identification of cosmographical instruments.

As important as any of these sources to Sempill, however, was surely the cosmographical tradition native to his adopted homeland

---

54 G. Mercator, *Atlas sive cosmographicae meditationes de fabrica mundi et fabricati figura* (Duisburg: Albert Buyss, 1595); see also P. van der Krogt, 'Gerard Mercator and His Cosmography: How the *Atlas* Became an Atlas', *Archives internationales d'histoire des sciences*, 59 (2009), pp. 465–83.

55 J. Stoeffler, *Cosmographicae aliquot descriptiones ... De Sphaera Cosmographica, hoc est de Globi terrestris, artificiosa structura* (Marburg: Eucharius Cervicornus, 1537). See Sempilius, *De Mathematicis Disciplinis*, p. 279.

56 See, for the full list of instruments it discusses, G. P. Gallucci, *Della fabrica et uso di diversi stromenti di Astronomia et Cosmografia* (Venice: Ruberto Meielti, 1597), sig. b4r. For the citation, see Sempilius, *De Mathematicis Disciplinis*, p. 284.

of Spain. In the sixteenth century cosmographers and professors of cosmography, identified and salaried as such, were appointed by Philip II to serve at the Casa de Contratación in Seville, its governing body, the Council of the Indies, and the Royal Mathematical Academy in Madrid.[57] The Jesuits of the Imperial College of Madrid, where Sempill taught, were quite literally the heirs to the cosmographic practice of the latter institution, which had been established in the 1570s to provide instruction in the mathematical sciences. After the death in 1625 of its last professor of cosmography, Juan Cedillo Díaz, responsibility for delivering its lectures, and subsequently the classes themselves, were transferred to the Jesuits' *Collegio*.[58] Sempill acknowledged the relationship in the dedication of his work to Philip IV, writing that the twelve books had been conceived in the sand of the Royal Academy, 'amidst the sphere and cylinder of Archimedes'.[59]

Cosmography in Iberia in the long sixteenth century was closely concerned not only with coordinate-based mapping, but also with mathematical techniques of navigation. Thus, cosmographers at the Casa de Contratación (or House of Trade) in Seville were charged with maintaining and improving the *padrón real*, the master map used to produce navigational charts for ships bound for the Indies. They were also responsible for producing and checking those derivative charts, constructing and certifying navigational instruments, and training ships' captains and pilots in navigational techniques. And they played a part, alongside the Casa's chief pilot, in examining those who aspired to occupy these crucial shipboard roles.[60] Cosmography was in this way institutionalised in Spain as a cluster of activities, dependent upon mathematical expertise, that were intimately associated with the practical problems of travelling to, and managing, overseas territories. The subject was similarly construed in Portugal, Spain's principal rival in the early years of the European expansion, but also a source of cosmographic expertise for

---

57 For a convenient summary, see M. E. Piñeiro, 'Los cosmógrafos del Rey', in A. Lafuente and J. Moscoso (eds.), *Madrid, ciencia y corte* (Madrid: Consejo Superior de Investigaciones Científicas, 1999), pp. 121–33. Other treatments include the essays collected in U. Lamb, *Cosmographers and Pilots of the Spanish Empire* (Aldershot: Variorum, 1995); and M. M. Portuondo, *Secret Science: Spanish Cosmography and the New World* (Chicago: University of Chicago Press, 2009).

58 Piñeiro, 'Los cosmógrafos del Rey', p. 133.

59 Sempilius, *De Mathematicis Disciplinis*, sig. *2r: 'inter sphaeram Archimedis & cylindrum'.

60 Piñeiro, 'Los cosmógrafos del Rey', pp. 123–5.

the Spanish court, especially during the Iberian Union from 1580 to 1640.[61] It is hardly surprising, therefore, that Sempill, writing not only as a Jesuit mathematician but also as a scholar embedded in the mathematical culture of Iberia, should have associated hydrography and instruments such as the mariner's compass with the practice of cosmography.

Because individuals could be and were appointed to cosmographical offices in Iberia in the long sixteenth century, a wider range of texts and instruments that they produced could perhaps be identified as cosmographical, both then and now, than is the case for some mathematicians working in other parts of Europe. Many Iberian cosmographers identified themselves as such in their writings as well as on their maps, although given the Spanish Crown's proprietary attitudes to knowledge that might be of use to its European rivals, these texts and charts were not printed as frequently as similar items created elsewhere.[62] Cosmographers in the service of the monarch produced translations of works by mathematical authorities, such as Euclid, treatises on the sphere, works on the art of navigation, sailing instructions (rutters), and *repertorios de los tiempos* – texts that treated calendrics and meteorology in forms intended to be particularly useful to mariners.[63] And they also wrote on particular instruments, including globes, cross-staves, mariner's astrolabes and compasses. Naturally, Sempill referenced many Spanish and Portuguese cosmographers in his indices of authors, including Pedro Nuñez, Alonso de Chaves, and Rodrigo Zamorano.[64]

Iberian cosmographers cited by Sempill even included references to dials in their works. Alonso de Chaves, for example, devoted a chapter of the second book of his manuscript *Quatri partitu en cosmografía práctica* to discussion of the universal dial of Regiomontanus, a close cousin to the dial incorporated by Apian into the pages of the *Cosmographicus liber*.[65] Sempill's inclusion of 'sundials'

---

61 V. N. Brótons, 'Astronomy and Cosmography 1561–1625: Different Aspects of the Activities of Spanish and Portuguese Mathematicians and Cosmographers', in L. Saraiva and H. Leitão (eds.), *The Practice of Mathematics in Portugal* (Coimbra: Imprensa da Universidade de Coimbra, 2004), pp. 225–74.

62 This is one of the themes of Portuondo, *Secret Science*.

63 See U. Lamb, 'The Teaching of Pilots and the *Chronographía o Repertorio de los tiempos*', in her *Cosmographers and Pilots of the Spanish Empire* (Aldershot: Variorum, 1995), pp. 1–17.

64 Sempilius, *De Mathematicis Disciplinis*, pp. 279, 291, 292.

65 P. C. Delgado, M. C. Domingo, and P. H. Aparicio (eds.), *Quatri partitu en cosmografía práctica, y por otro nombre, Espejo de navegantes* (Madrid: Instituto de Historia y Cultura Naval, 1983), pp. 160–2.

alongside 'other cosmographical instruments' is, therefore, not so difficult to comprehend given his access to the Iberian tradition and his manifest eclecticism. Yet the fact remains that his particular construal of the category of 'cosmographical instrument' is unusual. The point that Sempill's text appears to support, therefore, is not that sundials *were* cosmographical instruments in the long sixteenth century, but that they *could be* considered so – and the same applies to many of the other devices that historians have generally come to refer to as mathematical. Some of these objects were considered cosmographical by *some* makers, and by *some* authors and (we must assume) consumers of instrument literature, but the Renaissance category of cosmography was sufficiently protean to admit of many variant usages. Cosmography was not one thing, but many, and one and the same practice or instrument could be considered 'cosmographical' or not by different scholars, depending upon their personal preferences, the context in which they were operating, and the tradition or traditions with which they were familiar.

The Decline of Cosmography Revisited

The claim that cosmography disappeared shortly after 1600 has become established in the scholarly literature treating cosmography in the long sixteenth century. But this claim has also been advanced, and then repeated, without sufficient attention being paid to the full variety of cosmography's forms. Perhaps the clearest articulation of the thesis of cosmography's demise has been offered by Frank Lestringant.[66] He, however, was thinking particularly of the encyclopaedic form of the subject, as represented by the cosmographies of Münster, François de Belleforest, and André Thevet. While his account acknowledges the existence of one other genre of cosmographic authorship, the atlas, it neglects the mathematical textbook tradition and at the same time characterises the practical application of cosmographical learning to navigation and chartmaking, as witnessed particularly in Iberia, as somehow so technically difficult as to have rendered it incoherent and therefore unstable.[67] Encyclopaedic

---

66 F. Lestringant, 'The Crisis of Cosmography at the End of the Renaissance', in P. Desan (ed.), *Humanism in Crisis: The Decline of the French Renaissance* (Ann Arbor: University of Michigan Press, 1991), pp. 153–79. Lestringant's analysis finds an echo in the closing chapter of Portuondo, *Secret Science*, entitled 'Cosmography Dissolves'.

67 Lestringant, 'The Crisis of Cosmography at the End of the Renaissance', pp. 159–67.

universal cosmographies do seem to have fallen out of fashion after 1600, but perhaps not so abruptly as Lestringant supposes or for the reasons he supplies. In his analysis, the expiry of descriptive cosmography was especially associated with the hubris of his particular subject, the French royal cosmographer André Thevet, in whose *Cosmographie universelle* of 1575 eye-witness claims both clashed with received religious truths and stretched to breaking point his readers' credulity.[68] Despite this 'crisis', the last early modern edition of Münster's vast *Cosmographia* was published as late as 1628, while Peter Heylyn's eminently encyclopaedic *Cosmographie in Four Bookes* was published in London in 1652 and again in 1657.[69] Encyclopaedic cosmography persisted, therefore, at least until the middle of the seventeenth century.

Other forms of cosmography lasted even longer. Mercator's *Atlas* of 1595 was just the first of a series of cosmographic atlases published late into the seventeenth century in the Low Countries, by Jodocus Hondius, Johannes Janssonius, Willem Janszoon Blaeu, and their heirs and successors.[70] Vincenzo Maria Coronelli embraced the identity of 'cosmographer', and published works including the *Atlante Veneto* under the auspices of an Accademia cosmografica degli Argonauti – not so much a learned society as a way of publishing by subscription – in the 1690s.[71] A Kosmographische Gesellschaft was set up in Nuremberg in the late 1740s, again with the objective of facilitating the publication of various cartographical products.[72] A similarly named Cosmografiska Sällskapet, or Cosmographical Society, was established in Uppsala in 1758 and prioritised the production of geographical textbooks. Immanual Kant drew upon a German translation of one of these, Tobern Bergman's *Physik beskrifsning öfver jordlokot* (1766), in the lectures on physical

---

68 Lestringant, 'The Crisis of Cosmography at the End of the Renaissance', pp. 168–72. See also Lestringant, *André Thevet*, pp. 231–5.

69 P. Heylyn, *Cosmographie in Foure Bookes Contayning the Chorographie & Historie of the Whole World, and All the Principall Kingdomes, Provinces, Seas, and Isles, Thereof* (London: Henry Seile, 1652). Heylyn's work is discussed in R. J. Mayhew, '"Geography is Twinned with Divinity": The Laudian Geography of Peter Heylyn', *Geographical Review*, 90 (2000), pp. 18–34.

70 See P. van der Krogt, *Koeman's Atlantes Neerlandici*, 4 vols. to date ('t goy-Houten: HES & De Graaf, 1997–), vols. I–III.

71 M. Milanesi, *Vincenzo Coronelli Cosmographer (1650–1718)* (Turnhout: Brepols, 2016), especially pp. 317–42.

72 E. G. Forbes, 'Mathematical Cosmography', in G. S. Rousseau and R. Porter (eds.), *The Ferment of Knowledge* (Cambridge: Cambridge University Press, 1980), pp. 417–48.

geography that he gave at the University of Königsberg.[73] Globes, as well as texts, can be associated with these Enlightenment forms of cosmography: Coronelli, of course, was the pre-eminent globemaker of his day; the Nuremberg Gesellschaft produced globes, including a lunar globe designed by Tobias Mayer; and, from 1762 onwards, Anders Åkerman published globe pairs as part of the Uppsala cosmographical enterprise.[74]

Both the textbook and the Iberian traditions in cosmography also persisted long after the subject's supposed demise. Cosmographers continued to be appointed at the Casa de Contratación until the early eighteenth century, and Juan Baptista Mayor was appointed *cosmografo mayor* of the Indies as late as 1770.[75] More surprisingly, textbook treatments of cosmography continued to be produced in the nineteenth and even the twentieth centuries, particularly in France, Portugal, Spain, and former Spanish colonies. Mostly aimed at schoolchildren or recipients of a technical education, rather than those in higher education, the common starting point of these works was the celestial sphere and its projection onto the surface of the Earth in order to generate the terrestrial divisions of longitude and latitude. Examples include Auguste Tissot's *Précis de cosmographie* (1869), Manuel Burillo Stolle's *Elementos de cosmografía y nociones de física del globo* (1903), and António Barbosa's *Elementos de cosmografia* (1926).[76] The latter, a Portuguese-language text addressed to secondary-school educators, is especially noteworthy,

73 A. Buttimer and T. Mels, *By Northern Lights: On the Making of Geography in Sweden* (Aldershot: Ashgate, 2006), p. 19; and C. W. J. Withers, 'Kant's Geography in Comparative Perspective', in S. Elden and E. Mendieta (eds.), *Reading Kant's Geography* (Albany: SUNY Press, 2011), pp. 47–65, especially p. 58.

74 Milanesi, *Vincenzo Coronelli Cosmographer*, pp. 47–181; G. Oestmann, 'Der Mondglobus der Tobias Mayer', *Der Globusfreund*, 47/48 (1999), 221–8; and E. O. Bratt, 'Anders Åkerman: Ein schwedischer Globenmacher des 18. Jahrhunderts', *Der Globusfreund*, 9 (1960), 8–12. For an example of an Åkerman instrument, see the 1766 celestial globe in the collections of the Sjöhistoriska Museet, Stockholm which bears a cartouche reading 'Globus Coelestis . . . Cura Soc. Cosmogr. Upsal. delineatus ab Andrea Akerman'; images are available at https://digitaltmuseum.se/021025649620/himmelsglob.

75 J. Pulido Rubio, *El piloto mayor de la Casa de la Contratación de Sevilla: Pilotos mayores, catedráticos de cosmografía y cosmógrafos* (Seville: Escuela de Estudios Hispano-Americanos de Sevilla, 1950), pp. 981–3; and N. B. Martín, 'Juan Bautista Muñoz y la Sevilla del siglo XVIII', *Anales de la Real Sociedad Económica de Amigos del País de Valencia* (2001), pp. 902–9, on p. 903.

76 A. Tissot, *Précis de cosmographie* (Paris: Victor Masson et fils, 1869); M. Burillo Stolle, *Elementos de cosmografía y nociones de física del globo* (Madrid: Jaime Ratés, 1903); and A. Barbosa, *Elementos de cosmografia* (Coimbra: Imprensa da Universidade, 1926).

partly because its author recommended the construction and use of instruments as the best way to teach cosmography to pupils in the seventh grade.[77] Amongst the devices he discussed were the cross-staff, the mariner's astrolabe, and, indeed, the sundial. Barbosa explicitly invoked Portugal's long history of cosmographic excellence and the Renaissance textbook tradition in his work, citing both the *Tratado da sphera* (1537) of the Portuguese *cosmógrafo-mor* Pedro Nuñez, and a Spanish-language edition of Apian and Frisius's *Cosmographia* from 1575.[78] Twentieth-century school texts in *some* languages and cultures, therefore, not only presented cosmography in ways that seem remarkably similar to some sixteenth-century treatments of the subject, but might even have done so consciously.

Historians have not been entirely wrong to claim that cosmography declined after 1600, or to suggest that the category was increasingly displaced by those of astronomy and geography.[79] But the decline was not absolute, and it proceeded at different rates with respect not only to the various genres of cosmographic work, but also to different languages. In English, it seems clear, the terms 'cosmography', 'cosmographer', and 'cosmographic' were much less frequently employed after 1700 than cognate terms such as 'astronomy' and 'geography', and were rarely used post-1800 except by historians.[80] But in French, Spanish, and Portuguese, the category retained some currency well past this threshold. The eighteenth century was something of a transitional period, in which groups that sought to identify themselves with a cartographic tradition uniting astronomical and geographical techniques and products continued to do so under the aegis of 'cosmography' in Italian, German, and Swedish. Yet this was probably a deliberate decision, authorised by and alluding to the practices of the past, rather than merely a reflection of enduring preferences within those language

---

77 Barbosa, *Elementos de cosmografia*, pp. v–viii.
78 Barbosa, *Elementos de cosmografia*, p. 27.
79 Bennett's objection to this idea, that astronomy was *already* established as an independent discipline prior to the rise and decline of cosmography, can be answered by pointing to the convergence of cosmography and *spherical* astronomy noted above. See Bennett, 'Sundials and the Rise and Decline of Cosmography in the Long Sixteenth Century', p. 9.
80 A crude quantitative analysis is possible via the corpora *Early English Books Online* and *Eighteenth Century Collections Online*, accessed using the JISC Historical Texts interface. For eighteenth-century usage, for example, I searched ECCO and ECCO II for 'cosmograph*', and found 1,790 instances, with just three distinct works using the search term in their title. For 'astronom*' the same process resulted in 15,312 hits and 670 titles; for 'geograph*' there were 35,883 hits and 1,024 titles. These searches were undertaken on 15 June 2018.

Figure 3.7 The 'English Globe' designed by the Earl of Castlemaine and Joseph Moxon, 1679: a terrestrial globe set stationary above a celestial planisphere. Image © Whipple Museum (Wh.1466).

communities. Cosmography remained, as it had been in parts of Europe in the long sixteenth century, indeed, a discipline and practice that individuals might *choose* to identify with if their allegiance to it was not already determined by their context.

This characteristic of 'cosmography' helps to explain why *prima facie* analysis of the qualities of particular classes of instrument can be such an unreliable predictor of their categorisation by our historical subjects. Globe-pairs, and devices like the 'English' or Castlemaine globe devised in 1679 (Figure 3.7), which set a terrestrial globe over a celestial planisphere, might seem quintessentially cosmographical, combining as they do representations of the Earth, divided by longitude and latitude, and depictions of the heavens. And indeed, paired celestial and terrestrial globes were sometimes produced by makers who identified as cosmographers, such as Mercator and Coronelli, and were described in avowedly cosmographical texts. At least as frequently, however, globe-pairs were identified as celestial and terrestrial, or astronomical and geographical, without invoking cosmography explicitly as a category.[81] One and the same

---

81 See, for example, T. Hood, *The Use of Both the Globes, Coelestiall and Terrestriall* (London: Thomas Dawson, 1592). On sig. Kr of this text, Hood refers to 'Ptolemee, and the ancient Cosmographers', but he uses the categories of geography and astronomy much more liberally elsewhere in the text; he was therefore familiar with the concept of 'cosmography' but did not employ it as an overarching category to frame his treatment of globes.

instrument might therefore be understood as cosmographical or not, depending on the context on which it was produced and discussed. The Earl of Castlemaine, for example, introduced the globe he devised with Joseph Moxon without reference to cosmography; but Coronelli considered this *globo inglese* in his *Epitome cosmografica* of 1693.[82] Like sundials, therefore – and authors of globe manuals often emphasised the capacity of globes to solve problems in dialling – such devices *could be* considered cosmographical instruments, but weren't necessarily labelled as such, and this situation persisted for many years after the long sixteenth century.[83]

Conclusions

The historical usage of words cannot always be sharply distinguished from either historians' uses or the recoinages of present-day practitioners. In the nineteenth century, Alexander von Humboldt, who had published on the history of the New World discoveries and nautical astronomy, asserted in the French edition of his *Kosmos* that 'cosmography' was the proper title for this work.[84] Evidently, this was a use of the term informed by Humboldt's knowledge of the category's deep history. But the more-or-less transparent etymology of 'cosmography', readily understood to mean the description or depiction of the universe as a whole, has also allowed it to be periodically reintroduced without apparent reference to the past.[85]

82 [R. Palmer] Earl of Castlemaine, *The English Globe* (London: Joseph Moxon, 1679); and V. Coronelli, *Epitome cosmografica* (Cologne: Andrea Poletti, 1693), pp. 325–30. The Castlemaine globe is discussed, in its English context, in K. de Soysa, 'On the Use of the Globe: The Earl of Castlemaine's English Globe and Restoration Mathematics', unpublished MPhil thesis, University of Cambridge (2000).

83 On globes as cosmographical problem-solving devices, see E. Dekker, 'The Doctrine of the Sphere: A Forgotten Chapter in the History of Globes', *Der Globusfreund*, 49/50 (2002), pp. 25–44. Problem-solving using globes in seventeenth-century England has also been treated in K. de Soysa, 'Using Globes and Celestial Planispheres in Restoration England', unpublished PhD thesis, Cambridge University (2004), pp. 35–62 and Appendix 1, pp. 192–227.

84 A. von Humboldt, *Examen critique de l'histoire de la géographie du Nouveau Continent et du progrès de l'astronomie nautique aux quinzième et seizième siècles*, 4 vols. (Paris: Gide, 1836–9); and A. von Humboldt, *Cosmos, essai d'une description physique du monde*, trans. H. Faye, 4 vols. (Paris: Gide et Companie, 1846), vol. 1, p. 67: 'l'ouvrage que je publie devrait avoir la titre de Cosmographia'. That this passage is absent from the earlier German edition is itself suggestive of the greater currency of 'cosmography' in French than in German.

85 See, for example, S. Weinberg, *Gravitation and Cosmology: Principles and Applications of the General Theory of Relativity* (New York: Wiley, 1972),

Such fresh uses might themselves be considered part of a complete history of cosmography.

Problems arise, however, when scholars choose to convert historical categories into terms of their art, not in order to *capture* past usages but so as to overwrite them as a matter of analytic convenience. Such was the case when, for example, some historians chose to treat 'cosmography' simply as a synonym for 'geography' – overlooking, thereby, the substantial component of Renaissance cosmography that was, in fact, astronomical or navigational in kind. 'Geography' could then displace 'cosmography' as the object of study of such historians, contributing to the myth of the latter category's demise.[86] Such too would be the difficulty with Matthew Edney's attempt to rehabilitate 'mathematical cosmography' to capture the relations between astronomy, geography, and surveying in British cartography of the late eighteenth and early nineteenth centuries.[87] The category may have been virtually defunct in British scholarly discourse of this era but, as we have seen, it retained its currency elsewhere. Students of European cartography would, if this proposal were adopted, struggle unnecessarily to distinguish between the usages of their historical subjects and those of historians.

Of course, the writing of history involves translation of past concepts and terms into ones that can be comprehended by modern-day audiences. It would therefore be counterproductive, even were it possible, to insist that historians only ever used actors' categories as the actors did themselves.[88] But if we wish to understand the full range of meanings that past individuals associated with the subject and practice of cosmography, we must be sufficiently respectful of our subjects' uses of 'cosmographer', 'cosmography', and 'cosmographical' to notice that their application of these terms varied individually, by context, by tradition, by genre, and by language. At *every* point in its post-1400 history, the category of 'cosmography' was used alongside,

pp. 407–9; and M. Visser, 'Cosmography: Cosmology without the Einstein Equations', *General Relativity and Gravitation*, 9 (2005), 1541–8.

86 Historians of geography, of course, have been particularly prone to treat 'cosmography' in this way, just as historians of the mathematical sciences have tended to overlook its more encyclopaedic forms. See, for example, G. Kish (ed.), *A Source Book in Geography* (Cambridge: Harvard University Press, 1978), p. 350: 'cosmography (the term commonly used in the early 1500s to describe geography)'.

87 M. Edney, 'Mathematical Cosmography and the Social Ideology of British Cartography, 1780–1820', *Imago Mundi*, 46 (1994), pp. 101–16.

88 See N. Jardine, 'Uses and Abuses of Anachronism in the History of the Sciences', *History of Science*, 38 (2000), pp. 251–70, on p. 262.

sometimes as an umbrella term for, and sometimes in competition with, the alternative categories of 'astronomy' and 'geography'. Histories of *all* of these disciplines need to acknowledge the overlap in content of these categories and in the texts and the instruments that their practitioners used, discussed, and produced. But, if we elide them in the process we lose all sight of the different traditions in which our subjects worked, and the conscious choices they made about the disciplinary classifications they employed.

Jim Bennett's suggestion that sundials were 'cosmographical instruments' in the long sixteenth century was motivated, in part, by a desire to demonstrate that our understanding of dials' cultural significance is impoverished, and needs to be enhanced. By speculating that dialling was heir to the rich and interesting tradition of Renaissance cosmography, he offered a motivation for paying closer attention to the theory and use of these instruments. As we have seen, cosmography itself persisted long enough to be its own post-Renaissance heir. Nevertheless, Bennett's thesis retains much to commend it. Sundials *could be* considered 'cosmographical' in the long sixteenth century – and it may yet transpire that the treatment of sundials and dialling within cosmographical texts is even better documented in literature produced after 1650 than it is in that of the Renaissance.[89] Dialling may not be what the astronomical component of cosmography became. But histories of cosmography need to acknowledge the presence of sundials in *some* accounts of the subject. And the richer history of dialling that Bennett has envisaged will also need to accommodate cosmographical writings as one forum for the treatment of sundials, even after 1600. Dialling and cosmography are disciplines with *intertwined* histories, in other words, and there is much to be gained by studying them as such, provided that the distinctions between them are not lost in the process. Sundials were indeed *sometimes* considered cosmographical devices. But so too were many other species of mathematical instrument. Appreciating that fact will help historians to better understand the rich mathematical culture of the early modern period and, in turn, help curators to display and interpret their collections in ways that better communicate the significance of such objects to museum visitors.

---

89 For some further examples, see G. Gordon, *An Introduction to Geography, Astronomy, and Dialling* (London: A. Bettesworth, 1726), pp. 158–88, which describes itself as a 'Compendium of Cosmography' on p. 1; and C. Cornet, *Cosmographie et navigation, 1: Programme de capitaine et de l'élève de la marine marchande* (Paris: Gauthier-Villars, 1950), p. 75.

4   'That Incomparable Instrument Maker': The Reputation of Henry Sutton

JIM BENNETT

The London mathematical instrument maker Henry Sutton (c. 1624–65) has been recognised since his own time as one of the most skilled engravers in his trade in seventeenth-century England. His versatility allowed him to work directly on brass or on wood and also in reverse on a copper printing plate. Thus much of his surviving oeuvre is bound into books, although a number of his printed instruments have survived as single printed sheets, applied to a brass plate or more usually a wooden board.

The instruments of his preserved at the Whipple Museum are among those generally cited by collectors, curators, and instrument historians to justify a reputation that has continued to the present. Sutton's reputation is the theme of this chapter: how it was promoted and established in his lifetime, and how it survived him for a century or so, not simply for connoisseurs but for mathematical practitioners. The pioneering chronicler of these practitioners, Eva Taylor, offered a very fair assessment: 'one of the best known engravers of scales, quadrants, etc., of his day, was renowned for his accuracy and was in demand for drawing diagrams for mathematical books'.[1] Engraving skill, accuracy, and books were pillars of Sutton's work, and this account of the renown it achieved will be intertwined with a consideration of his instruments, specifically the horary quadrants.

Sutton made a great variety of mathematical instruments, and seems to have relished those requiring sets of engraved projection lines, such as astrolabes, types of horary quadrant, and William Oughtred's 'horizontal instrument'. Most of the well-known museums containing seventeenth-century instruments have a few in their collections, with the Whipple Museum holding a particularly rich and varied selection. At thirteen instruments, the Whipple's collection of Sutton material may be the largest of any museum.

---

1 E. G. R. Taylor, *The Mathematical Practitioners of Tudor & Stuart England, 1485–1718* (Cambridge: Cambridge University Press, 1970), p. 220.

Further, the Whipple has been a site for scholarship on Sutton's instruments. David Bryden's work showed a particular interest in Sutton,[2] and latterly Boris Jardine has produced a number of in-depth studies that have shown the benefits of looking at his instruments in detail.[3]

The Whipple Museum has four quadrants by Sutton, one in brass (using Gunter's projection) and three printed on paper (using the projections associated with Sutton and considered below). Of the printed instruments, a smaller one is applied to only one side of a brass quadrant shape, whereas two larger ones each have two sides and are mounted on wood.[4]

To begin with one of his instruments that is not in Cambridge but which can introduce the theme of reputation, the History of Science Museum in Oxford has a large universal astrolabe by Sutton, constructed on an orthographic (Rojas) projection and dated 1659.[5] There are a great many, very regularly engraved lines, in Sutton's

2 D. J. Bryden, *The Whipple Museum of the History of Science, Catalogue 6: Sundials and Related Instruments* (Cambridge: Whipple Museum of the History of Science, 1988), nos. 221, 282, 288A; D. J. Bryden, 'Evidence from Advertising for Mathematical Instrument Making in London 1556–1714', *Annals of Science*, 49 (1992), pp. 310–36, p. 319 and n. 89; and D. J. Bryden, 'The Instrument-Maker and the Printer: Paper Instruments Made in Stuart London', *Bulletin of the Scientific Instrument Society*, no. 55 (December 1997), pp. 3–15, *passim* but especially pp. 4–5, 8–9, 13–14. As Curator of the Whipple Museum, Bryden was responsible for acquiring the dialling scale 'In usum Euclidis Speidell Angli', Wh.2255.
3 C. Eagleton and B. Jardine, 'Collections and Projections: Henry Sutton's Paper Instruments', *Journal of the History of Collections*, 17 (2005), pp. 1–13; B. Jardine, 'Reverse-Printed Paper Instruments (with a Note on the First Slide Rule)', *Bulletin of the Scientific Instrument Society*, no. 128 (March, 2016), pp. 36–42; and B. Jardine, 'Henry Sutton's Collaboration with John Reynolds (Gauger, Assayer and Clerk at the Royal Mint)', *Bulletin of the Scientific Instrument Society*, no. 130 (September 2016), pp. 4–7.
4 These instruments have, respectively, Whipple Museum numbers Wh.0738, Wh.5831, Wh.2754, and Wh.6644. See Bryden, *The Whipple Museum of the History of Science, Catalogue 6*, catalogue numbers 282 and 288A; J. A. Bennett, *A Decade of Accessions: Selected Instruments Acquired by the Whipple Museum of the History of Science between 1980 and 1990* (Cambridge: Whipple Museum of the History of Science, 1992), catalogue number 5. On Sutton's quadrants, see M. Lowne and J. Davis, 'A Horizontal Quadrant of 1658 by Henry Sutton', *British Sundial Society Bulletin*, 23 (2011), pp. 8–13, 45–8; M. Lowne and J. Davis, 'The Stereographical Projection and Quadrant by Henry Sutton', *British Sundial Society Bulletin*, 24 (2012), pp. 8–15; Mike Cowham, *A Study of the Quadrant: Horary Quadrants, Sundial Making Quadrants, Surveying Quadrants, Astronomical Quadrants* (Cambridge: M. J. Cowham, 2014), pp. 30–4; and Hester Higton, *Sundials at Greenwich: A Catalogue of the Sundials, Nocturnals and Horary Quadrants in the National Maritime Museum, Greenwich* (Oxford: Oxford University Press, 2002), pp. 348–51.
5 History of Science Museum, University of Oxford, inventory no. 51786.

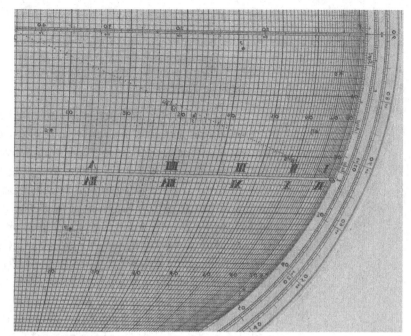

Figure 4.1 Detail of a print taken directly from an astrolabe by Sutton. Image © History of Science Museum, University of Oxford (inventory no. 56420).

typical manner and demonstrating his celebrated skill. There is also a remarkable companion: the Museum has an early print taken directly from this instrument, which must have been inked as though it were a printer's plate.[6] There has been discussion over the intention behind this pull and other direct prints from instruments, including others made by Sutton.[7] Being in reverse, they have no obvious use as instruments. Might they, for example, have been kept for record in the workshop or used to encourage future customers? Is it possible that the reverse print was an intermediate step to a counter-print? All these ideas have been canvassed but, whatever the intention, this astrolabe print is proof of quite exceptional skilled practice. It is much more difficult to notice any flaws in the brass plate, but any untidiness or unevenness of line will immediately be revealed by the print. While there is some leeway with a figurative print, here there is nowhere to hide faults in a challenging and detailed projection. As a demonstration of Sutton's accuracy, skill, and command, the print is extraordinary (Figure 4.1). Making such a print was an act of bravado, while preserving it was a statement of success. If Sutton

6 History of Science Museum, University of Oxford, inventory no. 56420.
7 Eagleton and Jardine, 'Collections and Projections', pp. 1–13; and Jardine, 'Reverse-Printed Paper Instruments', pp. 36–42.

himself made this print, as seems very likely, it surely reveals a concern for reputation.

Sutton's Quadrants

That concern is shown in a better-documented manner around the same time, one that gives us a more nuanced view of reputation in contemporary mathematics. In 1658 a book by the mathematician John Collins was published in London under the title *The Sector on a Quadrant, or, a Treatise Containing the Description and Use of Three Several Quadrants.*[8] Its author was described as 'Accountant, and Student in the Mathematiques', the printer was J. Macock, and the book was to be sold by two booksellers, George Hurlock and William Fisher, and also by 'Henry Sutton Mathematical Instrument Maker, at his House in Thredneedle street behind the Exchange.' The engraver of the plates is given unusual prominence on the title page and the relationship of the plates to the book is stated as 'With large Cuts of each Quadrant, printed from the original Plates graved by *Henry Sutton*, either loose, or pasted upon Boards'. Thus began a complex bibliographical sequence, where the content remained much the same, but was introduced by a variety of title pages. As early as 1659 it had been decided that there were in fact four quadrants, not three, though this was not altered in the preface.[9] The new title page insisted that there were 'Two small ones and two great ones' and then repeated that each was 'rendred many wayes, both general and particular'. The author was now described as 'Accountant Philomath'. A third bookseller, Thomas Pierrepont, was added and the description of the printed instruments altered to 'With Paper Prints of each Quadrant, either loose or pasted upon boards; to be sold at the respective places aforesaid'.[10] This reduced the former emphasis on the work of engraving and the prominence given to Sutton. There are copies containing both these title pages.

The relationship between printed instruments bound into books and separate paper instruments pulled from the same plate is a topic of interest to instrument historians, bearing as it does on the nexus of connections between engraving and the production of

8 J. Collins, *The Sector on a Quadrant, or, a Treatise Containing the Description and Use of Three Several Quadrants* (London, 1658).
9 J. Collins, *The Sector on a Quadrant, or a Treatise Containing the Description and Use of Four Several Quadrants* (London, 1659).
10 Collins, *The Sector on a Quadrant* (1659), title page.

instruments and books. The wording on the title pages seems to imply that the instruments are supplied separately from the book, the options mentioned being 'either loose or pasted upon boards', but this does not preclude their inclusion in the bound book, and in fact one plate includes the instruction 'Place this next after the Title page.' This is one of the smaller quadrants, and neither side of either of the larger quadrants is ever found in the book, although the scales and their uses are explained there. They would have had to have been folded down to fit within the usual quarto format.

Books such as this and their plates need to be studied in parallel with the printed instruments in collections such as that in the Whipple Museum. The bound prints and those pasted onto boards were produced from the same engraved plates. Museums with collections of early books or with associated libraries have perhaps been remiss in ignoring instruments by makers represented in their collections, simply because they happen to be bound into books. In Sutton's case the bound examples were probably not intended to have much use as instruments (though a solar declination scale, for example, could find ready applications). In the previous century, however, many such prints in books were certainly working instruments, a status emphasised by the inclusion of rotating discs and strings for reading scales.

In the sixteenth century, the instrument designer, and perhaps even the head of the workshop or print-shop, might also be the author. By the middle of the seventeenth century this was not at all common, and *The Sector on a Quadrant* brings a new collaborator into play, and a new relationship that would shape the venture. This book was a new departure for John Collins, who had previously published only some tables for currency exchange between England and Flanders, and England and France, and a short textbook on accountancy, *An Introduction to Merchants Accounts* in 1653. He had a sudden flowering in the field of mathematical instruments, with three books appearing in the late 1650s: *The Sector on a Quadrant* (1658), *Navigation by the Mariners Plain Scale New Plain'd* (1659), and *Geometrical Dyalling* (1659). All were linked to Sutton in some way, either as a stockist and seller, or as one of the publishers, and as the engraver for all three books.

The bibliography of these books is complex and requires more space and skill than are available here, but of the three titles it seems that *The Sector on a Quadrant* was the first to appear, which adds to the interest of an account of the genesis of the book, explained in unusual detail in a preface by Collins. It relates to Collins's entry into this field, as well as to Sutton's relationships with his clients and his quest for reputation. 'Thou hast in this Treatise', says Collins, 'the Description and Uses of three several Quadrants, presented to thy View and Acceptance; and here I am to give thee an account of their Occasion and Original.'[11]

The account is that a mathematical friend of Collins, Thomas Harvie, had worked out an idea for a quadrant, which he drew out on paper. It was a new design with a novel projection in the context of an horary quadrant, that would yield the time and the solar azimuth from the customary quadrant measurement of solar altitude. Harvie wanted to have one in brass for his own use, and approached Sutton, as an instrument maker. Having been told the general idea, Sutton agreed to make the instrument, and Harvie said he would come back in two weeks with the projection drawn out for him to copy in brass. Before this could happen, Collins tells us that

> M. Sutton having very good practise and experience in drawing Projections, speedily found out the drawing of that Projection, either in a Quadrant or a Semicircle, without the assistance of the promised directions, and accordingly, hath drawn the shape of it for all Latitudes, and also found how the Horizontal Projection might be inverted and contrived into a Quadrant without any confusion, by reason of a reverted tail, and let me further add, that he hath taken much pains in calculating Tables for the accurate making of these and other Instruments, in their construction more difficult then any that ever were before.[12]

Sutton asked Collins to write a few sheets on the use of the quadrant, for him to give to customers, when he supplied them with instruments. Once again matters were overtaken by Sutton's enthusiasm. He became dissatisfied with the idea of a few sheets and since, as Collins says, he 'very well understood the use as well as the making'[13] and had found many uses for his quadrant, he persuaded Collins to write a much fuller treatise. Sutton had also come up with further designs, and he continued to press ahead, engraving the plates after Collins had written the treatise and making some changes from the drawings from which Collins had been working. This meant that the text and plates, and therefore the instruments, did not quite coincide. In particular, whereas Collins had used right ascensions from current star tables, Sutton had calculated those on the

11 Collins, *The Sector on a Quadrant* (1658), sig. A2r.
12 Collins, *The Sector on a Quadrant* (1658), sigs. A2r–A2v.
13 Collins, *The Sector on a Quadrant* (1658), sig. A2v.

quadrant for a slightly later epoch, so as to lengthen the useful life of the instruments.

We learn a great deal from this preface about the relationships (almost certainly not typical) between the client, Harvie, the instrument-maker, Sutton, and the mathematical practitioner Collins. It is not insignificant that Harvie took his commission to Sutton in the first place. This tells us something about Sutton's reputation: he was not restricted to the standard designs, but would make a bespoke instrument on an original pattern. We learn of Sutton's very active engagement with the process, something that might easily have gone unrecorded. That it *was* recorded was also surely at Sutton's instigation. It is hard to see that Collins himself had anything to gain by the publication of this preface, though through the project itself, centred around Sutton's engraving of some very fine plates, he did achieve a successful book. Through this and the other titles, we know that he and Sutton were in a broader productive collaboration around this time.

The importance of Sutton's initiative survived in Collins's memory, when he wrote as follows in a later letter to John Wallis:

> At the request of Mr Sutton I wrote a despicable treatise of quadrants. His design was to demonstrate himself to be a good workman in cutting the prints of those quadrants, and thereby to obtain customers.[14]

We should not set too much store by the word 'despicable'. Aware of his humble origins and lack of formal education, Collins was inclined to refer to his work with excessive modesty, especially in writing to the renowned Professor of Geometry at Oxford. It is clear from his substantial book that Collins was thoroughly engaged with *The Sector on a Quadrant*.

Collins begins his account by offering two ways of thinking about the two projections to be used in the quadrants. His first way of thinking is related, he says, to how the projections 'may be demonstrated'. In the future we may expect a more general demonstration from Harvie, but for now the projections can be thought of as deriving either from Stoeffler's astrolabe (as he calls the ordinary astrolabe, and referring in particular to the projection of a latitude plate or tympan), or (in the second projection) from the horizontal instrument of William Oughtred.

---

14 Letter from Collins to Wallis, 28 February 1665/6, see P. Beeley and C. J. Scriba, *The Correspondence of John Wallis* (Oxford: Oxford University Press, 2005), vol. II, pp. 193, 460–2, quotation p. 462.

In the former projection the circles on the quadrant are the projections of the lines of altitude and azimuth, the point of projection is the south celestial pole, and the plane of projection contains the equator. The lines on the quadrant are, unlike the astrolabe, the projection of the altitude and azimuth below the horizon as far as the tropic of Capricorn, which, however, we are then advised to call the Tropic of Cancer. This is the projection used in three of the quadrants, the first three described by Collins, two of which are generally included as prints, although this cannot be assumed and neither can the positions of the prints in the text.

In the latter projection the circles to be projected are those of declination and right ascension (or hour lines), the point of projection is the observer's nadir, and the plane of projection contains the horizon. On the quadrant the projected lines are the arcs of these circles below the horizon and the user has to adopt a similar reversal in nomenclature between the tropics. This projection is used only for the fourth quadrant, which is never found as a print in the book.

As Collins admits, this is rather an unhelpful and counter-intuitive way of thinking about the two projections, but its purpose seems to be to relate the projections to the established work of Stoeffler and Oughtred. For Collins this constitutes a form of 'demonstration': these projections can be taken as established and something that simply extends them to cover a differently delimited area of the celestial sphere also partakes of that status. Surprisingly perhaps, he says that he gives this view 'for the accommodation of Instrument makers, to whom this Derivation may seem most suitable',[15] implying that they are the group who will want to see these new instruments within the established canon of projections of the sphere.

Collins then offers what he calls 'a more immediate account' of the projections, a view it seems was more readily understood. Now for the former projection he says that the point of projection is the north celestial pole, the plane of projection is equatorial, and the projected arcs are those of altitude and azimuth above the horizon and falling between the tropics, with the Tropic of Cancer being outermost on the projection and Capricorn innermost, that is, the reverse of 'Stoeffler's astrolabe'. For the horizontal projection, he now places the point of projection at the observer's zenith, projecting the lines of solar declination and right ascension onto the horizontal plane. This is the reverse of Oughtred's projection.

---

15 Collins, *The Sector on a Quadrant* (1658), sig. 22v.

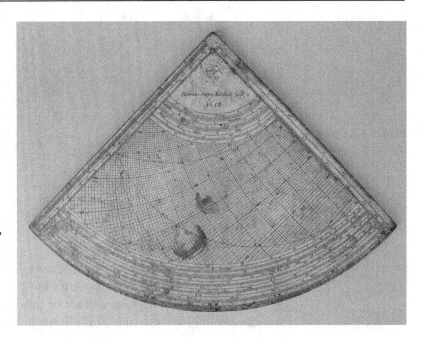

Figure 4.2 The 'great' (i.e. large) equatorial quadrant (i.e. having lines of solar altitude and azimuth, the ecliptic, and the horizon projected onto the plane of the equator). Image © Whipple Museum (Wh.2754).

There is confusion over the naming of these quadrants, through the uses of 'small' and 'universal'. Collins begins with an instrument he refers to simply as the quadrant. He describes its use at length, especially the lines and scales that allow proportional and trigonometrical calculations, i.e. the features that make this a 'sector on a quadrant', as many of these operations could be performed with a sector. Then he uses a further title page to announce *The Description and Uses of a Great Universal Quadrant: With a Quarter of Stoflers Particular Projection upon it, Inverted*, dated 1658 and describing the author as 'Accomptant, and Student in the Mathematiques'. What is meant by 'universal' here is not clear (and seems to be contradicted by 'particular'), as this instrument uses the same projection and is for a specific latitude (Figure 4.2). The previous instrument is now referred to as 'the small Quadrant'. Collins then describes the additional features on this larger quadrant and their use.

Both of these quadrants make use of what Collins calls the 'reverted tail'. This is a device he specifically attributes to Sutton, which is used to accommodate all the projection lines on the instrument, even though a portion as projected will fall outside the limits of a quadrant. The portions of the projected lines that fall beyond the 6 o'clock line to the north of the east or west point on the horizon (these points coincide on the 'folded' projection), needed for finding the time before 6 am and after 6 pm in summer, will lie outside the quadrant. However an equivalent, unused space arises from the sun being below

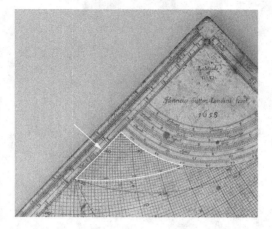

Figure 4.3 Detail of the great equatorial quadrant, with Sutton's 'reverted tail' indicated. Image © Whipple Museum (Wh.2754).

the horizon after 6 am and before 6 pm in winter (Figure 4.3). Sutton uses this space – smaller in area but with more closely packed lines – by adding the lines of negative altitude or depression and continuing the azimuth lines to the section of sky below the horizon and bounded by the Tropic of Capricorn and the 6 am/pm line.

To understand the use of the reverted tail, we must first be familiar with the normal operation for finding time. A bead slides friction-tight on a weighted thread suspended from the apex of the quadrant and must be adjusted ('rectified') for date (or solar declination). This is done by stretching the thread across the date point on the calendar scales close to the apex and setting the bead to the summer or winter section of the ecliptic line on the projection. The altitude is then measured by holding the quadrant vertical, aligning the edge sights with the sun, and noting the angle on the altitude scale at the limb. The bead is then placed on the equivalent line in the projection and the time found on the hour scales at the limb, where morning and afternoon hours run in opposite directions. This is very like the method of finding time with an astrolabe, here accommodated to a quadrant.

In the geometry of the projection the lines in the reverted tail are equivalent to those that would fall outside the quadrant area, but to use them for the absent dates and times the user must set the bead on the plumb-line to the winter ecliptic line even though the thread is stretched across a summer date, and must read the time from the 'wrong' hour scale on the limb – the afternoon hours in the morning and vice versa. Sutton's facility with projection allowed him to see this with ease, but his customers surely found it confusing.

Sutton seems to relish the opportunity to demonstrate his facility with projection in other ways in Collins's book. 'For varieties sake' he projects quadrants for different latitudes, illustrating the unexpected

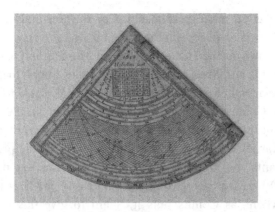

Figure 4.4 The 'small pocket quadrant'. Image © Whipple Museum (Wh.5831).

behaviour of the sun at a low latitude (Barbados, 13 °N) and a high one (Greenland, 75 °N). He also provides the projection of a full semicircle for the latitude of London, its advantages, he explains, being that there is no need for the operation of a reverted tail and that the projected area is narrower, not needing to accommodate the full range of solar altitudes at a given latitude within the space of a quadrant. Sutton wants it to be clear that he has not simply engraved a drawing projected by someone else: he signs the plate 'Henricus Sutton Londini deline= et sculp='.[16]

Collins then moves on to 'The Description of an Universal small Pocket Quadrant',[17] having scales on only one face. It can be small because the projection has lines only of solar altitude and the azimuth lines are not present. This allows Sutton to have summer and winter lines crossing each other, the same area of the quadrant being used for the northern and southern hemispheres and the outermost arc on the projection standing for either tropic. The Whipple Museum has an example unusually applied to a brass quadrant (Figure 4.4). A copy in the British Library of the relevant pages from Collins's book, with the printed plate, belonged to Robert Hooke.[18]

Finally, Collins describes the fourth quadrant, again a larger one and the only one based on his 'second' projection, which he describes as an inversion of the projection used for Oughtred's horizontal instrument. A final title page, dated 1658, which is also the date on all the plates, announces *The Description and Uses of a General Quadrant, with the Horizontal Projection, upon it Inverted,* and the instrument is referred

---

16 Collins, *The Sector on a Quadrant* (1658), pp. 32–3.
17 Collins, *The Sector on a Quadrant* (1658), p. 277.
18 British Library class mark 8561.a.27. Note the entry in 'Robert Hooke's Books', www.hookesbooks.com/wp-content/themes/hookesbooks/details_bh.php?id=2058 (accessed 15 May 2018).

to as 'the other great quadrant' or 'the horizontal quadrant'. Here again there is no plate with the book, but one surviving printed instrument is known, at the History of Science Museum in Oxford. It is not signed but is attributed to Sutton.[19] In Collins's book there follows a table of solar right ascension and declination calculated for 1666 by Sutton.

Collins may have written to Wallis in February 1665/6 that Sutton's intention for *The Sector on a Quadrant* 'was to demonstrate himself to be a good workman in cutting the prints of those quadrants', but the book itself shows that this was not the whole story. Otherwise there would be no reason for the unexpected preface, where, as Collins puts it, 'I am to give thee an account of their Occasion and Original'.[20] Sutton is explicit in his 'deline= et sculp=' inscription that he made the projection as well as the plate, so his reputation should encompass his facility with compass and rule, as well as with the burin.[21]

Sutton's Reputation

Sutton did achieve a substantial reputation, extending to the circle of the Royal Society. Of his death in the Plague of 1665, Sir Robert Moray wrote to Henry Oldenburg in October, 'wee all here are much troubled with the loss of poor [Anthony] Thomson & Sutton.'[22] Collins also wrote to John Wallis that on his return to London from Oxford, 'I found wanting Mr Anthony Thompson and Mr Henrie Sutton, two of the best Mathematicall Instrument Makers.'[23] We know of communication on dialling between Sutton and 'Doctor Richard Sterne',[24] who in all probability was the former master of Jesus College, Cambridge, who became Archbishop of York after the Restoration. Out of favour during the Commonwealth, he earned a living as a schoolmaster.[25]

19 Lowne and Davis, 'A Horizontal Quadrant of 1658 by Henry Sutton'.
20 Collins, *The Sector on a Quadrant* (1658), sig. A2r.
21 Collins mentions Sutton as someone who encouraged him to publish the dialling methods of Thomas Rice, which he had learned in turn from Gresham Professor of Astronomy Samuel Foster. See J. Collins, *Geometrical Dyalling, or, Dyalling Performed by a Line of Chords Onely* (London, 1659), preface.
22 A. R. Hall and M. B. Hall (eds.), *The Correspondence of Henry Oldenburg* (Madison: University of Wisconsin Press, 1966), vol. II, p. 561.
23 Beeley and Scriba (eds.), *The Correspondence of John Wallis*, vol. II, p. 189.
24 Collins, *Geometrical Dyalling*, p. 11.
25 For other evidence of Sutton's ingenuity and versatility, see Jardine, 'Henry Sutton's Collaboration with John Reynolds', pp. 4–7; and J. Bennett, 'Henry Sutton Thinking', *Sphaera*, no. 10 (1999), p. 6.

In 1668 the mathematical writer Robert Anderson published his *Stereometrical Propositions*, which he claimed would be useful for gauging, and announced that an instrument he recommended could be had from John Marke, who, he said, 'was formerly Servant to that incomparable Instrument maker Mr. Henry Sutton'.[26] In his letter to Wallis, Collins had mentioned Marke's succeeding to the business, 'We hope he may prove as good a Workeman as his deceased Master.'[27]

We shall see that Sutton's skill as an engraver was too valuable to allow his output to end with his death, but his reputation as a designer of instruments was less robust. In 1669 Robert Morden, an associate of Anderson and a maker of globes and seller of maps and instruments, published *A Description & Use of a Large Quadrant, Contrived and Made by H. Sutton*.[28] At this stage Sutton was understood not only to have made the quadrant but also to have contrived it. No author is credited and it is clear that this tract was meant to sit alongside the great quadrant, still available either from a stock of prints or pulled from the surviving plates. David Bryden mentions further early references to Sutton and his quadrant within the instrument trade.[29]

In 1703 there appeared the first edition of John Harris's *The Description and Uses of the Celestial and Terrestrial Globes; and of Collins's Pocket Quadrant*. The uses of the globes were a staple component of Harris's teaching, including his public lectures at the Marine Coffee House, and he explained that 'The Description and Use of Mr. Collins's Quadrant was occasioned by the Request of some Persons who would gladly know the best Uses of it, without being obliged to read over many Things which are little to their Purpose.'[30] Clearly *The Sector on a Quadrant* was no longer what was wanted. The quadrant Harris describes is the basic Sutton instrument with altitude and azimuth lines, but he neglects the verso, referring his readers to the 'large Account' in Collins's book.[31] Sutton is not mentioned in any

26 R. Anderson, *Stereometrical Propositions Variously Applicable, but Particularly Intended for Gageing* (London, 1668), p. 105; see also the edition of 1703, Robert Anderson, *Solid Geometry: or, Foundation of Measuring, of All Manner of Solid Bodies* (London, 1703), p. 105.

27 Beeley and Scriba (eds.), *The Correspondence of John Wallis*, vol. II, p. 189.

28 Robert Morden, *A Description & Use of a Large Quadrant, Contrived and Made by H. Sutton* (London, 1669); Taylor, *The Mathematical Practitioners of Tudor & Stuart England*, p. 237; and G. Clifton, *Directory of British Scientific Instrument Makers 1550–1851* (London: Zwemmer, 1995), p. 192.

29 Bryden, 'The Instrument-Maker and the Printer', pp. 13–14.

30 J. Harris, *The Description and Uses of the Celestial and Terrestrial Globes; and of Collins's Pocket Quadrant* (London, 1703), sig. A3r.

31 Harris, *The Description and Uses of the Celestial and Terrestrial Globes*, p. 53.

capacity. While there are four pages advertising books from the bookseller or publisher, readers are offered no advice on how to obtain a quadrant, which seems to imply that this was not difficult. The book went through a number of editions up to at least 1751.[32]

In 1710 the surveyor, dialist, and teacher of mathematics, John Good,[33] brought out a much-abridged account of the Sutton quadrants in a tract titled *The Description and Use of Four Several Quadrants, Two Great Ones, and Two Small Ones*. Sutton would not have been happy with the distribution of credit on the title page: 'Invented and Written by the Ingenious Mr. John Collins, and Engrav'd by the Curious Hand of Mr. Henry Sutton'.[34] The same view is repeated in the preface: all the instruments were 'invented by the Ingenious Mr. John Collins, and Engrav'd by that unparallel'd Artist Mr. Henry Sutton, Mathematical-Instrument-Maker'. Good explains that as the original book is 'now scarce and out of Print', he has 'drawn from it the usefullest Parts thereof'.

While Sutton (an instrument-maker) was ignored completely by John Harris (a successful clergyman, Royal Society fellow, and Boyle lecturer[35]), even for John Good Sutton's reputation rested on his engraving: he is 'that unparallel'd Artist'. Since, however, he is referred to as the engraver and since Good's text would have no purpose without the instruments, it is reasonable to assume that Sutton's plates had survived and that prints from them could be bought, perhaps from the promoters of Good's book. The successful chartmaker and bookseller Richard Mount was the publisher, in association with William Mount and Thomas Page. There was much acquisition of stock in books, maps, and plates between those engaged in mathematical commerce; Mount, for example, purchased the stock of instrument- and globe-maker Charles Price in 1706.[36] Price had been

32 *London Daily Advertiser and Literary Gazette*, Monday, 22 July 1751; Issue 121, 17th–18th Century Burney Collection Newspapers, accessed online 26 May 2018.
33 Taylor, *The Mathematical Practitioners of Tudor & Stuart England*, pp. 301–2; and E. G. R. Taylor, *The Mathematical Practitioners of Hanoverian England 1714–1840* (Cambridge: Cambridge University Press, 1966), p. 119.
34 J. Collins and J. Good, *The Description and Use of Four Several Quadrants, Two Great Ones, and Two Small Ones* (London, 1710).
35 For Harris, see L. Stewart, 'Harris, John (c. 1666–1719), writer and lecturer on science', in *Oxford Dictionary of National Biography* (Oxford: Oxford University Press, 2004; online edn, 2009), https://doi.org/10.1093/ref:odnb/12397 (accessed 27 May 2018).
36 E. G. Forbes, L. Murdin, and F. Willmoth (eds.), *The Correspondence of John Flamsteed, the First Astronomer Royal* (Bristol: Institute of Physics Publishing,

apprenticed to John Seller, and was in partnership for a time with-John Senex.[37] For men such as these a copperplate by Henry Sutton would be a valuable commodity, and his reputation as an outstanding engraver would have helped preserve such an item more effectively than if Sutton had been remembered as an inventor or designer.

Edmond Stone described Sutton's quadrant (the equatorial projection of altitude and azimuth lines, in two sizes) in his translation and edition of Nicolas Bion's *The Construction and Principal Uses of Mathematical Instruments*, published by John Senex in 1723. This was, Stone says, one of several different quadrants 'made by Mr. Sutton long since' and, while 'made by' is ambiguous, no other designer is mentioned, while Collins is referred to only as the author of the book where they are described.[38]

Good's book appeared again in 1750, published once more by Mount (now W. and J.) and Page, and, although the title page and text generally have been reset, the attributions to Collins and Sutton are unchanged.[39] Spelling is updated and grammar corrected, but examples that were updated from the year 1657 in *The Sector on a Quadrant* to 1709 in the 1710 edition are repeated unchanged in 1750. The Julian calendar is assumed, even though this is in the process of being abandoned in mid-century. Can we imagine that Sutton's plates survived still and that after a further forty years Mount and Page were still hoping to sell prints? It is hard to see why else they would have produced this revised edition, which continued to reference the work of the 'unparallel'd Artist'.

A second and augmented edition of Stone's Bion appeared in 1758, and several historians have noted the fulsome tribute paid to Sutton's quadrants in the introduction to Stone's 'Supplement'.[40]

1995–2002), vol. III, pp. 286, 288, 290; and Taylor, *The Mathematical Practitioners of Tudor & Stuart England*, pp. 276–7, 280.

37 Clifton, *Directory of British Scientific Instrument Makers 1550–1851*, pp. 223, 247–8.

38 N. Bion, *The Construction and Principal Uses of Mathematical Instruments*, trans. and ed. E. Stone (London, 1723), p. 197.

39 Bryden also mentions a reprint of 1723, Bryden, 'The Instrument-Maker and the Printer', p. 15.

40 A. J. Turner, 'Sutton, Henry (c. 1624–1665), maker of mathematical instruments', in *Oxford Dictionary of National Biography* (Oxford: Oxford University Press, 2004; online edn, 2009), https://doi.org/10.1093/ref:odnb/49540 (accessed 25 May 2018); Lowne and Davis, 'A Horizontal Quadrant of 1658 by Henry Sutton', p. 47; and Eagleton and Jardine, 'Collections and Projections', p. 5.

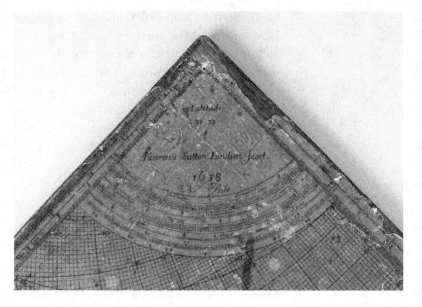

Figure 4.5 Detail of the great equatorial quadrant with a replacement solar declination scale or calendar, based on the 'New Stile', i.e. the Gregorian Calendar, which was officially adopted in England in 1752. Image © Whipple Museum (Wh.6644).

In justification for adding English instruments to Bion's account, the first instance Stone cites was as follows:

> I soon perceived that many *French* Instruments of Mr *Bion's* were excelled by some of ours, of the same kind in Contrivance; and as to Workmanship, I never did see one *French* Instrument so well framed and divided, as some of ours have been; for Example, Mr *Sutton's* Quadrants, made above one hundred Years ago, are the finest divided Instruments in the World; and the Regularity and Exactness of the vast Number of Circles drawn upon them is highly delightful to behold.[41]

Stone's account is to some extent historical, and we cannot infer from his description the availability of prints from any surviving plates. A recent acquisition by the Whipple Museum, however, does give us an unexpected coda to the history of at least one and probably two of the copperplates. In 2017 an example of the large version of the equatorial quadrant, with the two prints pasted onto a shaped wooden board in the usual way, was donated to the Museum. The verso has the customary scales, with the calendars, for example, unchanged, but the quadrant itself, on the front, has an unexpected feature. The date or solar declination scale, set out towards the apex in four quadrant sections, has been skilfully

---

41 N. Bion, *The Construction and Principal Uses of Mathematical Instruments*, trans. and ed. by E. Stone (London, 1758), *A Supplement*, sig. Yyy2.

replaced by one based on the Gregorian calendar and is inscribed 'New Stile' (Figure 4.5).[42] The legislation for adopting the Gregorian calendar was approved by Parliament in 1751 and the new calendar introduced the following year. The modification of Sutton's plate indicates that it had survived in a practical context for somewhere around a century.

'Sutton's Quadrant'?

Sutton wanted to have his engagement with geometry recognised alongside his skill as an engraver. However differently we may distinguish mathematical proficiency today, it is clear that, in Sutton's world, facility with projective technique counted as a species of mathematics. Sutton failed to achieve his dual ambition: by the eighteenth century he was not remembered as a competent geometer but as an 'unparallel'd Artist'.

In spite of eighteenth-century instances of naming the instrument after Collins, notably by John Harris, today the equatorial instrument, at whatever size, is generally referred to as 'Sutton's quadrant'. There are occasional reversions to Collins and even very occasional support for 'Harvey's quadrant', but Collins never claimed the instrument as his invention, and Harvey is surely too shadowy a figure and his connection too slightly documented to justify this name. The name itself might seem unimportant, but not if 'Collins' was introduced on the basis of a prejudice towards a mathematician and Fellow of the Royal Society over an instrument-maker. Collins wrote an account of the instrument that devoted more space to mathematical calculation than to instrumental astronomy and the odd title to his book, *The Sector on a Quadrant*, reflects this. It was mainly these sections that later writers stripped away.

In another sense as well, this quadrant began and remained as Sutton's. Unlike Gunter's quadrant, and in spite of surviving interest in published accounts, other makers did not take up Sutton's design with any enthusiasm. There are a very few instances, but nothing

42 Further examples are in the collections of the National Maritime Museum, see http://collections.rmg.co.uk/collections/objects/381692.html?_ga= 2.160740671.341929038.1537978801-1756226939.1514557254 (accessed 26 September 2018), and of the Science Museum, London, see Bryden, 'The Instrument-Maker and the Printer', pp. 13–14.

substantial,[43] and engraving the projection was a challenge. In its near-exclusive use by Sutton, both living and posthumous, the quadrant embodies his geometry and his engraving together, while nothing we have seen here suggests that Sutton himself would have made this distinction.

43 Bryden, *The Whipple Museum of the History of Science, Catalogue 6*, no. 289; Higton, *Sundials at Greenwich*, pp. 254–6; D. J. Bryden, 'Made in Oxford: John Prujean's 1701 Catalogue of Mathematical Instruments', *Oxoniensia*, 58 (1993), pp. 263–85; and Lowne and Davis, 'The Stereographical Projection and Quadrant by Henry Sutton', p. 14.

5 ⚘ Specimens of Observation: Edward Hobson's *Musci Britannici**

ANNE SECORD

The set of mosses in the Whipple Museum labelled *Musci Britannici*, bearing a title page dated 1818 declaring it to be *A Collection of British Mosses and Hepaticae, Collected in the Vicinity of Manchester, and Systematically Arranged with reference to the Muscologia Britanica, English Botany, &c, &c, &c*, is hard to define (Figure 5.1).[1] It belongs to a genre of publication involving specimens alone that arose out of reservations about the adequacy of drawings in those 'difficult divisions of the Flora' neglected by most botanists.[2] These sets of labelled specimens are known as exsiccatae (from the Latin for 'dried'). They are available in multiple copies, and typically consist of pressed plants all belonging to the same taxonomic group whose identification and arrangement follows that of the most established botanical authorities.[3] The specimens are usually mounted on loose sheets contained in covers or boxes.

The *Musci Britannici* is an early example of such a set of published specimens. It is also an object that, depending on its contexts of use and of preservation, can be seen as a book or as a collection. It thus highlights and straddles the modern division between libraries and museums. Spaces of science have been used to differentiate both practices and things, but the *Musci Britannici* challenges this

---

* I am very grateful to the Gifford family (Kinnordy Archive); the Trustees of the Natural History Museum, London; the Trustees of the Royal Botanic Gardens, Kew; the Herbarium Archive, Manchester Museum, University of Manchester; the West Yorkshire Archive Service, Calderdale; and the Archives of the New York Botanical Garden, for kind permission to quote from manuscripts in their collections.

1 Whipple Museum catalogue number Wh.4577. On the original title page 'Britanica' was spelled incorrectly; some copies of volume one, produced after a second volume was published in 1822, have an altered volume two title page with the correct spelling (see, for example, the copy in the Herbarium, Manchester Museum, University of Manchester).

2 F. Hanham, *Natural Illustrations of the British Grasses* (Bath: Binns and Goodwin, 1846), p. ix.

3 G. Sayre, 'Cryptogamae exsiccatae', *Memoirs of the New York Botanical Garden*, 19, nos. 1–3 (1969–75).

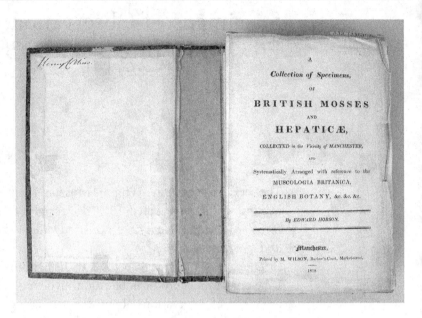

Figure 5.1 The first volume of Edward Hobson's *Musci Britannici* (Manchester, 1818), showing the casing and title page. This volume contains 119 sheets of specimens. Image © Whipple Museum (Wh.4577).

analytical framework. It also blurs any sharp divide between cabinet and field work, as well as between commerce and the established practice of gift exchange in natural history. As either book or collection, the *Musci Britannici* comes across as a 'black box', in that its scientific and technical work is made invisible by its own success at stabilising and making obvious the objects of scientific study – in this case, the species and genera of mosses and liverworts. It is regarded as both the product and the confirmation of botanical taxonomic practices.

Exploration of the production and distribution of exsiccatae – at a time when taxonomic systems were in formation and discoveries of rare and new species were still being made in certain groups of plants – indicates that, more than books or collections, they were instruments for seeing. The function of the *Musci Britannici* was to hone visual skills and calibrate observational powers. The aim was to produce a consensus about how mosses should be classified by providing the least ambiguous means of observing the basis on which they were ordered.

The *Musci Britannici* did so even for keen field botanists by providing them with the best and most complete specimens available. The importance of having dried plants of this quality was made clear by the Yorkshire botanist Benjamin Carrington, who complained in 1857 that some of his moss specimens were so scrappy that it was 'doubtful how far an opinion can be gained of a

species from such fragments'.[4] When precisely what was being seen
was at stake, specimens allowed readers to observe and judge for
themselves; they guided and trained the eye in the 'study and
collection' of plants.[5] Historians' lack of attention to the observational function of exsiccatae is due perhaps to the more obvious
utility and appeal of illustrations. But botanists interested in the
classification of contested and difficult groups of plants favoured
specimens, precisely because illustrations embodied theoretical decisions concerning which classificatory characters should be noticed.

Botanical Instruments

Descriptive botany remained the benchmark by which botanists
were measured well into the nineteenth century. When, after five
successive failures, Charles Darwin was finally elected a corresponding member of the Académie des Sciences in Paris on 5 August 1878,
he was surprised to find himself in the botany section rather than
zoology. 'It is funny', he wrote to a friend, 'the Academy having
elected a . . . member in Botany, who does not know the characters of
a single natural order.'[6] Despite his numerous botanical publications,
Darwin did not regard himself as a botanist because he engaged in
experimental physiological botany and had never done the taxonomic work regarded as fundamental to botanical expertise. Just a
year earlier, he had complained to the American botanist Asa Gray
that 'It is dreadful work making out anything about dried flowers;
I never look at one without feeling profound pity for all botanists,
but I suppose you are used to it like eels to be skinned alive.'[7]

The study of plant physiology depended upon intricate experimental set-ups involving apparatus of varying degrees of
sophistication. Darwin's son Horace, who undertook an engineering
apprenticeship from 1875 to 1878, and established the Cambridge
Scientific Instrument Company in 1881, devoted some of his earliest
efforts to making instruments for his father's botanical research.[8]

4 B. Carrington to M. J. Berkeley, 23 September 1857, Natural History Museum,
   London, Botany Library (hereafter NHM), Berkeley Correspondence, vol. 2.
5 Hanham, *Natural Illustrations*, p. vii.
6 C. Darwin to T. H, Huxley, 11 August [1878], *The Correspondence of Charles
   Darwin* (Cambridge: Cambridge University Press, 2018), vol. 26, pp. 343–4.
7 C. Darwin to A. Gray, 8 March 1877, *The Correspondence of Charles Darwin*
   (Cambridge: Cambridge University Press, 2017), vol. 25, p. 118.
8 M. J. G. Cattermole and A. F. Wolfe, *Horace Darwin's Shop: A History of the
   Cambridge Scientific Instrument Company 1878 to 1968* (Bristol: Adam Hilger,
   1987).

These instruments were designed to record specific movements in plants, and were inspired by reports of the precision equipment in Julius Sachs's botanical institute in Würzburg, where Horace's brother Francis carried out research over the summer of 1878. After seeing a klinostat, designed by Sachs to measure the effect of gravity on plant growth, Francis told his father that it was 'one machine we must have'. He also expressed his belief that Horace could design an instrument superior to Sachs's, which was 'far from well made'.[9] Francis's confidence was probably based on the expertise Horace had displayed in 1876, when he had built an auxanometer – a self-recording instrument invented by Sachs for measuring the growth of a plant (Figure 5.2).[10]

This emphasis on apparatus, experiment, and measurement seems far removed from the observational taxonomic work Darwin believed marked a true botanist. However, earlier in the century, when floras had yet to be fully catalogued and taxonomic systems based on artificial characters were being challenged by ones based on natural affinities, the classification of plants also required instruments and a variety of manual skills.[11] 'I am become a passionate admirer of the Natural Orders as far as I yet understand them', declared the botanist and future director of Kew Gardens William Jackson Hooker in 1816. Emphasising the 'immense application' that this study required, Hooker was also aware that he had an advantage over most other botanists: 'I may thank my good fortune in having begun Botany with the Cryptogamia, which has given me a habit of dissection that I find of the utmost importance in the analysis of the

9 F. Darwin to C. Darwin, [before 17 July 1878], *The Correspondence of Charles Darwin* (Cambridge: Cambridge University Press, 2018), vol. 26, pp. 295–8. Francis worked as his father's botanical assistant in their home in Down, Kent, from 1874 to 1882. The klinostat designed by Horace Darwin was described and illustrated in Francis Darwin, 'On the Power Possessed by Leaves of Placing Themselves at Right Angles to the Direction of Incident Light', *Journal of the Linnean Society (Botany)*, 18 (1881), pp. 449–55. While Francis believed Sachs's instruments were not well made, Sachs believed the Darwins' botany was wretched; see S. de Chaderavian, 'Laboratory Science versus Country-House Experiments: The Controversy between Julius Sachs and Charles Darwin', *British Journal for the History of Science*, 29 (1996), pp. 17–41.

10 In 1894, E. Hamilton Acton and Francis Darwin, then reader in botany at Cambridge University, stated in their *Practical Physiology of Plants* (Cambridge: Cambridge University Press, 1894), p. 140, n. 2, that the auxanometer constructed by Horace Darwin in 1876 was still being used in the Cambridge laboratory.

11 J. Endersby, *Imperial Nature: Joseph Hooker and the Practices of Victorian Science* (Chicago: University of Chicago Press, 2008), pp. 54–83.

Figure 5.2 A self-recording auxanometer for measuring plant growth, made by Horace Darwin in 1876. Image © Whipple Museum (Wh.2766).

flowers & fruits of the phænogamous plants.'[12] Unlike phaenero-gams (flowering plants), which were easy to classify using the artificial system of Linnaeus, cryptogams (non-flowering plants such as mosses, algae, and lichens) had long been regarded as some of the most complex groups of plants to order. Not only was their manner of reproduction puzzling and their family connections difficult to determine, but their minute size required the use of a microscope for the detection of the relevant characters by which their identity and affinities could be established.

In late 1816, Hooker was working with the Irish botanist Thomas Taylor on a monograph of British mosses, the *Muscologia Britannica*, which contained both written descriptions and illustrations of the plants at their natural size, with magnifications of the features by which they were classified (Figure 5.3). The skilful manipulation of a microscope, some artistic talent, and a competent engraver were essential to producing reliable information about these plants. But there was nothing easy or consistent about any of these stages. Not only did Hooker and Taylor drastically reduce the number of moss

12 W. J. Hooker to C. Lyell, 2 October 1816 (Kinnordy Archive).

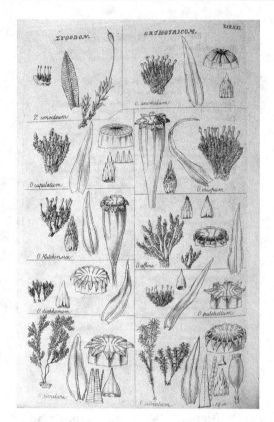

Figure 5.3 Plate 21 from W. J. Hooker and T. Taylor's *Muscologia Britannica* (London, 1818), showing the magnified features by which mosses of the genera *Zygodon* and *Orthotricum* were identified. Author's copy.

species, they also 'declined quoting' the illustrations in one of the standard floras of the period because they were so 'excessively bad'.[13] The variability in quality of how these plants had been figured by earlier botanists, and the fact that illustrations embodied theoretical decisions concerning which characters were thought to define a species, made the use of dried specimens preferable, especially before the classification of mosses was fully established. An 'admirably preserved & arranged' moss specimen 'is better distinguished than by the most elaborate figure', Hooker stated in a private communication, and he and Taylor also declared this publicly in their illustrated monograph: although they emphasised the 'utmost care' with which their figures of mosses had been drawn, they admitted that well-prepared specimens were far superior 'in point of accuracy to the best of plates'.[14] Hooker and Taylor did not refer to specimens in general but directed their readers' attention to the *Musci Britannici*.

13 W. J. Hooker to C. Lyell, 9 January 1817 (Kinnordy Archive).
14 W. J. Hooker to C. Lyell, 14 October 1821 (Kinnordy Archive); and W. J. Hooker and T. Taylor, *Muscologia Britannica: Containing the Mosses of Great*

Systematic botany has been characterised by Lorraine Daston as a process of identification and nomenclature, in which descriptions, illustrations, dried specimens, and actual plants are not interchangeable but interlocked. Descriptions and illustrations aimed to represent plants in general terms, emphasising only the essential characters that distinguished a species, while dried and growing plants conveyed the idiosyncracy of individuals, omitting none of their non-essential features. The interlocking of these elements of descriptive botany was essential both in the field and in the herbarium.[15] However, until a classification was stabilised through repeated and consensual observation, descriptions and illustrations were not regarded as reliable. The *Musci Britannici* was a key element in establishing the early-nineteenth-century order of mosses.

Making the *Musci Britannici*

The copy of *Musci Britannici* in the Whipple Museum must be one of the most unusual products of Eton College to end up in Cambridge. It was purchased in 1997 when the Eton College Natural History Museum sold this collection of mosses following the successful sale of several other sets of dried herbarium specimens.[16] The privileged provenance of this copy of *Musci Britannici* stands in stark contrast with the impoverished status of its maker, but in so doing it reflects the history of its production and distribution. It is one of about twenty-five sets made by Edward Hobson, a poor warehouseman in Manchester, in 1818. Hobson was born in Ancoats Lane, a working-class industrial area of Manchester, in 1782, but from the age of three was raised by an uncle in Ashton-under-Lyne following his father's death and his mother's subsequent alcoholism. First trained as a muslin weaver, by 1815 Hobson had become a warehouseman. From 1809, he had established friendships with other artisans in the area who collected plants,

Britian & Ireland, Systematically Arranged and Described; with Plates Illustrative of the Characters of the Genera and Species (London: Longman, Hurst, Rees, Orme, and Brown, 1818), pp. viii, x.

15 L. Daston, 'Scientific Objectivity with and without Words', in P. Becker and W. Clark (eds.), *Little Tools of Knowledge: Historical Essays on Academic and Bureaucratic Practices* (Ann Arbor: University of Michigan Press, 2001), pp. 271–4.

16 Maggs Bros Ltd, Catalogue 1224 (1997), 'Medicine, Science and Natural History', item 207.

but it was Hobson who stimulated an interest in mosses.[17] In order to identify the mosses that they found, Hobson visited Chetham's Library in Manchester to consult the most authoritative book on moss classification.[18] Unable to afford a microscope, the only instrument he had to examine his specimens was a 'common pocket lens'.[19]

When William Hooker and Thomas Taylor embarked on their monograph, mosses were regarded as fiendishly difficult – hard to see when growing, impossible to investigate without a microscope, and with no stable classification. Their study required exceptional powers of observation and, given how few botanists collected mosses, a dedication to obtaining specimens. In a botanical community consisting largely of private individuals, held together by correspondence, exchanges of specimens and information, and the bonds of friendship thus generated, the discovery of a keen observer, regardless of social class, was greeted with the same delight as the discovery of a rare plant. On hearing about a workingman whose particular skill lay in the ability to find mosses, Hooker therefore made a point of meeting Hobson for the first and only time in Manchester in 1815. Hobson, who had been allowed a couple of hours off from his work as a packer in a warehouse, delighted Hooker with 'some very excellent mosses' and by 'how well he had named his specimens'. 'I hardly ever saw a man possessed of more enthusiasm than this poor fellow', Hooker declared soon after the meeting.[20] By way of encouragement, he gave Hobson his Ellis aquatic microscope (Figure 5.4). The instrument had been Hooker's 'companion for many years', which

17 'Edward Hobson' in H. C. G. Matthew and B. Harrison (eds.), *Oxford Dictionary of National Biography: From the Earliest Times to the Year 2000*, revised edn, 60 vols. (Oxford: Oxford University Press, 2004) (hereafter *ODNB*).
18 J. Moore, 'A Memoir of Mr. Edward Hobson, Author of *Musci Britannici, &c*', *Memoirs of the Literary and Philosophical Society of Manchester*, 2nd series, 6 (1842), pp. 297–324, on p. 307. According to G. P. Greswell, *Bibliotheca Chethamensis: Sive Bibliothecae publicae Mancuniensis* (Manchester: J. Harrop, 1826), p. 113, Chetham's Library included J. Hedwig, *Descriptio et Adumbratio Microscopico-Analytica Muscorum frondosorum*, 2 vols. (Leipzig: In bibliopolio I. G. Mülleriano, 1787–93), a folio work consisting of illustrations and Latin descriptions of mosses. Chetham's Library, founded in 1653, was one of the very few public libraries in England before the Public Libraries Act of 1850.
19 Moore, 'A Memoir of Mr. Edward Hobson, Author of *Musci Britannici, &c*', p. 321.
20 W. J. Hooker to D. Turner, 14 October 1815, Royal Botanic Gardens, Kew (hereafter RBGK), 'Sir W. J. Hooker Letters', vol. 1, fols. 200–1, WJH/2/1.

Figure 5.4 An Ellis-type aquatic microscope, similar to the microscope given to Edward Hobson by W. J. Hooker in 1815. Image © Whipple Museum (Wh.1824).

allowed him to know exactly what could be seen through such a microscope.[21]

Hobson not only supplied Hooker with fine specimens of rare and new species of mosses, but also, with Hooker's encouragement and guidance, produced several sets of dried specimens for sale arranged according to Hooker and Taylor's monograph, which was also published in 1818. In early-nineteenth-century Britain, when botany was pursued mainly by independent individuals scattered across the country, often with little or no access to the few public collections of note, herbaria were largely private collections. Moss specialists in particular collected in the field as much as they prepared and studied dried specimens in their cabinets in order to build up their collections, even if they also employed collectors to travel further afield. There was therefore a market for exsiccatae. Hobson's *Musci Britannici* sold for £1, and was widely admired for its excellence and beauty. For Hobson, producing sets of specimens both enhanced his reputation and was a way of making some extra money.[22] The context of the making of the *Musci Britannici* clearly shows the interaction of patronage, commerce, polite exchange, and working-class participation in science.

---

21 Moore, 'A Memoir of Mr. Edward Hobson, Author of *Musci Britannici*, &c', p. 27. The 'Ellis' was a dissecting microscope with moving objective; see G. L'Estrange Turner, *The Great Age of the Microscope* (Bristol: Adam Hilger, 1989), p. 270.

22 £1 was probably more than Hobson's weekly wage (Moore, 'A Memoir of Mr. Edward Hobson, Author of *Musci Britannici*, &c', p. 322).

The production of sets of specimens for sale reveals a division of labour and distinctions in intellectual or social status. Apart from the initial identification and arrangement of specimens, gentlemen botanists regarded the preparation of exsiccatae as unremunerative and time-consuming work, undertaken only when essential for the benefit of science. Hooker, for example, rejoiced in the public interest in the Reverend Miles Joseph Berkeley's sets of fungus specimens, but regretted 'the great *manual* labor [*sic*] you have in collecting & preparing the specimens'.[23] In contrast, when the Scottish workingman Thomas Drummond began making extraordinary discoveries of mosses in Scotland, Hooker thought it entirely appropriate to encourage him, as he had Hobson, to prepare exsiccatae for sale. Aware that Drummond had a not very 'creditable' character, rather than send him money Hooker proposed to provide him with 'five pounds worth of neatly done up books' in which to fasten the specimens, and to take in return some copies of the work.[24] Drummond was later sponsored to collect in America with the aim of making exsiccatae, but his sudden death in Cuba in 1835 left Hooker feeling obliged to 'convert what specimens of plants are in hand into money' for the benefit of Drummond's family. To this end, Hooker recruited the help of the moss expert William Wilson in Warrington, who was willing to identify Drummond's mosses but not to prepare the exsiccatae. Instead, he considered hiring 'some neat handed female willing to work for 6^d or 1/– a day' to fasten down the specimens, before persuading his wife to do the work.[25]

Wilson's stress on neat-handedness in preparing exsiccatae is telling, and Hobson struggled more with the basic manual skills of laying down, ordering, and labelling specimens than might appear from his *Musci Britannici*. While Hooker acknowledged that he did

23 W. J. Hooker to M. J. Berkeley, 3 September 1836, NHM, Berkeley Correspondence, vol. 7. Emphasis in the original.

24 W. J. Hooker to Lyell, 16 November 1823 (Kinnordy Archive). Drummond possessed a 'fatal propensity for strong drink' (*ODNB*). He did, however, produce two volumes of *Musci Scotici; or, Dried Specimens of the Mosses That Have Been Discovered in Scotland; with Reference to Their Localities* in 1824 and 1825, and *Musci Americani; or Specimens of the Mosses Collected in British North America, and Chiefly among the Rocky Mountains* in 1828.

25 W. Wilson to W. J. Hooker, 15 November 1839, RBGK, Directors' Correspondence, vol. 13, letter 174; Wilson to Hooker, [16 March 1840], RBGK, Directors' Correspondence, vol. 15, letter 245. Drummond's mosses were issued in 1841 as *Musci Americani; or, Specimens of Mosses, Jungermanniae, &c. Collected by the Late Thomas Drummond, in the Southern States of North America*, with the title page stating that they were arranged and named by W. Wilson and W. J. Hooker.

'not know any Naturalist who has searched for Mosses more successfully than Hobson has done in their native stations, nor one who
has discriminated them more accurately', his efforts in bringing out
Hobson's work were directed largely to improving Hobson's manual
skills.[26] From the very start of their exchange, Hooker had urged
Hobson to take more care in drying specimens; he was still com-
plaining in 1818 that 'the specimens you have sent me if they were
ever so rare are hardly fit for my herbarium the leaves are so twisted
and muddled'.[27] Hooker had also criticised Hobson's preparation of
a specimen that had arrived 'so loaded with the earth on which it
grows that I can hardly distinguish the fructification nor fasten it
down in my herbarium'. In preparing exsiccatae, neatness was
essential. Hooker sent Hobson a published set of Swiss mosses to
act as a model, and suggested that Hobson

> make up a hundred good specimens ... & fasten them down
> *neatly* upon paper of the size & form of the Swiss ones ... There is
> no need for so very smart a cover as the one I send. But the whole
> should be got up *very neatly* ... Whatever you put in dry carefully
> & let me see specimens ... that I may confirm the names ...
> Observe not to dry *thick tufts* of specimens, but rather divide them
> & let them be slightly pressed, so that they may lie well between
> the papers.[28]

Hobson, acting on this advice, prepared a preliminary set of
mosses which 'much pleased' Hooker, but also produced another
spate of instructions. The paper must be thicker, the casings must
accommodate the number of pages exactly, the pages must be cut
'with an instrument at the Bookbinders', the ribbands with which the
casings were tied needed to be narrower, and the little bands of paper
used to fasten down some mosses should be as small as possible and
only used for woody stems. 'I have sent a list of 100 arranged &
named correctly', Hooker told Hobson, suggesting he add 'the places
of growth to such as are not very common'. Two days later Hooker
remembered to remind Hobson not to place his mosses in the same
place on every page but to vary their positioning so that the pages lay

26 J. Moore, 'A Memoir of Mr. Edward Hobson, Author of *Musci Britannici, &c*',
   [2nd edn] (Manchester: Simms & Dinham, and Samuel Boardman, 1843),
   title page.
27 W. J. Hooker to E. Hobson, 27 October 1816 and 1 August 1818, Herbarium
   Archive, Manchester Museum, University of Manchester, GB 2875 BAL/1
   (MANCH 595153), Edward Hobson correspondence (hereafter MM), pp. 153
   and [160].
28 W. J. Hooker to E. Hobson, 21 June [1817], MM, p. 155.

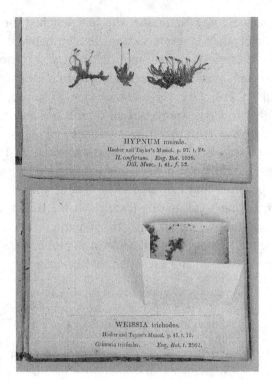

Figure 5.5 Pages from Edward Hobson's *Musci Britannici* showing (top) fixed moss specimens and (bottom) a small pocket containing loose specimens. Image © Whipple Museum (Wh.4577).

flat in the case, to fasten them with 'strong gum mixed with flour-paste', and to enclose very minute ones in little cases of paper (Figure 5.5).[29]

Then there were the instructions for the labels. If Hobson did not think he could get the labels printed, Hooker pointedly suggested that perhaps 'some friend' could 'write them in a good hand.'[30] Neatness was an attribute much valued and noted by botanists, and included the labelling of specimens. Good handwriting was thus another manual skill necessary for the maintenance of a well-ordered collection. It was for the herbariums of expert cryptogamists and genteel collectors that the specimens in *Musci Britannici* were destined. Hobson chose to have his labels printed.

29 W. J. Hooker to E. Hobson, 18 August 1817 and 20 August 1817, MM, pp. 157 and 158. The Whipple set no longer has the ribbons by which the loose sheets were secured in the case, but the inside of the case shows faint marks on the side, top, and bottom where they were positioned (see Figure 5.1).
30 W. J. Hooker to E. Hobson, 18 August 1817, MM, p. 157.

The Publication Circuit

The ability to produce multiple sets of the same plants was limited by the quantity of the rarest specimen.[31] It was therefore essential to build up stocks of specimens before embarking on the sale of exsiccatae. However, given the time-consuming labour of producing sets of specimens as well as the costs involved in printing labels and buying paper and cases, it was also important that Hobson, before starting work, acquired subscribers for the *Musci Britannici* to ensure that he made 'no more than are spoken for'.[32] 'I will do all I can (if you determine upon it) to recommend it', Hooker assured Hobson, '& will mention it in my Muscologia, which is now about to appear.'[33] Hooker and Taylor announced Hobson's intention to produce exsiccatae, pointing out how much more accurate and how much cheaper sets of specimens were than plates.[34] The orders began to flow in.

But this was not all that flowed in. Both specialists and enthusiasts began to offer Hobson mosses. The production and distribution of the *Music Britannici* thus reveals how even a commercial enterprise was dependent on the system of knowledge and specimen exchange built up through correspondence networks for mutual benefit of all participants. 'I shall be very glad at any time to supply you with any specimens in my power, that may be likely to be of service to you', the botanist and clergyman William Bree told Hobson after purchasing his copy of *Musci Britannici* and ordering two more copies for Warwickshire botanists.[35] Edinburgh botanist Robert Kaye Greville placed an order after he had seen his friend John Stewart's copy of Hobson's 'valuable work', offering at the same time a good stock of some specimens.[36] Greville continued to supply Hobson with specimens, and by 1820 hoped that what he sent might 'hasten the appearance of a second volume'.[37] Stewart, a botanical lecturer in

---

31 After distributing Drummond's mosses, Hooker and Wilson complained that his 'stock' of American liverworts was 'very meagre; scarcely sufficing for the 20 sets which are already sold' (W. Wilson to M. J. Berkeley, 28 January 1843, NHM, Berkeley Correspondence, vol. 11).

32 W. J. Hooker to E. Hobson, 21 June 1817, MM, p. 155.

33 W. J. Hooker to E. Hobson, 21 June 1817, MM, p. 155.

34 Hooker and Taylor, *Muscologia Britannica*, p. x.

35 W. T. Bree to E. Hobson, 23 July 1818, MM, p. 123.

36 R. K. Greville to E. Hobson, 28 June 1819, MM, p. 131. John Stewart issued *Hortus Cryptogamicus Edinensis* (exsiccatae) in 1819; see R. Desmond, *Dictonary of British and Irish Botanists and Horticulturists* (London: Taylor & Francis, 1994).

37 R. K. Greville to E. Hobson, 30 September 1820, MM, p. 133.

Edinburgh, also offered to help Hobson, and commented that anyone in Edinburgh acquainted with cryptogamic botany was 'quite delighted' with the *Musci Britannici*.[38] Hooker's close friend and keen muscologist Charles Lyell (father of the geologist) sent choice specimens to Hobson, and also hoped that the demand for the first volume would encourage Hobson to publish another volume very speedily.[39] Susannah Corrie of Woodville near Birmingham regretted she was prevented by illness from sending more specimens, while the plant collector Margaret Stovin of Derbyshire wondered how Hobson made the time 'with other necessary avocations to compleat so beautiful a work'.[40]

Time was the crucial issue. Hooker thought that preparing moss exsiccatae might be a way for Hobson to earn 'a few shillings', but acknowledged that Hobson alone could judge the 'value' of his time and whether it was worth undertaking such work.[41] As demand for the first volume of *Musci Britannici* grew, Hobson himself began to express concern that making up the volumes was so time-consuming that it left him little opportunity to collect mosses. It was only with the help of Hooker and Lyell in particular that Hobson was able to complete twenty-five copies of the first volume and then embark on twenty sets of a second volume.[42] However, progress was so slow that the naturalist John Edward Gray, then an assistant in the British Museum, wrote to the botanist Roberts Leyland of Halifax in July 1822 to enquire whether he knew '*Mr. Hobson* the author of the *Musci Brittanici*, a most excellent collection of British specimens of Mosses'. 'I have his first part & wrote directly for the second, but I have [not] heard any thing from him & have lost his Direction,' Gray explained to Leyland.[43] By this time, Hobson had, in fact,

---

38 J. Stewart to E. Hobson, 6 July 1818, MM, p. 179. When Edinburgh lecturer James Robinson Scott showed his class Hobson's *Musci Britannici*, his student William Jardine immediately placed an order for his own copy (Jardine to Hobson, 13 July 1818, MM, p. 165). Scott went on to issue *Herbarium Edinense* (exsiccatae) in 1820 (Desmond, *Dictonary of British and Irish Botanists and Horticulturists*).

39 C. Lyell to E. Hobson, 18 January 1819, MM, p. 170.

40 S. Corrie to E. Hobson, 18 April [1822] and 6 November 1823, and M. Stovin to Hobson, 12 April 1829, MM, pp. 125, 126, and 180. Susannah Corrie was the wife of the Unitarian minister John Corrie, who was president of the Birmingham Philosophical Society from 1812 to 1839.

41 W. J. Hooker to E. Hobson, 21 June 1817, MM, p. 155.

42 Draft of a letter from E. Hobson to W. J. Hooker, n.d., MM, p. 159; E. Hobson to C. Lyell, 3 February 1819 (Kinnordy Archive).

43 J. E. Gray to R. Leyland, 24 July 1822, West Yorkshire Archive Service, Calderdale, SH:7/JN/B/66/78.

begun preparing his second volume, and Hooker was one of the first to receive a copy in 1822. Declaring himself 'much pleased' with the 'very interesting volume', Hooker urged Hobson to supply as quickly as possible copies to the purchasers of the first volume who wished to buy the second.[44]

The publication of a second edition of Hooker and Taylor's *Muscologia Britannica* in 1827, detailing some newly discovered species, prompted Hobson to consider a third volume of *Musci Britannici*. At this point it was not time that Hobson lacked but specimens, as he explained to Hooker: 'In consequence of the Bankruptcy of my late Master ... I am now out of employment for some time and should have time to go on with a third Vol^{m}. of British Mosses &c if I had sufficient quantity of some species that are mentioned in the annexed list.'[45] On this occasion, however, Hooker was discouraging. He did not possess sufficient specimens himself and did not think Hobson could obtain adequate supplies to make up volumes 'without great delay'; instead he suggested that the volume be devoted to cryptogams more generally and also mentioned that Hobson could obtain Scottish mosses by 'entering into an exchange' with Hooker's Scottish protégé, Thomas Drummond.[46]

The production of Hobson's *Musci Britannici* shows that, even as a commercial object, it depended upon the networks of polite exchange. But it is important to recognise just what was being purchased. It was not the case that gentlemen like Lyell provided Hobson with specimens that were then sold back to them. The principle of gift exchange in natural history with respect to specimens and knowledge was not violated.[47] Rather, what was being paid for was the manual labour involved in making exsiccatae and the quality of the specimens included therein. This was especially the case with species that were difficult to find 'in fruit', that is with the capsules that were essential to identifying some species of moss. The difficulties of collecting sufficient fruiting plants, the time-consuming fixing of specimens, and the system of payment may have

---

44 W. J. Hooker to E. Hobson, 8 June 1822, MM, p. 162.
45 E. Hobson to W. J. Hooker, 20 June 1827, RBGK, Directors' Correspondence, vol. 8, fols. 32–3 (letter 22).
46 W. J. Hooker to E. Hobson, 20 March 1828, MM, p. 164. Hobson, in fact, persevered and a few copies of a third volume of *Musci Britannici* were produced (see, for example, the set in the Herbarium, Manchester Museum, University of Manchester).
47 For acceptance of these norms of exchange by all social classes, see A. Secord, 'Corresponding Interests: Artisans and Gentlemen in Nineteenth-Century Natural History', *British Journal for the History of Science*, 27 (1994), pp. 383–408.

made preparing exsiccatae unappealing to gentlemen botanists, but such work did not threaten the norms of exchange networks.

Conclusion: Exsiccatae Unbound

The role of different observational tools for seeing in botany is exemplified in the early career of William Hooker's son Joseph. When Joseph Hooker set off as assistant surgeon and ship's botanist on an expedition to the southern oceans and Antarctica, his ambitions included describing a genus of mosses for his first paper at the Linnean Society of London.[48] Conditions were hardly favourable. In rough icy seas often 'he & his microscope had to be lashed to the table from the rolling of the ship'.[49] Nonetheless, Joseph managed to produce copious drawings of highly magnified dissections that were essential for identification. His appreciation of the rationale behind the classification of this difficult group of plants had, however, been formed much earlier through exsiccatae.[50] In the calmer waters of Berkeley Sound, Falkland Islands, he received a reminder of what had inspired his love of mosses. His father had sent him, half way round the world, the recently published 'Memoir of Mr. Edward Hobson'.[51] Although Joseph regarded himself 'a born Muscologist' because both his mother and his father independently began their botanical studies with the mosses, his latent powers were, he claimed, stimulated 'by a book in my father's library ... by Edward Hobson, of Manchester'.[52]

48 J. D. Hooker to W. J. Hooker, 25 November 1842, RBGK, 'J. D. Hooker Correspondence 1839–45 from Antarctic Expedition', letter 72 (fols. 128–33), JDH/1/2.
49 W. J. Hooker to W. Wilson, [1843], RBGK, 'Letters from W. J. Hooker', fol. 90, WJH/2/8.
50 For more extensive discussion of exsiccatae as observational tools, see A. Secord, 'Pressed into Service: Specimens, Space, and Seeing in Botanical Practice', in D. N. Livingstone and C. W. J. Withers (eds.), Geographies of Nineteenth-Century Science (Chicago: Chicago University Press, 2011), pp. 283–310. Exsiccatae are still a common way for lichen taxonomists to convey and distribute their species concepts.
51 This copy of Moore, 'A Memoir of Mr. Edward Hobson, Author of Musci Britannici, &c' bears the inscription 'J. D. Hooker. R.N. | H.M.S. "Erebus" | Received Berkeley Sound | Falkland Isld^s | Novr. 23. 1842.' (RBGK, Library, P920.HOB).
52 L. Huxley, Life and Letters of Sir Joseph Dalton Hooker, 2 vols. (London: John Murray, 1918), vol. 1, pp. 3, 5; and 'Sir Joseph Hooker's Reminiscences of Manchester', Lancashire Naturalist, 1 (1907–8), pp. 118–20, p. 119, reprinted from Manchester Guardian, 30 March 1898, p. 10. Joseph's mother, Maria Hooker, was the daughter of Dawson Turner, who had studied and published on cryptogamic botany.

The *Musci Britannici* probably remained part of William Hooker's library until his death in 1865, when his cryptogamic collections, his private property up to this point, were sold to the Royal Botanic Gardens at Kew, where he had served as director from 1840.

The Yorkshire botanist and clergyman James Dalton, Joseph Hooker's godfather and William Hooker's close friend, probably kept his copy of Hobson's *Musci Britannici* in his library too. But this presented Dalton with a dilemma when he decided to donate his moss herbarium to the York Philosophical Society. He wished to include the mosses prepared by Hobson in his collection as they possessed 'the *authority* of a good Muscologist'. There was only one solution. Dalton hoped that Hobson would not be 'offended' by his 'begging to be considered a purchaser' of another set of specimens because he could not bear to break up the 'beautiful' set he had already received.[53] Moreover, for those actively studying mosses, dissection of specimens was often essential; for this reason the Irish botanist Thomas Taylor had asked for duplicates of Drummond's American mosses 'in order that he might be able to preserve the published spec^{ms}. from mutilation'.[54]

Hobson's *Musci Britannici* was an observational tool. Yet, from the perspective of the present, it is all too easy to regard it only as a self-explanatory taxonomic exercise showing how a particular group of plants was classified at a specific point in time. Hobson's *Musci Britannici* is thus taken to represent the end point of a collection rather than a stimulus to observation. Many of the copies in public institutions reinforce this notion. Where preserved in libraries, the scientific relevance of the *Musci Britannici* has dwindled to little more than a collection of specimens trapped in an obsolete taxonomic system. The most extreme case is the copy in Chetham's Library, Manchester, which has been bound as a book. In contrast, when found in herbariums, the pages of *Musci Britannici* are either dispersed among the larger collection of plants, or, if kept in their covers, reordered by later users who have arranged and renamed the specimens according to more recent

53 J. Dalton to E. Hobson, 20 March [1819], MM, p. 127. W. J. Hooker gave Joseph the second name of Dalton after James Dalton, and both editions of Hooker and Taylor's *Muscologia Britannica* are dedicated to Dalton.

54 R. Spruce to W. Wilson, [31 October 1843], Archives of the New York Botanical Garden, William Wilson Papers.

classifications.[55] Even those copies that remain in their original format relatively intact, like the copy in the Whipple Museum, no longer explicitly impart their function as a method for learning how to observe. It is by considering both production and consumption that the *Musci Britannici* shows its potential as an instrument of observation. The point of exsiccatae was not only to convey a systematic understanding of difficult groups of plants, but also to hone observational skills by guiding and training the eye. The publication of specimens labelled with their species names and arranged into genera provided a way for the botanical community to calibrate its vision and test new classifications.

55 For example, the Manchester Central Library set (BR 588.2 Ho 1) was later rearranged according to William Wilson's *Bryologia Britannica* (London, 1855), while the set in the Olney herbarium, Brown University, was reorganised according to P. Bruch, W. P. Schimper, and T. Gümbel's *Bryologia Europaea* (6 vols., Stuttgart, 1836–55).

6   Ideas Embodied in Metal: Babbage's Engines Dismembered and Remembered*

SIMON SCHAFFER

Memory, that treacherous friend but faithful monitor, recalls the existence of the past.

(Charles Babbage, *The Ninth Bridgewater Treatise: A Fragment*, 1837)[1]

In the last few years of his life, although his memory for general matters had become impaired, he still retained a perfect recollection of the details of his workshops.

('Mr Charles Babbage', *Athenaeum*, October 1871)[2]

I began to think of making a small piece of Calculating Machinery to embody the ideas of my father ... I wished, if I could, to justify the confidence he had shown in me by embodying some of his ideas in metal.

(Henry Babbage, *Memoirs and Correspondence*, 1915)[3]

It has often been said that museum collections embody memories, their artefacts and displays able to summon past experience through artful disposition. Complex assemblages of objects, texts, and people within and around museums allow the recall of what might otherwise seem lost or at least beyond reach. It is therefore perverse, if understandable, that such gatherings are so often used rather to evoke singular heroic individuals than to realise the extended webs

* Thanks for their generous help are due to Will Ashworth, Jenny Bulstrode, and Joshua Nall.

1 C. Babbage, *The Ninth Bridgewater Treatise: A Fragment*, 2nd edn (London: John Murray, 1838), p. 161.
2 'Mr Charles Babbage', *Athenaeum* no. 2296 (28 October 1871), p. 564.
3 H. Babbage, *Memoirs and Correspondence* (London: William Clowes, [?1915]), p. 225. The preface is dated August 1910; but the Science Museum Library copy (92 BAB) was inscribed to Henry's nephew Herbert Ivan, son of Benjamin Herschel Babbage, in April 1915. Other copies were also inscribed in 1915: see G. Tee, 'The Heritage of Charles Babbage in Australasia', *Annals of the History of Computing*, 5 (1983), pp. 45–59, on p. 47 n. 1; and I. Bernard Cohen, 'Babbage and Aiken', *Annals of the History of Computing*, 10 (1988), pp. 171–93, on p. 191.

of labour processes and social relationships embodied in each object. Museums in this sense risk becoming – or might deliberately aim to become – systems for the attribution and celebration of exclusive authorship and property.[4] Such highly charged dilemmas of collective and hagiographic memory and the prerogatives of ownership were especially marked in nineteenth-century exhibitions and galleries of science and art. Property relations, collective enterprise, and rights of labour were the very stuff of the political economy of display during the Age of Capital. A manifesto for the new Polytechnic Institution on London's Regent Street sent in 1839 to a nearby resident and supporter, the mathematician Charles Babbage, explained how the costly investment in 'its laboratory, its theatre and its splendid Gallery is well adapted for the display of scientific discoveries and were it truly in scientific hands, so that scientific discoveries were thrown off *hot* from the brain and before they had become public property by publication, sufficient novelty would be produced to excite public attention and to make it pay'.[5]

Dominant centres of scientific inquiry and accumulation, museums sustained vital if troublesome linkages between sites of artisan manufacture in urban workshops and the emergent factory system and sociable realms of theatre and consumption. Objects on show, turned into commodities, depended quite directly on hosts of workers, clients, and patrons elsewhere, while the status of technical knowledge embodied in such objects hinged on how they were publicly displayed.[6] Museums and galleries were made into repositories of historical narratives and travellers' tales. In a society cultivating a renewed obsession with sentimental and evocative relics and memorials, powerful fantasies of access to the past and the exotic were nourished by this set of relations, even though their realities

4 R. Lumley (ed.), *The Museum Time-Machine: Putting Cultures on Display* (London: Routledge, 1988); S. Crane (ed.), *Museums and Memory* (Stanford: Stanford University Press, 2000); and S. Vackimes, *Science Museums: Magic or Ideology?* (Almeria: Albedrio, 2008).
5 Cayley to Babbage, November 1839, British Library MS Add.37191, fol. 271 (stress in original); see B. Weeden, *The Education of the Eye: History of the Royal Polytechnic Institution* (Cambridge: Granta, 2008), pp. 12–13.
6 I. R. Morus, 'Sights and Sites: The National Repository and the Politics of Seeing in Early Nineteenth-Century England', in C. Berkowitz and B. Lightman (eds.), *Science Museums in Transition: Cultures of Display in Nineteenth-Century Britain and America* (Pittsburgh: University of Pittsburgh Press, 2017), pp. 87–107, on p. 89 (on the status of applied knowledge); and J. Bulstrode, 'The Industrial Archaeology of Deep Time', *British Journal for the History of Science*, 49 (2016), pp. 1–25, on pp. 16–23 (on exhibits and embodiment).

were often effaced.[7] Public museums were stocked with loot from antiquarian and oriental sites brought to the metropole through military and rapacious expeditions. They hosted technical equipment designed to signal the recoverable past and conjectural future of a politically and economically unstable society.

No doubt the fraught connections that flourished in Victorian capitals between state agencies, private accumulation, and commercial projects gave such displays significance. Commenting on the display of newfangled calculating engines at the 1862 South Kensington international exhibition, the medical statistician William Farr of the General Register Office explained that 'there are besides the thousands of machines in the clouds of inventors' brains, many ingenious and beautiful machines in exhibitions of no practical use whatever. How can the spectator know whether they will execute genuine work at all?' Judgment depended on objects' track records and their makers' promises. For a candidate to be judged a discovery or invention, expert public communities had somehow to go back over traces of labour and material manipulation in a kind of retrospective inquiry: the exhibitions helped nourish these genealogical exercises and were subject to radical criticism from artisan activists keen to redistribute the property rights of inventors and masters.[8] In priority disputes and labour conflict, museological memory became a matter of material politics. Connections with the accumulated records and imagined future of labour and materials made such shows resemble devices that might somehow move through time, through the reconstruction, conservation, and show of their culture's antecedents and subjects. This was an indispensable aspect of what has been called the museums' 'uncanny social technology'. Reflection on museological memory thus highlights themes such as the hard labour of salvage and reconstruction and the commemorative practices of nostalgia and piety.[9]

---

7 D. Lutz, *Relics of Death in Victorian Literature and Culture* (Cambridge: Cambridge University Press, 2015), pp. 8–9; and A. Craciun, *Writing Arctic Disaster: Authorship and Exploration* (Cambridge: Cambridge University Press, 2016), pp. 34–7.

8 W. Farr, *English Life Table* (London: Longman, 1864), p. cxxxix. See W. J. Ashworth, 'England and the Machinery of Reason 1780–1830', *Canadian Journal of History*, 35 (2000), pp. 2–36; C. Pettitt, *Patent Inventions: Intellectual Property and the Victorian Novel* (Oxford: Oxford University Press, 2004), pp. 88–110; and C. Macleod, *Heroes of Invention: Technology, Liberalism and British Identity 1750–1914* (Cambridge: Cambridge University Press, 2007), pp. 153–80.

9 For 'uncanny social technology' see D. Preziosi, 'Brain of the Earth's Body: Museums and the Framing of Modernity' (1996), cited in T. Baringer, 'Victorian

Themes that became especially significant included the exhibition of technology in museums. The display of machinery helped turn such museums into something like memory devices. This transformation was especially marked because of the vexed issue of the ownership of technology. In London between 1820 and 1833 Babbage was engaged in producing a calculating engine to manufacture mathematical tables for fiscal and navigational purposes. Significant public cash was invested in the engine, its assembly dependent on pugnacious and mutable relations with skilled labour in the city's machine-tool workshops. His scheme had to turn memory into mechanism in intricately indispensable features of the engine's operation. He worked out the mechanical principle for his difference engine to govern each of its figure wheels during carriage, the process in which wheels were compelled to pass from nine back to zero. Memory was embodied both in the machine and in the relations established by the machine's display. Writing of what he called Babbage's proposition 'to substitute an automaton for a compositor', the industrial publicist and science lecturer Dionysius Lardner observed that this vital principle involved 'in effect a *memorandum* taken by the machine of a carriage to be made'. In a timely polemic about industrial expositions, intellectual property, and the calculating engines written for the Great Exhibition, Babbage himself claimed that 'there is in this mechanism a certain analogy with the act of memory'. During such carriage, a lever was pushed back, 'the equivalent of the note of an event made in the memory', then a spiral arm would restore the lever and register the new number, a movement which 'in some measure resembles the endeavour made to recollect a fact'.[10]

Dependent on workers' skilful gear-cutting and draftsmanship, these processes of mechanised memory formed part of a practical culture of labour and performance, systematically embodied in the

Culture and the Museum', *Journal of Victorian Culture*, 11 (2006), pp. 133–45, on p. 133. Compare R. Altick, *The Shows of London* (Cambridge: Belknap Press, 1978), pp. 375–89; and I. R. Morus, 'Manufacturing Nature: Science, Technology and Victorian Consumer Culture', *British Journal for the History of Science*, 29 (1996), pp. 403–34.

10 [D. Lardner], 'Babbage's Calculating Engine', *Edinburgh Review*, 59 (July 1834), pp. 263–327, on p. 297 (my stress); C. Babbage, *The Exposition of 1851*, 2nd edn (London: John Murray, 1851), p. 182. Compare C. Babbage, *Passages from the Life of a Philosopher* (London: Longman, 1864), p. 62: 'the mechanical means I employed to make these carriages bears some slight analogy to the operation of the faculty of memory'; see W. J. Ashworth, 'Memory, Efficiency and Symbolic Analysis: Charles Babbage, John Herschel and the Industrial Mind', *Isis*, 87 (1996), pp. 629–53, on pp. 649–52; and M. L. Jones, *Reckoning with Matter: Calculating Machines, Innovation and Thinking about Thinking from Pascal to Babbage* (Chicago: University of Chicago Press, 2016), pp. 47–55.

urban milieux of display. The difference engines' fate was largely governed by the complex class topography of artisan and managerial enterprise. The Lambeth journeyman Richard Wright, who started as Babbage's valet in the late 1820s before touring the principal northern engineering works and becoming a master artisan for Babbage's mechanics projects, explained how paperwork and metal-work had to be combined in the calculating engine: 'a journal carefully connecting the different parts might serve the future as a perfect guide for its completion'.[11] The decisive experience of young workmen such as the Manchester artisan and lathe-maker Joseph Whitworth, who worked for the master engineer Joseph Clement in south London on the calculating engine project in 1830–1, con-firmed Wright's judgment that the project's reputation as a training system was significant: 'a man who has worked at it has a greater chance of the best work', Wright told Babbage from Manchester. Though an indispensable component of drives to automate and mechanise production, the machine tools designed by Clement and Whitworth demanded ever more intense craft skills in forging, assemblage, and maintenance; and the valorisation of such artefacts relied on reputation and the memory of workers' skill.[12]

Paper and metal memoranda passing between centres of calcula-tion and of work defined this topography's troubles. Conflicts about the disciplinary and labour systems of the naval dockyards, involving several of the engineers who would form Babbage's closest collabor-ators, had already intensely raised the basic spatial and political problems of inscribed plans, artisan skill, and managerial control.[13] Babbage's relations with his government patrons and with Clement's team hinged significantly on the physical and social distance between Lambeth workshops, Whitehall offices, and Babbage's domestic quarters in fashionable Marylebone. The machine eventually became memorable for its notoriously disjointed makers and dismembered materials. It was said 'Mr. Babbage made Clement. Clement made Whitworth. Whitworth made the tools.' 'When I first employed

11 Wright to Babbage, 25 September 1859, British Library MS Add.37197, fol. 440.
12 N. Atkinson, *Sir Joseph Whitworth, the World's Best Mechanician* (Stroud: Sutton, 1996), pp. 26–7; Wright to Babbage, 18 June 1834, British Library MS Add.37188, fol. 390. See R. Samuel, 'Workshop of the World: Steam Power and Hand Technology in Mid-Victorian Britain', *History Workshop Journal*, 3 (1977), pp. 6–72, on pp. 39–40.
13 P. Linebaugh, *The London Hanged: Crime and Civil Society in the Eighteenth Century* (London: Penguin, 1993), pp. 371–401; and W. J. Ashworth, 'System of Terror: Samuel Bentham, Accountability and Dockyard Reform during the Napoleonic wars', *Social History*, 23 (1998), pp. 63–79.

Clement', Babbage bitterly recalled, 'he possessed one lathe (a very good one) and his workshop was in a small front kitchen. When I ceased to employ him he valued his tools at several thousand pounds and he had converted a large chapel into workshops.'[14] Whitworth and Clement both occupied sacred places in Samuel Smiles's pantheon of heroic nineteenth-century engineers: eventually, Clement's workshop was broken up by his nephew Wilkinson, and the relics of the calculating-machine project held there were dismantled and melted down.[15] As relations with Clement soured in the early 1830s, the new chief draftsman Charles Jarvis privately advised Babbage that 'the plan I wish to recommend is that the designs and drawings be all made on your premises and under your immediate inspection'. The eminent engineer Marc Brunel agreed: the engine must be built at a site 'close to your own garden', certainly not where the new Whig administration proposed, in a workroom at the British Museum. The Treasury even suggested that Babbage might be persuaded to 'take a residence nearer to the Museum'.[16]

Though the difference engine was never fully completed, various of its components and relics were in fact destined to spend time nearer and often inside museums and showrooms. Once on display, the calculating machine was welded to histories of its development and fate: memoirs of its construction, funding, ownership, and disassembly always formed part of every attempt to explain its function. The principal material realisation of the addition and carriage mechanisms, completed in late 1832, was kept for display in his Marylebone drawing room and shown to Babbage's house guests in re-enactments of the engine's philosophical and economic lessons about mind and matter.[17] Other models of the difference engine were proposed. In 1834 Babbage's publicist Lardner toured northern England and Scotland lecturing on the engine's capacities, reportedly drawing vast crowds and cash. He hired the Charing Cross instrument-maker Francis Watkins, former apparatus curator at University College London, where Lardner briefly occupied the

14 Babbage memorandum, 9 November 1869, British Library MS Add.37189, fol. 499.
15 S. Smiles, *Industrial Biography: Iron Workers and Tool Makers* (London: John Murray, 1863), pp. 253–7; and H. Babbage, 'Babbage's Analytical Engine', *Monthly Notices of the Royal Astronomical Society*, 70 (1910), pp. 517–20, on p. 518.
16 Jarvis to Babbage, 25 August 1833, British Library MS Add.37188, fol. 39; and Brunel to Babbage, 11 January 1831, British Library MS Add.37185, fol. 439. See Jones, *Reckoning with Matter*, pp. 206–7.
17 H. Babbage, *Memoirs and Correspondence*, p. 89; Babbage, *Passages from the Life of a Philosopher*, pp. 425–6; Babbage, *The Ninth Bridgewater Treatise*, pp. 32–43.

natural philosophy chair until 1831. Over the winter of 1833–4, in consultation with Babbage and aided by drawings from Babbage's eldest son Benjamin Herschel Babbage, Watkins produced two steel models of the carriage mechanisms and the printer of the difference engine.[18] Lardner then sought to use the models in Manchester, Liverpool, and Sheffield and at the Royal Institution and the British Association, grumbling whenever deprived of the apparatus for his performances. Organisers complained, in turn, that the theme was 'too hard and too scientific': 'I find I have too much disregarded and looked down upon the exhibition of apparatus.' Babbage contacted allies such as Alexander von Humboldt in Berlin, and the Paris science writer and educator Charles Dupin, who went to one of Lardner's shows, to recruit interest in the tour and extend it Europe-wide.[19]

The lectures coincided with further abortive attempts to make demonstration models of the engine. In early 1834 the young Harvard-trained physician Henry Ingersoll Bowditch, who spent 'nearly a whole day and the greater part of the night' at Dorset Street examining the engine, begged Babbage for guidance: 'expense would be to me of little moment, could I hope to shew in America a work of art, which might excite in many a mind trains of thought, which might otherwise remain dormant ... It is the hope of producing some such beautiful result that I wish to present to the scientific world of America, and the mechanics of Boston, a model of the calculating machine.' Bowditch planned to recruit the American instrument-maker Joseph Saxton, colleague of Watkins at London's premier mechanics showroom, the Adelaide Gallery, to produce the device. Saxton promised to visit Babbage to examine the possibilities. Just as negotiations between Babbage and Clement began to collapse, it had become evident that demonstration of the principles of the

18 Watkins to Babbage, 31 December 1833 and 15 January 1834, British Library MS Add.37188, fols. 119 and 160; Lardner to Babbage, [1833], British Library MS Add.37188, fol. 203. See B. Gee, *Francis Watkins and the Dollond Telescope Patent Controversy*, ed. A. McConnell and A. D. Morrison-Low (Farnham: Ashgate, 2014), pp. 276–83.
19 Babbage to Dupin, 30 December 1833, British Library MS Add.37188, fol. 117; Babbage to Humboldt [1833], British Library MS Add.37188, fol. 123; and Lardner to Babbage, 3 January, 11 January, 23 January, 16 February, and 16 October 1834, British Library MS Add.37188, fols. 140, 154, 176, 208, and 494. See J. N. Hays, 'The Rise and Fall of Dionysius Lardner', *Annals of Science*, 38 (1981), pp. 527–42, on p. 529. In 1839 Babbage and Lardner became embroiled in fierce disputes about railway gauges on I. K. Brunel's Great Western Railway.

calculating engine required what Lardner called 'the trickery of the lecture table by the introduction of apparatus'.[20]

When finally abandoned, the relics of the engine, so Babbage recommended, 'should be kept in a warm well-ventilated room', perhaps as a guard against rust and decay, and 'placed where the public can see it, for example the British Museum'. There, he intended, 'it would form a beautiful specimen of the state of the mechanical arts at the time when it was made', a souvenir of skill embodied in mechanism. By summer 1843 it was decided to shift it from Marylebone to the Museum at King's College on the Strand, and the brass, gun metal, and steel parts associated with it were crudely valued for scrap. 'The property should remain in the Government, in the event of its being at any time hereafter required for public use.'[21] Though excluded from the Crystal Palace in 1851, in 1862 it was shifted to South Kensington under the management of the railway engineer William Gravatt, as part of the subsequent international exhibition of industry. To Babbage's fury, King's College refused its return.[22]

In 1872, the year after Babbage's death, his eldest son Benjamin, by then a South Australian engineer, surveyor, and enthusiastic wine-grower, composed a small guidebook for the model machine to aid museum visitors. Benjamin added a drawing he'd made of the

20 Bowditch to Babbage, 18 February 1834, British Library MS Add.37188, fol. 212; V. Y. Bowditch, *Life and Correspondence of Henry Ingersoll Bowditch*, 2 vols. (Boston: Houghton, Mifflin, 1902), vol. 2, pp. 267–8; and Lardner to Babbage 29 March 1834, fol. 288. For Saxton and Watkins see I. R. Morus, *Frankenstein's Children: Electricity, Exhibition and Experiment in Early-Nineteenth-Century London* (Princeton: Princeton University Press, 1998), pp. 83–92. Bowditch was introduced to Babbage through his father, the eminent Boston mathematician Nathaniel Bowditch, actuary, nautical almanac-maker, and translator of Laplace: see Henry Bowditch to Nathaniel Bowditch, 13 December 1833, in Bowditch, *Life and Correspondence of Henry Ingersoll Bowditch*, vol. 1, pp. 68–9: 'How can one tell the effect which the examination of such a machine might produce upon the minds of some of our young and intelligent mechanics?'
21 Babbage to Milne, 1842; Milne to Babbage 5 June 1843; Milne to Babbage 20 July 1843, British Library MS Add.37192, fols. 224, 326, and 381; and A. Filipoupolitti, 'Premises for Exhibition and Use', *Museums History Journal*, 4 (2011), pp. 11–28, on p. 21.
22 Babbage, *Passages from the Life of a Philosopher*, 147–67; *International Exhibition of 1862: Illustrated Catalogue of the Industrial Department, British Division* (London: HM Commissioners, 1862), vol. 2, p. 46 (no. 3012); and L. Purbrick, 'The Dream Machine: Charles Babbage and His Imaginary Computers', *Journal of Design History*, 6 (1993), pp. 9–23, on p. 12. Gravatt also tried to assemble 'a number of separate parts' of the difference engine given him by Babbage: see Gravatt to Jelf (KCL Principal), 7 November 1861, British Library MS Add.37198 fol. 258.

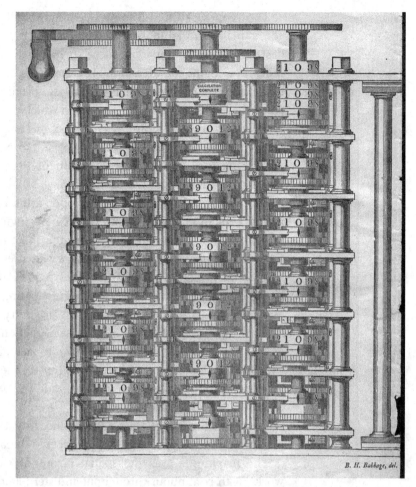

Figure 6.1 Benjamin Herschel Babbage's drawing of the fragment of the difference engine on show at South Kensington in 1872. From *Babbage's Calculating Machine or Difference Engine* (London: HMSO, 1872). Image © Whipple Museum (Wh.2339).

relic's three principal axes and number wheels, an image subsequently treated as the definitive rendering of the device (Figure 6.1). In 1876, under the aegis of the Devonshire Commission on scientific instruction, and its energetic secretary Norman Lockyer, once again a major exhibition was staged at South Kensington of both current and historic apparatus as a means to weld contemporary scientific hardware and education to the material memory of their development. The Babbage model, with its carriage mechanism and ambitious design given due attention as 'a machine for manufacturing tables', went on display and was occasionally put to work; and has stayed in South Kensington ever since.[23]

23 B. H. Babbage, *Babbage's Calculating Machine or Difference Engine* (London: HMSO, 1872); *Catalogue of the Special Loan Collection of Scientific Apparatus at the South Kensington Museum*, 3rd edn (London: Eyre and Spottiswoode, 1877),

The repute of this machine, seemingly safely ensconced as an admirable memorial in its museum, has nevertheless somehow been destabilised by its apparent successor. As his scheme for the first difference engine foundered, Babbage projected a much more ambitious analytical engine, a general-purpose machine never fully to be completed, which demanded a 'store' in which operations and results to be performed by a 'mill' could be mechanically stashed and recovered when needed. In 1835 Babbage told Nathaniel Bowditch, Henry Ingersoll's father and pre-eminent Boston mathematician, that during carriage 'in the old Engine when addition takes place a memorandum is made' and 'the proper arms then pick up these memoranda in *succession*', while in 'the new Engine after addition all the carriages are affected at once (at the *same instant*) the engine *foreseeing* if necessary when a carriage will itself cause carriage to several nines above'. The difference engine's successive carriage would be replaced by a much faster anticipating carriage and thus an accelerated calculation mechanism. Memory must be complemented by foresight. The mechanical function of memory was geared to the capacity to know how to act ahead of time. He explained in his summary memorandum of late 1837 that, 'if the mechanism which carries could be made to *foresee* that its own carriage of a ten to the digit above … would at the next step give notice of a new carriage, then a contrivance might be made by which, acting on that knowledge, it should effect both carriages at once'.[24]

The inspiration was the spatial layout of steam-driven textile works, with continuous throughput and stern labour discipline. Karl Marx's 1860s London writings on industrial capital used Babbage as key evidence that 'the factory is still described in English as a "mill"', and stated that 'Babbage treated large-scale industry from the standpoint of manufacture alone.' In his analysis of steam engineering and the factory system, Babbage explained that 'whenever the individual operation demanding little force for its own performance is to be

pp. 6–7 (no. 23); and R. Bud, 'Responding to Stories: The 1876 Loan Collection of Scientific Apparatus and the Science Museum', *Science Museum Group Journal*, 1 (spring 2014), http://dx.doi.org/10.15180/140104 (accessed 13 December 2018).

24 Babbage to Bowditch, 2 August 1835, in M. R. Williams, 'Babbage and Bowditch: A Transatlantic Connection', *Annals of the History of Computing*, 9 (1988), pp. 283–90, on p. 287 (stress in the original); and C. Babbage, 'On the Mathematical Powers of the Calculating Engine' (December 1837), in H. W. Buxton (ed.), *Memoir of the Life and Labours of the Late Charles Babbage*, ed. A. Hyman (Cambridge: MIT Press, 1988), p. 187 (stress in the original). See Ashworth, 'Memory, Efficiency and Symbolic Analysis', pp. 650–1.

multiplied in almost endless repetition, commensurate power is required'.[25] This multiplication of power, so evident in the inter-actions of steam technology and textile manufacture, required ordered control. Artisan skill, so it was intended, would be system-atically disciplined and confined through reduction of degrees of freedom in the programmed operations of the engine. The 'beautiful contrivance' of pasteboard or metal cards used to manage a Jacquard loom provided both model and control system for the engine. Alongside store and mill, therefore, 'the Analytical Engine will pos-sess a library of its own. Every set of cards once made will at any future time reproduce the calculations for which it was first arranged.'[26] The crucial innovation of the analytical engine was exactly this division of labour, prompted by the automation of the anticipatory system of carriage and of managerial reproduction and control, between the tasks of calculation in the mill and those of recall from the store and library.

Babbage turned this system of mechanical memory and anticipa-tion into a moral cosmology. In a *Treatise* composed during his analytical engine project and tellingly subtitled 'a fragment', he meditated on the relation between memory, immortality, and fame. In a distinctly autobiographical passage, he foresaw 'the approaching dawn of that day' when 'more highly endowed' minds would 'exchange the hatred they experience from the honest and dishonest intolerance of their contemporaries for that higher homage, alike independent of space and of time, which their memory will forever receive'. It was not simply that in some millenarian future Babbage and his ilk would at last be rewarded with deserved memorials. Rather, 'memory seems to be the only faculty which must of neces-sity be preserved in order to render a future state possible'. The very existence of prospective punishment and reward depended on material preservation of individual memory. The designer of the difference and analytical engines had a candidate mechanism for memory preservation: the embodiment of voice and movement in air and water. Like the calculating engine, 'the air itself is one vast library', he argued, 'the never-failing historian of the sentiments we

25 K. Marx, *Capital: A Critique of Political Economy, Volume 1* (Harmondsworth: Penguin Books, 1976 [1867]), pp. 468–70; and C. Babbage, *On the Economy of Machinery and Manufactures*, 4th edn (London: Charles Knight, 1835), pp. 49–50. See M. Berg, *The Machinery Question and the Making of Political Economy 1815–1848* (Cambridge: Cambridge University Press, 1980), pp. 182–97.
26 Babbage, *Passages from the Life of a Philosopher*, p. 119.

have uttered'. Particles in motion worked like mechanical records in the engine, forever available to recall to presence the traces of past actions, 'the eternal witnesses of the acts we have done'.[27]

In Babbage's mix of tough materialism about the labour process and ruthless immaterialism about the dominance of thought over mechanics and of memory over time, the movement between factory, mind, and machine was explicit: 'the analogy between these acts and the operations of mind almost forced upon me the figurative employment of the same terms'.[28] Babbage himself may well have treated the relics of the machines as so many gifts, material fragments of his own memorable achievements and his aims at a legitimate afterlife. He presented fragments of the Difference Engine to his friend Harry Buxton, and other components of the machine survived in family possession, being handed on to Nevil Francis Babbage, Benjamin's great-great-grandson. The latter are now in the Macleay Museum at the University of Sydney, one of the largest surviving collections of machine components from the original project of the 1820s.[29] Babbage's calculating engines thus embodied mnemotechnics. They aimed at the economical mechanisation of memory and were caught up with mechanisms of the Victorian commemorative economy. They were at least at home in showrooms as workshops, while their notoriety within official memory long depended on cautionary parables about their failure ever to be completed, and the ironies of the subsequent histories of automatic computing they allegedly spawned.[30]

The Whipple Museum holds a remarkable segment of an addition and carry mechanism of a difference engine put together around 1879 (Figure 6.2), eight years after his father's death, by Babbage's

27 Babbage, *The Ninth Bridgewater Treatise*, pp. 54, 112; and Babbage, *Passages from the Life of a Philosopher*, p. 405. See J. Picker, *Victorian Soundscapes* (Oxford: Oxford University Press, 2003), pp. 15–17; and W. Schivelbusch, 'World Machines: The Steam Engine, the Railway and the Computer', *Log*, 33 (winter 2015), pp. 54–61, on p. 61.

28 Babbage, *Passages from the Life of a Philosopher*, p. 119; and Babbage, 'On the Mathematical Powers of the Calculating Engine', p. 216.

29 The Buxton material is at the Museum of the History of Science, Oxford, no. 94229. The Macleay Museum, University of Sydney, object no. 1993.3, holds the components presented by Nevil Francis Babbage. Thanks are due to Jude Philp at the Macleay Museum for her help.

30 D. Swade, *The Cogwheel Brain: Charles Babbage and the Quest to Build the First Computer* (London: Little, Brown, 2000), pp. 308–14; and Purbrick, 'The Dream Machine', pp. 14–20. For Babbage's own retrospection and reconstruction of his own role and repute, see M. Fisch, 'Babbage's Two Lives', *British Journal for the History of Science*, 47 (2014), pp. 95–118.

Figure 6.2
Demonstration model of a calculating segment of the difference engine with five cages assembled by Henry Prevost Babbage in Bromley in 1879 and sent to Cambridge in 1886. Image © Whipple Museum (Wh.2339).

youngest son Henry, a former Indian army officer. Not the only such model then assembled by Henry, it was almost certainly the largest, constructed from precisely machined components of brass, steel, and a nickel–copper alloy described as German silver. These were material relics of the original difference engine project managed by Babbage and Clement. They were accompanied by a brief sheet of instructions from Henry, together with his elder brother Benjamin's descriptive exhibition pamphlet. The Whipple Museum's device was sent to Cambridge in December 1886. It has a pair of principal columns set up to carry series of figures vertically on their axes, three squat cylinders on one and two on the other. These five metal rings, with digits marked on them, are now papered over. The rings were originally to be concealed by screens set up to hide unwanted digits, which 'require slight fitting'. Indeed, the expectation was that key components of the carefully boxed-up machine sent from Henry's Cheltenham retirement home would then be assembled in

Cambridge. 'The German silver rings with the numerals engraved on them', he wrote, 'have been made recently and never fitted on.'[31]

The hand-turned axes interlock to perform addition by rotating the columns in series. The principle of such engines was that successive addition could generate the exact numerical values of the terms of a range of power series by the addition of a constant difference. They could also calculate terms of algebraic series such as logarithms where, though the differences were not constant, this inaccuracy could be ignored over large ranges of values. In such cases, Benjamin's pamphlet explained, 'the greater the number of difference columns that is worked out, the nearer a constant difference is approached'. This demonstration device also significantly exhibits the elegant carriage mechanism, a rudimentary version of its automated successive memory. The device has a supplementary axis through which, when a figure wheel travelled from nine to zero, numbers were to be sequentially carried over to the next column of digits after moving what, ever keen to move between machine and mind, Babbage called a warning lever. 'A warning of carriage will be heard when the carriage from the wheel below is being picked up, which warning will be followed almost immediately by the actual carriage on the wheel above.'[32]

As part of a long and obsessive campaign of filial piety and earnest technical enterprise, it might have seemed apt that Henry's gift went to Cambridge, where his father had been a precocious undergraduate (who nevertheless sat no examinations) and erstwhile professor (who gave no lectures). It was not the first token Henry gave the University in his father's memory. Back in May 1871, the new Cambridge professor of experimental physics, James Clerk Maxwell, then preoccupied with his *Theory of Heat*, set out to obtain a seventeenth-century Florentine glass thermometer that Charles Babbage had got from the director of the Florence museum, Vincenzo Antinori. When Charles died in autumn 1871, Henry at once contemplated transferring the instrument to Cambridge. 'Send the thermometer and letters to Professor Clerk Maxwell,' Henry was advised by his father's friend Frederick Pollock, lawyer and man of letters. Maxwell 'has charge of the collection of Philosophical Instruments, among which I apprehend it will be placed in the new building now about to be erected', the Cavendish Laboratory on

31 H. Babbage, Whipple Museum MS 2339, December 1886, sheet 3 ('note'). See the appendix to this chapter.
32 B. H. Babbage, *Babbage's Calculating Machine or Difference Engine*, p. 7.

the New Museums Site. Babbage's thermometer reached Cambridge in summer 1872.[33]

It was in this collection of philosophical instruments, too, that Henry's model difference engine was housed from 1887. The recently appointed head of the Cavendish, J. J. Thomson, reported that, alongside the acquisition of the whole of Maxwell's scientific library, the laboratory had also been presented by Henry with 'a portion of the very interesting Difference Machine invented by his father'.[34] The model's home remained the New Museums site, tracing a pathway between university labs and their pedagogical exhibitions. Despite subsequent claims about the influence of Babbage's projects on the modern development of computation, Cambridge institutional memory of the 'very interesting machine' and its associations seems to have faded in the interim during the regimes of Thomson and his successor Ernest Rutherford. In 1936 the University's Mathematics Faculty Board backed the establishment of a Mathematical Laboratory for computing, in the wake of the construction of a Meccano version of a differential analyser for calculating molecular wave functions, prompted by the chemistry professor John Lennard-Jones and aided by the Manchester numerical analyst Douglas Hartree. Lennard-Jones was *pro tem* Laboratory director, and the Cavendish researcher Maurice Wilkes was charged with working on the proposed machines. Wartime mobilisation halted university plans while intensifying state and industrial investment in automatic computation. Only in autumn 1946 was Wilkes at last placed at the head of the Mathematics Laboratory, based on the eastern side of the New Museums Site, with computer research as his

33 Maxwell to Tait, 25 and 27 May 1871, in *The Scientific Letters and Papers of James Clerk Maxwell*, ed. P. M. Harman (Cambridge: Cambridge University Press, 1995), vol. 2, pp. 645, 648; Pollock to Henry Babbage, 17 July 1872, British Library MS Add.37199, fol. 576; Henry Babbage to Power, 7 August 1872, fol. 581; and J. C. Maxwell, 'Report on the Cavendish Laboratory', 14 April 1875, in *The Scientific Letters and Papers of James Clerk Maxwell*, ed. P. M. Harman (Cambridge: Cambridge University Press, 2002), vol. 3, p. 213. This thermometer now resides in the Whipple Museum, accession no. Wh.1116.

34 J. J. Thomson, 'Experimental Physics', in 'Museums and Lecture Syndicate Annual Report for 1886', *Cambridge University Reporter*, no. 688 (26 May 1887), p. 749. For the role of historical instruments and collections in the early Cavendish Laboratory projects see B. Jardine, 'The Museum in the Lab: Historical Practice in the Experimental Sciences at Cambridge, 1874–1936', in B. Jardine, E. Kowal, and J. Bangham (eds.), *How Collections End: Objects, Meaning and Loss in Laboratories and Museums*, BJHS Themes Vol. 3 (Cambridge: Cambridge University Press and British Society for the History of Science, 2019).

brief, and his ally Hartree as the new Cambridge mathematical physics professor.[35]

That same autumn an explosive argument erupted in the correspondence columns of *The Times*, provoked in part by publicity given a speech by Lord Mountbatten as president of the Institution of Radio Engineers. The speech was a response to Alan Turing's work on automatic computers, which the noble Admiral, soon to become the very last Viceroy of India, reckoned showed it possible 'to evolve an electronic brain'. Hartree wrote from the Cavendish to protest against Mountbatten's phrase, explaining that while such devices, even purely mechanical ones, might exercise a form of judgment, this was 'no substitute for thought', the prerogative of human operators. At this point in November 1946 the polymathic Rupert Gould, naval officer, horologist, and broadcaster, intervened to refresh *The Times* readers' memories of Charles Babbage's analytical engine, a mechanical device capable, as Gould put it, of 'memorizing in its store for future use the results of its calculations'.[36]

It was Gould's letter that prompted Hartree to consult Babbage's autobiography in which he learnt at last of the nineteenth-century mathematician's doctrines about the mechanisation of judgment and memory. Wilkes remembered Hartree delightedly passing round a copy of Babbage's book in the Mathematics Laboratory. Hartree published a summary of the analytical engine project in 1949, while Wilkes himself went to South Kensington to consult the Babbage notebooks Henry Babbage had given to the Science Museum. It was thanks to Hartree that Henry's difference engine model fragment held at the Cavendish was then remembered, and moved in about 1950 from west to east across the site to the Mathematics Laboratory, where Wilkes used it to lecture on the principles of automatic addition.[37] Wilkes recalled that at that point 'an object of this sort was not so highly regarded', and could scarcely be seen as 'epoch-making' in comparison with the analytical engine. This was a period of intense interest in automatic memory and the psychological and

35 M. Croarken, 'The Emergence of Computing Science Research and Teaching at Cambridge, 1936–1949', *Annals of the History of Computing*, 14 (1992), pp. 10–15.
36 A. Hodges, *Alan Turing: The Enigma* (London: Vintage, 1992), p. 347; D. Hartree, 'The electronic brain: a misleading term', *The Times*, 7 November 1946, p. 5; D. Hartree, 'The electronic brain', *The Times*, 22 November 1946, p. 5; and R. Gould, 'The electronic brain', *The Times*, 29 November 1946, p. 5.
37 M. V. Wilkes, *Memoirs of a Computer Pioneer* (Cambridge: MIT Press, 1985), pp. 195–9; D. Hartree, *Calculating Instruments and Machines* (Urbana: Illinois University Press, 1949), pp. 69–72; and Cohen, 'Babbage and Aiken', p. 189.

moral implications of recall and mechanisation. Even if the difference engine model was not 'epoch-making', it nevertheless went on show at the Science Museum's 1976 exhibition 'Computers Then and Now', after which, through the lobbying of the Whipple curator David Bryden, who noted his Museum's possession of other 'Cambridge firsts' in the history of calculating instruments such as Oughtred's circle of proportion, it was deposited in the Whipple. In 1980 Wilkes also passed Henry's accompanying notes and Benjamin Herschel Babbage's pamphlet to the Museum.[38]

It is thus uncharacteristically appropriate that the Babbage fragment is held on the Cambridge site and put on public show there – the location and display are congruent with the artefact's original purpose as demonstration model and its lengthy afterlife. Henry Babbage's commemorative enterprise incorporated laborious manufacture of a range of models; significant publicity initiatives; and the systematic distribution of material mementos, somewhat akin to the actions of traditional impresarios of saintly relics.[39] This enterprise of assemblage and display was thus never entirely a fanciful vision of a dimly predictable future in which the calculating engines would at last occupy their proper place; rather, it was a deliberate exploitation of a highly crafted past, bringing souvenirs of artisan workshops and hardware of the Age of Reform back to life amidst the pomp and circumstance of Victorian *fin-de-siècle* shows and salons. For Babbage, memory was precisely a moral and material assemblage of traces and relics that somehow might survive, despite their transience and fragility.[40] The Whipple fragment is thus aptly and tenuously positioned between uncertain afterlife and patriarchal provenance. It was as much pious resuscitation as prophetic vision.

Embodied in that fragment is a complex and telling relation between memories of Henry's upbringing and the labour relations of the engines' manufacture under his father's direction. 'Now what

38 Bryden to Wilkes, 1 March 1977 and Wilkes to Bryden, 4 March 1977, Whipple Museum Archive file P 009 and 011. Thanks to Joshua Nall for these materials. For the technoscience of automatic memory in the 1950s through the 1970s see A. Winter, *Memory: Fragments of a Modern History* (Chicago: University of Chicago Press, 2012).

39 P. Geary, 'Sacred Commodities', in A. Appadurai (ed.), *The Social Life of Things* (Cambridge: Cambridge University Press, 1986), pp. 169–92; S. Stewart, *On Longing: Narratives of the Miniature, the Gigantic, the Souvenir, the Collection* (Durham: Duke University Press, 1993), pp. 133–9; and A. Walsham, 'Relics and Remains', *Past and Present*, supplement 5 (2010), pp. 9–36, on pp. 31–2.

40 Fisch, 'Babbage's Two Lives', pp. 115–16 on the artifice and tragedy of Babbage's memory.

was your first serious mistake in life, Henry?', Charles allegedly once asked his son. 'I answered on the instant, "Alas, my choice of a parent!"' Up to his enrollment at University College London in 1840–2, Henry lived with his paternal grandmother Elizabeth on Devonshire Street in Marylebone, rather than round the corner at his father's house and workshop in Dorset Street. 'I feared him, and often left the house to avoid meeting him.'[41] Encounters were almost entirely mediated through the labours of the workshop and the showmanship of Babbage's fashionable soirées. The difference engine had by this time become a party piece for display to prestigious visitors at Dorset Street, where its tricks of automated memory and foresight would be used to telling effect. Charles also made the young UCL student come to his drawing office two or three times a week, where he was taught to 'handle tools and to draw machinery'. After leaving college in summer 1842, Henry spent several days a week in the workshops, training with the Lambeth engineer William Garton in lathe-work, metal forging and 'a little hardening and tempering steel'. He also studied mechanical drawing, 'such as clock work', with Jarvis, who 'made all the beautiful drawings for the analytical engine and knew something of mathematics'.[42]

These same months, while Henry encountered the collective and individual skills on which the engine enterprise depended, also saw the termination of the difference engine project, damned by the Astronomer Royal George Airy and axed by Robert Peel's government in November 1842 on grounds of cost, and the inauguration the next year of the machine's display at King's College London. 'It is amazing', Henry later reminisced, that the government 'did not see the advisability of having the calculating part completed. A few hundred pounds would probably have been sufficient for this . . . a Difference Engine might have existed.'[43] As was common in Henry's cohort of engineers and servicemen, and true of both his elder brothers, colonial employment eventually provided a career. By the end of 1842 it had been decided Henry would enter the East India Company military under the patronage of his father's friend William

41 H. Babbage, *Memoirs and Correspondence*, p. 93.
42 H. Babbage, *Memoirs and Correspondence*, pp. 9–11. Henry mentions 'a workman called Garton'; William Garton, engineer, is listed in the 1841 census at Anderson's Walk, Lambeth, then aged fifty, with son Charles, apprentice toolcutter.
43 H. Babbage, 'Conclusion' (1888), in H. Babbage (ed.), *Babbage's Calculating Engines* (London: Spon, 1889), pp. 339–42, on p. 340. For Airy's role see Swade, *The Cogwheel Brain*, pp. 134–54.

Plowden, one of the Directors. He left for India, at the age of eighteen, in spring 1843. Seven decades later, memory of cool separation stayed vivid: 'my father bade me good-bye in his library at 1 Dorset Street. He did not see me into the cab.'[44]

Henry's early Indian service was a characteristic Company combination of strenuous disciplinary training and intense if irregular violence. He was rapidly promoted Lieutenant in the Bengal Infantry, studied Hindi and Bengali, and became an interpreter. He led fighting both in the vicious Company war with Sikh forces in the Punjab in 1846 and in military raids in Assam against the Adi hill peoples in 1848, aggression which, Henry long remembered, taught the value of well-armed punitive expeditions rather than any attempt at permanent fortified settlement along the imperial frontiers.[45] He married the much younger Irishwoman Mary Bradshawe, nicknamed Min, whose father, an officer in the Company's army, had allegedly been poisoned with diamond dust on the orders of the ruler of Awadh. Henry returned to London on leave in 1854–6. The thirty-year-old military veteran and family man, accompanied by wife, baby daughter, and Indian nurse, engaged in a period of committed work and transformation in his relations with his father and the calculating engines. Decisively, Henry and his family settled at his father's house in Dorset Street. Charles installed a mirror in the dining room 'so that he could see Min in the glass without looking in her direction'. Henry reflected that 'I met him on more equal terms … the wish to merit his approval since I had grown up had always been a strong motive with me, but was now strengthened and endured to the end.'[46]

This was a period of intense labour. Henry worked on mathematics, cryptography, and practical arts such as photography and electroplating; he joined his father on exhibition tours and visits to engineering sites such as Isambard Brunel's *Great Eastern*. It also coincided with important developments in the calculating engine projects, in the wake of the Great Exhibition and the polemics and

---

44 H. Babbage, *Memoirs and Correspondence*, 11–12. Charles Babbage recalls obtaining Henry's East India Company place through Plowden, in Babbage to Alexander Dallas Bache, 8 August 1854, Yale University Bienecke Library General MSS 1322, box 1, folder 4. Thanks are due to Alexi Baker for this source. For the difference engine's transfer to King's College see Milne to Babbage, 5 June 1843, British Library MS Add.37192, fol. 326.

45 H. Babbage, 'Expeditions against the Abors', *The Times*, 24 March 1894, p. 10; and H. Babbage, 'The operations in 1848', *The Times*, 15 September 1911, p. 3.

46 H. Babbage, *Memoirs and Correspondence*, p. 94.

publicity raised about new engineering and its capital effects. In summer 1852 Charles Babbage had sought unsuccessfully to get government funds and engineers' commitment for a second, more efficient version of the difference engine.[47] At the same time, William Farr vainly proposed that Babbage's difference engine be used to verify joint life tables for his General Register Office. Babbage also began negotiating with former Clement employee Whitworth, now the pre-eminent Manchester engineering master. Whitworth's repute as manufacturer of machine tools had been established at the 1851 Great Exhibition and marked in the notorious engineers' lockout against his own workforce over union demands to abolish piece-work and against deskilling the following year. In summer 1855 Whitworth began abortive discussions with Babbage about constructing some version of the analytical engine, the ultimate embodiment of automatic skill.[48]

Meanwhile, the Swedish printers Georg Scheutz and his son Edvard brought to London their own version of a difference engine, inspired by Lardner's earlier accounts of Babbage's machine. They visited Charles and Henry at Dorset Street over two days in November 1854, and with the aid of Babbage and his engineer colleague Gravatt began to publicise their device for computing and printing mathematical tables with a bronze framed machine on a steel base plate and silver-plated number wheels with figures engraved in black enamel. The engine was praised for its simplicity, its cheapness, and because 'it can easily be taken to pieces and examined if need be'.[49] Eventually, the government took the Swedish engine for Farr's Office, though by the 1870s Charles Thomas de Colmar's arithmometer, a device whose repute was also established through successive industrial and technical exhibitions, was beginning to displace the grander Scheutz device. Both for the French machine and for the Swedish one, issues of labour skill and reliability were decisive. Farr refused to put the Scheutz machine on show at South Kensington in 1862 alongside Babbage's difference engine, since 'its work had to be

---

47 H. Babbage, *Memoirs and Correspondence*, pp. 80–3; and Babbage, *Passages from the Life of a Philosopher*, pp. 100–107.

48 Farr to Babbage, 2 September 1852, British Library MS Add.37195, fol. 135; A. Hyman, *Charles Babbage: Pioneer of the Computer* (Oxford: Oxford University Press, 1982), p. 231; and Atkinson, *Sir Joseph Whitworth, the World's Best Mechanician*, pp. 172–90.

49 M. Lindgren, *Glory and Failure: The Difference Engines of Johann Müller, Charles Babbage and Georg and Edvard Scheutz* (Cambridge: MIT Press, 1990), pp. 184–92; and H. Babbage, *Memoirs and Correspondence*, p. 83.

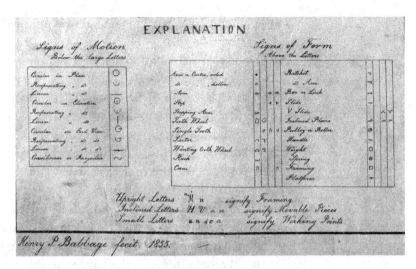

Figure 6.3 Henry Babbage's explanation of the mechanical notation of the Scheutz engine, drawn in London in summer 1855. By permission of the Science and Society Picture Library (Science Museum Library MSR/0012).

watched with anxiety and its arithmetical music had to be elicited by frequent tuning and skilful handling'. He insisted that 'it consists of a multitude of pieces and some of these occasionally get deranged ... it is not infallible, except in very skilful hands'.[50]

Because of such ineffable skill and dangers of showmanship, the scheme designed by the Swedish father and son had somehow to be rendered legible to its audience so as to be successfully marketed to the British state. It seemed important to display the distinction between Babbage's difference engines and the Scheutz layout, especially its carriage mechanism, which adopted a completely contrasting mode of automatic memory. Babbage worked closely with his own son, Henry, to achieve this legibility. In spring 1855 Henry trained himself in copying the technical notation of the mechanical drawings for his father's second version of the difference engine, distracted somewhat briefly by the birth of a son in late March. Then he was taken to Somerset House to inspect the Swedish version, and applied the same mechanical notation to this new machine (Figure 6.3). His vast diagrams of gear and wheel trains, ranging from eight to over thirteen feet in length, were pasted onto calico sheets at Dorset Street by Min, Henry, and Charles. 'They were really a work of art and would have done credit to a professional draughtsman,' Henry boasted. Somewhat to Henry's surprise, his father treated commodification literally, and formally bought the drawings

---

50 Farr, *English Life Table*, pp. cxl–ii; and Lindgren, *Glory and Failure*, pp. 216, 224–5, 285. For the arithmometer see S. Johnston, 'Making the Arithmometer Count', *Bulletin of the Scientific Instrument Society*, 52 (1997), pp. 12–21.

from him for £50, even though 'I always intended them for him, and considered them his.'[51]

The mechanical notation helped make Henry's repute as a drafts-man of memory, however transiently. He went to the British Associ-ation at Glasgow in September 1855 to lecture on the notation's panoptic virtues: 'we can demonstrate the practicability of any con-trivance and the certainty of all its parts working in unison *before a single part of it is made*'. Thus not only had the language of signs in some way compensated for the labour and financial uncertainties that plagued the engine-construction projects, but Henry's paper-work had become an integral part of the substitution of mechanism for weak memory: 'it would be beyond the powers of the human mind to master and retain the details of the complicated machinery'. At just the same time, at the Clydeside shipyard where the *Great Eastern*'s paddle shaft was being forged, Henry witnessed the power of controlled machinery and labour on metal: 'masses of about two tons were welded in one operation under the blows of a steam hammer'. The family inheritance of embodied skill in manufacturing paper imagery became part of their domestic and industrial legacy. Charles and Henry took the drawings to the Institution of Civil Engineers in Westminster, insisting that the notation allowed even 'the most fleeting movements' to be captured forever. 'It had, as it were, *photographed the footsteps of time*', Charles told the Civils; 'it had conferred fixity and permanence on the swiftest motion.'[52]

Alongside the inscription system that would somehow make the engine's functions visible and permanent, at the very same moment in late 1855 Charles and Henry also tried an experiment in model building to illuminate the carriage and addition mechanisms in the Dorset Street difference engine and help establish their independ-ence and originality. From the workshop stores they 'fitted and selected' five sets of number wheels and gears for their model. Henry 'cleaned them up and did what was necessary, and put them together, two cages in one column and three cages in the other

---

51 H. Babbage, *Memoirs and Correspondence*, pp. 85–6; and H. Babbage, 'Conclu-sion', p. 341.

52 H. Babbage, 'On Mechanical Notation as Exemplified in the Swedish Calculating Machine of Messrs. Scheutz' (1855), in H. Babbage (ed.), *Babbage's Calculating Engines* (London: Spon, 1889), pp. 246–7, on p. 246 (my stress); C. Babbage, 'Scheutz's Difference Engine and Babbage's Notation' (1856), in H. Babbage (ed.), *Babbage's Calculating Engines* (London: Spon, 1889), pp. 248–57, on p. 257 (my stress); H. Babbage, *Memoirs and Correspondence*, pp. 87, 89; and Lindgren, *Glory and Failure*, p. 192.

column'. Over a few weeks Henry learnt how to work the mechanism for carriage by hand. Significantly, this was precisely the same layout of three wheels on one column and two on the other as was also evident in the Whipple Museum's model engine, built almost a quarter of a century later. In 1856 Henry and Min returned to India and the violent crisis of the first Indian war of independence, leaving children, drawings, and the new demonstration model as part of the paternal inheritance. Over a century later, Maurice Wilkes was so struck by this episode in the history of the machines and the family that he turned it into a play, performed as a Christmas entertainment in 1982 at Boston's Computer Museum.[53]

Embodiment is also expropriation: to claim that what was fixed in metal's body was pure intellect was to assert the primacy of the sole inventor's abstract thought. Charles Babbage unambiguously connected his authority to determine the calculating engine's legacy with the claims of pure reason over mere artisan labour and brute engineering. During the fight with Clement, Babbage told the erstwhile Prime Minister, the Duke of Wellington, that 'my right to dispose as I will of such inventions cannot be contested, it is more sacred in its nature than any hereditary or acquired property, for they are the absolute creations of my own mind'.[54] Yet the ailing Charles had made no disposition for the engines' fate as hereditary property when Henry again returned from India in spring 1871. It was only a few days before Charles's death in October 1871, with advice from his brother-in-law, the senior Indian lawyer Edward Ryan, that he drew up a will leaving Henry 'for his own absolute use and disposal his calculating machines and the machinery, tools, models and drawings of every kind relating thereto and all the contents and materials of his work-rooms'.[55]

53 H. Babbage, *Memoirs and Correspondence*, pp. 88–9. See M. V. Wilkes, 'Pray Mr Babbage: A Character Study in Dramatic Form', *Annals of the History of Computing*, 13 (1991), pp. 147–54; and [Boston] *Computer Museum Report* (Spring 1983), p. 15.

54 Babbage to Wellington, 23 December 1834, British Library MS Add.37188, fol. 525; and Hyman, *Charles Babbage*, 134. In September 1838 it was reported from the Newcastle BAAS by the Harvard lawyer Charles Sumner that Babbage planned to build the analytical engine in the United States, even though Sumner reckoned 'our Government ... would no more give that sum for that purpose than keep a hunting pack of hounds': Sumner to Henry Bowditch, 28 September 1838, in Bowditch, *Life and Correspondence of Henry Ingersoll Bowditch*, vol. 1, p. 109.

55 H. Babbage, *Memoirs and Correspondence*, p. 181; the provisions of the will are printed in the *Pall Mall Gazette*, no. 2134 (15 December 1871), p. 7.

Figure 6.4 Charles Babbage's brain, presented to the Hunterian Museum by Henry Babbage in October 1871. From Victor Horsley, 'Description of the Brain of Mr Charles Babbage', *Philosophical Transactions of the Royal Society,* series B, **200** (1909), pp. 117–31, plate 9. Image courtesy of Cambridge University Library.

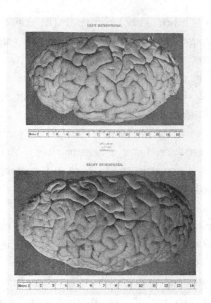

Henry took his father's cerebral and material powers very seriously, believing them worthy of posthumous show. The terminology of intellectual labour was then rapidly changing: in 1871 novelists invented 'brainwork' and psychical researchers started investigating 'brain-waves'; in a lecture on 'body and mind' in 1874 the UCL mathematics professor William Clifford argued that humans were simply conscious automata; from 1877 the *Brain* got its own eponymous journal; and in 1878 Clifford's ally Joseph Hooker used the term 'brain-power', while the author of a text on *The Hygiene of Brain and Nerves* began referring to 'brain-workers'.[56] The day following his father's death, Henry arranged for Charles's brain to be deposited and preserved in the Hunterian Museum at the Royal College of Surgeons (Figure 6.4). This was a public experiment on Babbage's doctrine of the materiality of the afterlife. The specimen was not to be anonymised: 'his character is known by his deeds and his published works and the brain should be known as his'; and it was to be understood as a paternal gift: 'I have but one standard to guide me, my thoughts of what could be the judgment of my father.' A few weeks later Henry sent the Royal College his father's portrait to accompany the brain.[57]

---

56 *Oxford English Dictionary*, 3rd edn (2011), s.v. 'brain'.
57 Henry Babbage to Paget, 19 October 1871, Archives of the Museum of the Royal College of Surgeons, Letters Book, cited in J. Agar, 'Bodies, Machines and Noise', in I. R. Morus (ed.), *Bodies/Machines* (Berg: Oxford, 2002),

The commemorative bequests launched the campaign to embody Charles Babbage's 'absolute creations' in metal. Henry spent much of 1872–3 settling his father's affairs. His initial aims were to write Charles's biography and to construct 'the Mill or actual working part of the Analytical complete in metal'.[58] Commemoration, politics, and engineering were entwined. In May 1872 Henry took his family to hear Clifford deliver a Royal Institution lecture in Babbage's memory. The young mathematician and evolutionist spoke on the plans for the difference and analytical engines, foresaw their completion when 'machines must be made all over the country', and explained how Babbage's life had become 'embodied in [his] workshop'. He also devoted a section of his lecture to Babbage's doctrines of foresight and immortality, the mechanical means through which apparent miracles could be programmed into the machine by its designer, and through which the traces of past actions could be materially preserved. All this, Clifford urged, showed the 'deeper law' of 'evolution as the true statement of the world's history'. Henry admired Clifford's enthusiasm about the analytical engine, but did not think he put enough stress on Charles's views about 'the never ending effects of our words and actions'. His father's doctrine of immortality was fundamental for Henry: an extant copy of the 1837 *Treatise* in which these views were developed still has Henry's annotations insisting on the scientific significance of such a view of memory and a future state.[59] Reminiscence and foresight were commonplaces in testimonies to Babbage's legacy. In his eulogy to Babbage at the Statistical Society of London, William Farr prophesied the completion of the analytical engine to calculate 'those still more complicated coefficients and variables which, it is easy to foresee, will be in requisition when future State problems are dealt with scientifically by *a political Newton*'. Henry promised Farr in January 1872 that he would indeed finish a working section of the

pp. 197–220, 297, 215; and Paget to Henry Babbage, 12 January 1872, British Library MS Add.37199, fol. 574.

58 Henry Babbage to Pollock, 8 January 1872, British Library MS Add.37199, fol. 558.

59 'Professor Clifford's notes for his lecture', in Lucy Clifford to Henry Babbage, 26 July 1879, University Library Cambridge MS Add.8705 no. 33; and H. Babbage, *Memoirs and Correspondence*, p. 184. Henry's copy of the *Treatise* is Harvard University Houghton Library *74 434: see Picker, *Victorian Soundscapes*, 158 n. 7. For Clifford's views see S. Cook, *The Intellectual Foundations of Alfred Marshall's Economic Science* (Cambridge: Cambridge University Press, 2006), pp. 189–91.

Figure 6.5 The Babbage family in the early 1870s: (from the left) Mary (Min), Sophie (b. 1860), Mary (b. 1857), Henry, Georgiana (b. 1852), and Harry (b. 1855). By permission of the Science and Society Picture Library (SSPL 10300411).

great engine, 'sufficient to perform those calculations which from their unmanageable extent baffle human skill'.[60]

In the event, it was the construction of the analytical engine's mill that would prove baffling and almost unmanageable. Henry made a brave start, hoping to finish the Mill to tabulate the reciprocal of π (defined as 7/22) to twenty-five significant figures. During 1872 he completed a set of drawings of the design; retained the services of the elderly engineer Richard Wright, whose four decades' experience on the calculating engines might prove invaluable; hired a new work-man named Dancaster to 'push on the machine work'; commissioned a new workshop and house in suburban Bromley; and sold the house at Dorset Street. Almost all the equipment there was auctioned off: forges, fly presses, lathes, steam engines, and metal scrap. Henry retained but a little hardware for his own enter-prise: 'a lathe or two, a small planing machine and some smaller tools'. These plans in place, he then returned to India for a final two years of military service and promotion to Major-General, setting up permanently back in England from early 1875 (Figure 6.5).[61]

60 W. Farr, 'Inaugural Address', 21 November 1871, *Journal of the Statistical Society of London*, 34 (December 1871), pp. 409–23, on p. 415 (my emphasis); and Henry Babbage to Farr, 20 January 1872, British Library MS Add.37199, fol. 564.

61 H. Babbage, *Memoirs and Correspondence*, pp. 183–4; and Fuller, Horsey, Son & Co., *Catalogue of a Collection of Engineers' Tools and Plant Used by the Late Mr Babbage* (1 March 1872), Erwin Tomash Library B65, Charles Babbage Institute, University of Minnesota.

Henry was committed to his ultimate ambitious goal, the analytical mill. In early 1878 he compiled detailed manuscript notes on the best method for using the mill to compute the powers of the reciprocal of π by experimental methods using cross-multiplication, and in 1880 organised the casting of the mill's frame plates, 'not as part of the larger machine which my father proposed but something which would be practically useful by itself in the hands of a skilled operator'.[62] At the same time, with Henry's assistance and spurred by Farr's arguments, a British Association committee led by Clifford and managed by the civil servant Charles Merrifield considered the feasibility of completing the analytical engine, this 'marvel of mechanical ingenuity and resource'. They expressed understandable scepticism about the analytical plan, wondering 'until it leaves the inventor's hands in the finished state whether it really represents what is meant to be rendered in metal'; and they reported that by 1878 the analytical engine, designed to demonstrate the principle of anticipation, existed only as a series of gunmetal wheels and cranks on steel shafts, though to keep costs down most had been pressure-moulded in zinc-hardened pewter. At all events, despite the considerable possible utility of such an engine, the committee calculated its construction would cost up to £40,000 (£4,000,000 in modern terms), apparently a prohibitive sum, though perhaps some limited component might be built as 'a simple multiplying machine'. The mathematicians also cautioned that in cases of calculation by finite differences the 'specialization of the difference engine would probably give it an advantage over the more powerful engine'.[63]

Henry did not dissent from this somewhat frosty judgment – but his interests were subtly different, less concerned with precise calculations of the engines' future viability, more focused on commemorative vindication of paternal and mechanical virtues. The BAAS mathematicians dwelt on the fundamental principle of digital, or what they called 'discontinuous', machines, which they associated with the mechanism of 'millwork and clockwork'.[64] Henry certainly shared their interest in the economic and material distinctions

62 H. Babbage, 'Computations of the powers of π', University Library Cambridge MS Add.8705 no. 34 (entries dated January 1878); and H. Babbage, 'Babbage's Analytical Engine', p. 518.

63 C. Merrifield, 'Report of the Committee Appointed to Consider the Advisability of Constructing Mr Babbage's Analytical Machine' (1878), in H. Babbage (ed.), *Babbage's Calculating Engines* (London: Spon, 1889), pp. 323–30. Clifford's lead role in the committee is mentioned in Henry Babbage, *Memoirs and Correspondence*, p. 184.

64 Merrifield, 'Report of the Committee Appointed to Consider the Advisability of Constructing Mr Babbage's Analytical Machine', p. 323.

between continuous and discontinuous calculating devices. This concern emerged dramatically in his attack on a lecture given to the Institution of Civil Engineers by the new Liverpool engineering professor Henry Selby Hele-Shaw, who praised mechanical integrators and harmonic analysers, analogue devices that continuously traced the surfaces of areas to be computed 'in a way that could not be effected by mere trains of wheel-work, such as form the mechanism of some kinds of calculating machines'.[65] The professor's higher valuation of planimeters and mechanical integrators and his implied dismissal of the difference and analytical engines' wheelwork enraged Henry: in his father's calculating machines 'there was absolute accuracy of result, and the same with all operators, and there were mechanical means for correcting, to a certain extent, slackness of the machinery'. His memory of the calculating engines and their mastery of mechanical memory showed 'all except the simplest planimeters would become obsolete'. Hele-Shaw responded in kind: 'all efforts to employ mere combinations of trains of wheelwork for such operations as were required in continuous integrators had hitherto entirely failed', he claimed. The exchange about the virtues of discontinuous devices showed Henry how crucial it was to resuscitate the repute of Babbage's versions of the calculating engines.[66]

The conflict with the engineers took place after Henry had spent ten years at his Bromley house and workshop from 1875, during which he set out to publicise and, if possible, model the components of the simpler difference engine using his Dorset Street inheritance. He set out to combine the 'waste metal' left by his father with the museum machine on public show described in Benjamin's recent pamphlet. 'There was nearly, if not all of the Difference Engine . . . enough to put together the calculating part of the machine.' What was missing were gears and frame plates, many of which had been cut up for his father's experiments. Henry commissioned new frame plates as well as 'a new driving gear which answered perfectly'.[67] He sought to use the north London instrument-maker Robert William Munro, head of a prestigious engineering firm then building printing machines for the Bank of England and harmonic analysers, efficient examples of continuous integrators, designed by the eminent Glasgow professor William

65 Henry Hele-Shaw, 'Mechanical Integrators', *Minutes of the Institution of Civil Engineers*, 82 (1885), pp. 75–143, on pp. 76–7.
66 'Correspondence', *Minutes of the Institution of Civil Engineers*, 82 (1885), pp. 163–4.
67 H. Babbage, *Memoirs and Correspondence*, pp. 224–5; and H. Babbage, 'Babbage's Analytical Engine', p. 518.

Thomson for the Meteorological Office.[68] Henry's own memories of
what happened next are confused, but are at least consistent in
recording that in the 1870s he soon decided the project of building
the difference engine's entire calculating section was beyond his
limited means and those of his collaborators and peremptorily 'sent
the whole to the melting pot'. He also learnt 'about the same time' of
the destruction of the relics of the engine project held by Clement.
Henry's reminiscences differ about whether he heard of Clement's
workshop's burnt offering *after* he'd destroyed most of his own
difference engine components, which would then be a tragic decision
to be regretted ('I might have kept what I had'). Or perhaps he learnt
of the fate of Clement's workshop *beforehand*, in which case his own
metallurgical meltdown was apt ('I decided to dismantle the work').
Whatever the reliability of his memory, Henry's enterprise was
decidedly backward-looking. 'I still hold to the opinion that the
calculating part of the difference engine might have been completed
at the time the Government gave it up for £500.'[69]

From then on, the difference engine was unambiguously consigned
to history. This was a history with a melancholy moral. The Babbages'
family friend Frederick Pollock, who had advised on the despatch of
Babbage's thermometer to Cambridge, reminisced after a visit to
Dorset Street that 'it was a strange fortune for a man to have eclipsed
himself, as it were, in this way, and the deserted work benches, lathe
and tools presented a dreary and melancholy spectacle'. The brilliant
mathematician and patent lawyer John Fletcher Moulton recalled a
visit there in May 1869 as 'one of the sad memories of my life'.
Commenting on the Merrifield report, Moulton added that 'not only
had [Babbage] constructed no machine, but the verdict of a jury of
kind and sympathetic scientific men who were deputed to pronounce
upon what he had left behind him either in papers or mechanism was
that everything was too incomplete to be capable of being put to any
useful purpose'.[70] Henry also often adopted the same tone in his own

68 H. Babbage, 'Computations of the powers of π' (entries dated August 1895). For
   Munro see https://collection.sciencemuseum.org.uk/people/ap26820/r-w-munro-
   ltd (accessed 2 March 2018).
69 H. Babbage, *Memoirs and Correspondence*, pp. 224–5; H. Babbage, 'Babbage's
   Analytical Engine', p. 518; H. Babbage, 'Conclusion', p. 341; and H. Babbage,
   'Computations of the powers of π' (entries dated August 1895). See Cohen,
   'Babbage and Aiken', p. 187; and Swade, *The Cogwheel Brain*, p. 316.
70 F. Pollock, *Personal Remembrances*, 2 vols. (London: Macmillan, 1887), vol. 2,
   pp. 9, 206; and J. F. Moulton, 'The Invention of Logarithms, Its Genesis and
   Growth', in C. G. Knott (ed.), *Napier Tercentenary Memorial Volume* (London:
   Longmans, Green, 1915), pp. 1–32, on pp. 19–20. See G. Williams, 'Engine

memoirs. He told the British Association that this history 'is sufficient to damp the ardour of a dozen enthusiasts', and added in his entry on calculating machines for *Chambers's Encyclopaedia* that 'the engagement was to the inventor a disaster'.[71]

Glum reflection was balanced by enthusiastic exhibition. During 1876, when the calculating engine was on show and occasionally working at the South Kensington exhibition of scientific apparatus, Henry lent his father's examples of much older calculating devices such as those of Samuel Morland (1660s) and Charles Stanhope (1770s) to be juxtaposed with his father's machine. The following year, the distinguished railway engineer William Prime Marshall lectured on the history of Babbage's machine at the Birmingham Philosophical Society, describing the engine as a 'mechanical treat' at last put on proper public display.[72] A similar performance took place at the Manchester Society of Chartered Accountants when its president, the statistician and Liberal politician Edwin Guthrie, evoked the South Kensington display of calculating devices during a lecture on the history of numeration. Guthrie's Manchester lecture was accompanied by an exhibition of model calculating machines, including Thomas de Colmar's arithmometer, slowly becoming standard if not always reliable issue in the major insurance firms, and, once again, Babbage's example of Stanhope's calculating machine 'which had never been in the provinces before'. Importantly, Henry also lent the Manchester accountants 'a small original portion of a machine' designed 'to exhibit the principle of the Difference Engine'.[73] This reference is suggestive, since as the example of the Whipple engine shows, there were indeed several fragmentary models of the mechanism of the difference engine assembled in 1878–9 and in circulation during the 1880s through Henry Babbage's enterprise. Significantly for the politics of memory and

Noise and Artificial Intelligence: Babbage's London', in J. Q. Davies and E. Lockhart (eds.), *Sound Knowledge: Music and Science in London 1789–1851* (Chicago: University of Chicago Press, 2016), pp. 203–25, on pp. 203–4.

71 H. Babbage, 'Paper Read at Bath, 12 September 1888', in H. Babbage (ed.), *Babbage's Calculating Engines* (London: Spon, 1889), pp. 331–7, on p. 337; and [H. Babbage], 'Calculating Machines' (1888), *Chambers's Encyclopaedia*, new edn (Edinburgh: Chambers, 1901), pp. 633–4. Henry's proof copy of the *Chambers* entry is at Powerhouse Sydney, Museum of Applied Arts and Sciences MS 97/186/1-5/1/31, p. 14.

72 *Catalogue of the Special Loan Exhibition*, pp. 5–6; and W. P. Marshall, 'Babbage's Calculating Machine', *Proceedings of the Birmingham Philosophical Society*, 1 (1879), pp. 33–48, on p. 45.

73 E. Guthrie, 'The Development of the Art of Numeration', *The Accountant*, 10, no. 518 (8 November 1884), pp. 7–15, on pp. 12, 14. The arithmometer's career is described in Johnston, 'Making the Arithmometer Count'.

display, they were part of his active campaign to exhibit and recall the historic principles on which his father's engine had worked.

During 1878–9, Henry at last recalled the project he and Charles had briefly tried back in 1855, the construction of a working fragment of the carriage and addition mechanisms of the difference engine. Once again, his memory of the details was inconsistent. Emulating the managers of saintly relics, Henry decided to make a whole series of such small models, eventually presenting some as gifts, though the exact number is not clear. In a manuscript note of December 1884 he referred to 'six specimen pieces'. In a similar document of December 1886 he mentioned 'seven fragments'. In a printed editorial of October 1888 he listed six 'sample pieces', and in his 1910 *Memoirs* he counted 'five small separate pieces'.[74] Apart from the Whipple Museum fragment, only three others assembled by Henry are extant. One was despatched to Harvard, a destination that might have recalled the Harvard physician Bowditch's visionary plans for a 'work of art' half a century earlier. The Harvard device was sent in December 1886 at the same time as that for Cambridge University. Henry also presented a piece to University College London, his alma mater and erstwhile base for Lardner and Watkins, a model transferred to the Science Museum in 1967 (Figure 6.6). A similar sample acquired by Henry's nephew and Benjamin's son Charles Whitmore Babbage, surveyor, clerk, convicted fraudster, and farmer in South Australia and New Zealand, has since 1996 been held by the Powerhouse Museum in Sydney. In 1888 Henry also mentioned a gift to Owens College Manchester, though it is possible that there was some confusion with the model shown at Guthrie's October 1884 lecture in the city or perhaps through some connection with the College's great patron Whitworth, who died in 1887. In any event, no model of the difference engine survives anywhere in Manchester.[75]

---

74 H. Babbage, 'Note on Specimen Piece of Babbage's Difference Engine' (December 1884), in Tee, 'The Heritage of Charles Babbage in Australasia', p. 59; H. Babbage, 'History of This Fragment' (December 1886), in Cohen, 'Babbage and Aiken', p. 185; H. Babbage, 'Conclusion' (October 1888), in H. Babbage (ed.), *Babbage's Calculating Engines* (London: Spon, 1889), p. 341; and H. Babbage, *Memoirs and Correspondence*, p. 225.

75 Inventory numbers of extant machine fragments are Harvard Collection of Historical Scientific Instruments 1991-1-0001a; Science Museum London 1967-70; and Sydney Museum of Applied Arts and Sciences 96/203/1. For Bowditch see Williams, 'Babbage and Bowditch', p. 285; for Charles Whitmore Babbage see S. O. Reader, *The Vision Splendid* (Canberra: National Library of Australia, 2011), pp. 116–21; and Tee, 'The Heritage of Charles Babbage in Australasia', pp. 51–53. Thanks are due to James Sumner and Erin Beeston for help with Manchester sources.

Figure 6.6 Henry Babbage's model of the difference engine mechanism with three cages assembled in 1879 and presented to University College London. By permission of the Science and Society Picture Library (Science Museum 1967–70).

Like the Whipple piece, the fragments at Harvard and in Sydney are also accompanied by Henry's instructions, which vary but little among themselves, though, in contrast to the Cambridge notes' indication of the novelty of some of the components, the notes for Harvard mention that 'the whole is dull with time and dust' and that the lever might not sit on the teeth of the crown wheel 'from dirt and weakness of the spring', while the Sydney notes uniquely refer to the whole engine's calculating mechanism as composed of no fewer than eighteen figures in the result column. Henry's *Memoirs* are also somewhat confused about the details of the models he made and sent as gifts. He misremembered that the Harvard model, and the fragment for his own use, were 'of five cages each, to show two figures added to three', while all the others showed the addition of one figure to two figures. Yet in fact, like those in London and Sydney, the Harvard model carries only three cages; the sole extant model with as many as five cages is that in the Whipple Museum.[76]

Just before despatching model fragments of the machine to British and North American universities, the sixty-year-old Henry and his family moved away from Bromley to Cheltenham in 1885. He took with him the materials for a small workshop, including the lathe originally acquired from Clement.[77] He never again did any work on the difference

---

76 Cohen, 'Babbage and Aiken', p. 185; Tee, 'The Heritage of Charles Babbage in Australasia', p. 59. There are also early difference engine fragments at the Museum of the History of Science, Oxford, no. 94229, which were given by Charles Babbage to Harry Buxton; and at the Macleay Museum, University of Sydney, object no. 1993.3, components presented by Nevil Francis Babbage, great-grandson of Benjamin. Thanks are due to Jude Philp at the Macleay Museum for her help.

77 H. Babbage, *Memoirs and Correspondence*, pp. vi–vii, 228. The Clement lathe is presumably Science Museum London object no. 1878–89.

engine. The models he sent out in 1886 were thus a form of terminal valediction. From then on, it was the mill of the analytical engine that absorbed attention. In September 1888 he travelled from Cheltenham to Bath to lecture the British Association on the analytical engine, and the following month added he the text of this lecture to a large printed collection put together when in Bromley, consisting of papers by his father and other protagonists of the engineering project designed to offer a synoptic history of the calculating engines. He told his readers that it was because of his father's dying bequest of the 'calculating machines and all that belonged to them at my absolute disposal . . . and not to any special fitness for the task that it has fallen to me to complete it'. In several respects, notably its stress on the power of recall and surveillance offered by the system of mechanical notation, and the elegance of memory and anticipation embodied in the analytical mill, the 1888 lecture and publication were to be read as a long riposte to Merrifield's pessimistic BAAS report of a decade earlier, though Henry conceded finally that he saw 'no hope of bringing any profit to its constructor'.[78]

Nevertheless, he did renew his enterprise to build the mill. During 1888 he once again contacted Munro at Tottenham, whose firm worked with Henry's advice to complete a cast iron bed for the mill together with the rudiments of the steel cam mechanisms for number carriage. Visitors such as the ingenious South Kensington physics lecturer Charles Vernon Boys, who designed his own machine to effect carriage and was an authority on office calculating machines and arithmometers, visited Cheltenham to see the mill and learn how simultaneous carriage, with its elegant principle of anticipation, was achieved. The mill project lasted eight years, at the end of which Henry was discouraged by systematic errors in the smooth running of the machine when calculating multiples of π, halted the work and deposited the mill in the South Kensington Museum (Figure 6.7).[79] At the same time, in 1896, the elderly retired Indian general travelled up from Cheltenham to the offices of the best-selling *Strand Magazine* in London, a journal then otherwise best

78 H. Babbage, 'Preface' and 'Paper Read at Bath, 12 September 1888', in H. Babbage (ed.), *Babbage's Calculating Engines* (London: Spon, 1889), pp. 331–7, on p. 337.

79 H. Babbage, *Memoirs and Correspondence*, p. 227; H. Babbage, 'Babbage's Analytical Engine', pp. 518–19; and Charles Boys in 'Meeting of the Royal Astronomical Society, 8 April 1910', *The Observatory*, 33 (1910), pp. 191–201, on pp. 195–6. Thanks are due to Joshua Nall for this reference. For Boys and arithmometers, see A. Warwick, 'The Laboratory of Theory', in M. Norton Wise (ed.), *The Values of Precision* (Princeton: Princeton University Press, 1995), pp. 311–51, on p. 334.

Figure 6.7 The analytical engine mill completed by R. W. Munro for Henry Babbage in 1906–10. By permission of the Science and Society Picture Library (Science Museum London 1896–0058).

known for its publication of Conan Doyle's detective stories, to advise the journalist William Fitzgerald on appropriate coverage of the display of the difference engine model and the analytical mill at the Museum.[80] From 1906, Henry would pay further regular visits to the Museum to inspect the model, devise more alterations, notably to the anticipating carriage, and devise a new printing mechanism. 'He was not a mathematician', so the Science Museum's mathematics curator David Baxandall remembered, but Henry 'looked upon it as a duty to try and complete some little portion of the analytical engine which his father had designed but only partially constructed'. The same year Henry decided to resuscitate his work with Munro's workshop, which eventually organised the analytical printer to generate impressions of a sequence of multiples of π to twenty-eight places.[81]

In 1910–11 Henry staged his last set of exhibitions of the calculating engines. In April 1910 he took the printouts from the mill to a meeting of the Royal Astronomical Society at Burlington House, where he gave a talk to the astronomers on the analytical engine and showed a photograph of the device: he lauded the aim to use mechanical calculators to unveil 'the laws of the Cosmos', referred to the sacred trust granted by his father's legacy, and explained that he had 'endeavoured to complete the work as far as the instructions in

80 W. Fitzgerald, 'The Romance of the Museums, Part 5', *Strand Magazine*, 11 (January 1896), pp. 710–15, on p. 713.
81 D. Buxton, 'Charles Babbage and His Difference Engine', *Transactions of the Newcomen Society*, 14 (1933), pp. 43–65, on p. 60.

the hands of the workmen permitted'. Charles Boys was at the meeting to express his continuing puzzlement about the analytical mill's mechanisms for recall and anticipation. The Cambridge astronomy professor Robert Ball reminisced, if 'a treacherous memory is not deceiving me', that as late as 1865 Charles Babbage had 'young men specially trained as highly skilled mechanics for the purpose of remarking on the construction of the analytical engine'. Embarrassingly, in the 1910 display mistakes in the value of π fed into the machine, and weakness in the printer's springs led to errors in the output.[82]

Undeterred by this publicity setback, Henry then arranged for the mill to be put on show both at the Astronomical Society's soirée and more publicly at the prodigious entrepreneur Imre Kiralfy's glamorous and crowded exhibitions at White City, the British–Japanese exhibition of 1910 and the Coronation exhibition of 1911. The mill and printer were placed in the British Science section near Lyons restaurant and the Wood Lane underground station: Henry grumbled that there was 'no one to explain it on these occasions and no great interest was taken in it. After this it went back to the Museum.'[83] By summer 1914, on the eve of war, the devices appeared in Edinburgh in a display of calculating machines marking the tercentenary of Napier's invention of logarithms, but merely as a Science Museum photograph alongside other images of the difference engine and a commemorative portrait of Charles Babbage himself.[84] In his notes for the display, the Dublin accountant Percy Ludgate summed up the achievements of the Babbage enterprise, before then announcing his own development of a completely newfangled analytical engine. The Babbage machines were now consigned to a past of original invention, with a singular and heroic inventor and tragic aftermath.[85]

82 H. Babbage, 'Babbage's Analytical Engine', p. 518; H. Babbage, 'Meeting of the Royal Astronomical Society', pp. 195–7; and H. Babbage, 'Errata', *Monthly Notices of the Royal Astronomical Society*, 70 (June 1910), p. 645.

83 H. Babbage, *Memoirs and Correspondence*, p. 228; and *Coronation Exhibition Official Guide and Catalogue* (Derby: Bemrose, 1911), p. 120. See Swade, *The Cogwheel Brain*, p. 315.

84 E. M. Horsbugh (ed.), *Modern Instruments and Methods of Calculation* (London: Bell, 1914), p. 27.

85 P. Ludgate, 'Automatic Calculating Machines', in E. M. Horsbugh (ed.), *Modern Instruments and Methods of Calculation* (London: Bell, 1914), pp. 124–7 describes the Babbage machines; for his own engine see B. Randell, 'Ludgate's Analytical Machine of 1909', *Computer Journal*, 14 (1971), pp. 317–26.

The fragment of the model difference engine on display in the Whipple Museum had a much humbler fate than the analytical mill in South Kensington: Wilkes had told Bryden that to call the Whipple model 'epoch making' was to confuse it with the analytical engine.[86] The relationship between such projects and their makers' repute has become even more marked since the assemblage of a working version of the second of Babbage's difference engine schemes at the Science Museum in 1985–91, launched in an exhibition called *Making the Difference* alongside corresponding hagiographies both of machine and of individual author: 'Museum revives Georgian genius's technology'.[87] All these devices share significant patriarchal, technical, and economic histories, different versions of a plan for a world-machine managed by what Farr once called a 'political Newton'. Babbage's difference engine was proposed and funded as a device for the manufacture of printed tables, and its function showed the close if perhaps surprising relationship between calculation and measurement. The disciplinary organisation of precision measures in the workshop was very closely related to the reliability of the calculation of successive terms in a mathematical series. While the former seemed to be a matter of judgment and skill in gearcutting and forging, proper to the engineering workshop, the latter was surely merely a question of following a rule, appropriate for the student's study and the actuary's desktop. The fate of the Babbage engines showed this contrast was illusory: to organise computation was always also to organise labour. At successive public exhibitions in South Kensington, Manchester, and White City, the calculating engines were displayed alongside impressive tables and humbler desktop calculators. The arithmometers that, unlike the difference engine, did come to dominate government and private calculation offices could become sites of skill and conflict; and by 1910 were not used to make tables but to replace them by performing computation directly. Displays of the model difference engines should juxtapose and connect them with these modest and indispensable contemporaries, rather than persistently seeking to find the ancestry of the modern electronic computer somewhere buried inside Henry Babbage's device.[88]

---

86 Wilkes to Bryden, 4 March 1977, Whipple Museum archive P 011.
87 Purbrick, 'The Dream Machine', pp. 15–19; and Swade, *The Cogwheel Brain*, p. 279.
88 Warwick, 'The Laboratory of Theory', pp. 331–6; Johnston, 'Making the Arithmometer Count'; and Schivelbusch, 'World Machines', pp. 59–61.

   That device is a relic, an object deliberately designed to evoke
certain souvenirs of filial piety alongside the ingenious manipulation
of storage and recall. When Charles and Henry Babbage assembled
their very first version of the five-cage difference engine model in
1855, London was agog with the display of the relics of the Franklin
expedition to the Arctic and its evidently tragic fate. Dismembered
equipment, machinery, and more mundane materials from Babbage's friend John Franklin's catastrophic voyage to the Beaufort
Sea were put on show just round the corner from Dorset Street at the
Polytechnic Institution. In this presentation of melancholy polar
detritus at the Institution where 'scientific discoveries thrown off
hot from the brain' were normally displayed, components of a
technically sophisticated mid-Victorian steam-driven scientific and
political enterprise became a means, through an exhibition, of
imagining the salvage of a project that was nevertheless decisively
lost. The project became intensely identified with the persona of the
solitary hero, martyred by an unforgiving establishment. The analogy with the calculating engine enterprise is instructive – an assemblage for public display of a set of relics of an endeavour that might,
at least in the imagination, somehow be rescued from oblivion. The
souvenir and the machine played crucial roles in this relic cult and its
complex motivations.[89] Babbage himself had a model of how such
memories and relics might survive forever, through the traces left in
the atmosphere. It might be apt to accompany the museum display
of the difference engine model with a recording of the noisy warning
it issues during carriage – listeners might then just catch the traces of
all the memorable histories embodied in its workings.

89 Craciun, *Writing Arctic Disaster*, pp. 45–50.

Appendix

Henry Babbage's notes on the model fragment of the difference engine
sent to Cambridge in December 1886: Whipple Museum MS 2339
[Sheet 1: Contents]
Contents of this box
A piece of 5 figures. Diff. Engine
5 German silver rings with numbers 0 to 9 engraved. These have
to be mounted.
4 screens. These require a little fitting, and a mark made on each
for the index guide (see print). They also want screws to keep them
in position when raised to hide the figure.
A pamphlet prepared for S. Kensington but useful here.
H P Babbage
Dec 1886

—

[Sheet 2: Instructions]
To work the Machine
The Axes have the numbers I, II and III attached to them.
The axis No. II should be so placed that the sliding bolts fixed to it
are perpendicular to the front or face of the Machine. In this
fragment this must be adjusted by hand: this having been done
Axis No. I must be turned once which will shoot the sliding bolts
in all cases except where the figure is 0.
Axis No. II should then be turned a half circle:- this will do
addition, the figure on the left hand column being added to that
on the right hand one, the first remaining, the second of course
changing and if a 'carriage' has become necessary the warning lever
will have moved (giving a 'click')
Axis No. III should now be turned and the 'carriage' will be made
and the calculation completed.
Where any difference is *minus* the complement is added.
H P Babbage
Dec 1886.

—

[Sheet 3: Note]

Note

The german silver rings with the numerals engraved on them have been made recently and never fitted on. The figures run reverse ways on the adjacent columns. Five are sent. In fitting them on care should be taken that the Index of the figure wheel is about mid way between 0 and 9 when the carriage warning lever is released.

The index is on the screen in front: four screens are sent and require slight fitting. On each of them should be made a mark thus ♦ (see print) which is the Index or guide to the figure and when the figure is not wanted in the calculation or for any reason it is wished to hide it, the screen can be raised and so brought over the figure. The print in the S. K. pamphlet will be found useful in making these arrangements.

H P Babbage

7 ❧ Galvanometers and the Many Lives of Scientific Instruments*

CHARLOTTE CONNELLY AND HASOK CHANG

Introduction: Into the Black Box

Electrical measuring tools now epitomise 'black-boxed' technologies. Since the second half of the nineteenth century, ammeters and voltmeters have been developed that users could apparently simply connect up to their electrical circuitry, allowing them to read off a number giving them a measure of current or voltage. It seems that a clear majority of those who use electrical measuring instruments today lack any clear or detailed understanding of the theoretical principles and material designs inside the black boxes which they rely on so readily.

In this chapter we want to illustrate what we can learn by getting inside such black boxes. In this enterprise, museum collections play an essential role, complementing written records, because they constitute tangible traces of how what is now black-boxed has developed. By analysing these instruments carefully, we will interrogate the craft of both instrument maker and user, some of the different types of user whose practices are embodied in the instruments, and the lessons we can learn from a close look at instruments and collecting practices. We will look into the mechanisms and historical contexts of selected pairings of electrical measuring instruments from the Whipple Museum's collection, with a focus on galvanometers. In fact, Robert S. Whipple himself paid considerable attention to the history of galvanometers, publishing an informative paper on the subject in 1934, and galvanometers were a major line of instruments produced and marketed by the Cambridge Scientific Instrument Company, where Whipple worked from 1898, rising eventually to the position of Managing Director and then Chairman.[1]

* Thanks are due to BT Connected Earth for funding Charlotte Connelly's research internship at the Science Museum in 2009. This chapter drew extensively on material gathered during the placement.
1 R. S. Whipple, 'The Evolution of the Galvanometer: Report of a Discourse Given at the Twenty-Fourth Annual Exhibition of the Physical Society', *Journal of Scientific Instruments*, **11** (1934), 37–43.

Most of what we now know as electrical measuring instruments were made possible by the spectacular constellation of developments in the first third of the nineteenth century that enabled the monitoring of electromagnetic effects. Key steps in these developments included the invention of the battery and the electrical circuit by Alessandro Volta, which was publicised in 1800; the discovery of the magnetic action of electrical currents by Hans Christian Ørsted in 1820; and the establishment of the famous law relating voltage, current, and resistance by Georg Simon Ohm in the late 1820s.

Galvanometers were instruments designed to measure the strength of current going through a wire, whose various designs evolved in interesting ways. Different understandings of what electricity was, and the different ways it might be measured, were part of the changing theoretical landscape in which galvanometers developed. Shifts in instrumentation arrived in tandem with shifts in modes of thought about electricity and understandings of the different ways in which electrical phenomena could be interrogated. When combined with other theoretical and physical trappings, galvanometers were at the core of almost all electrical measurements. By the time electrical measurements had become well-established in the late nineteenth century, the handiest method of voltage measurement was to use a galvanometer to measure a current going through a resistor of known resistance, relying on Ohm's law to deduce the voltage from the current. And the easiest method of measuring resistance was to apply a power source of a known voltage to the resistor and measure the current that flows through it, calculating the resistance from the current, again using Ohm's law. So we can see that the galvanometer was the instrumental heart of electrical science and technology, while Ohm's law was the theoretical heart.

Given the paramount importance of electrical technology and science in the European and European-influenced civilisations of the world from the second half of the nineteenth century onwards, it would not be too much of an exaggeration to say that the galvanometer was *the* defining measuring instrument of pre-electronic modern society. Whipple opens his historical account of galvanometers as follows: 'There are few scientific men, and presumably no electrical engineers, who have not been called upon to use a galvanometer at some period of their lives.'[2]

---

2 Whipple, 'The Evolution of the Galvanometer', p. 37.

Prologue: The Continuity and Stability of Electrometers

Our discussion will focus on nineteenth-century galvanometers; however, as many histories of electrical measurement begin with tools for measuring static electricity, it is instructive to start with a brief look at a much older and more basic type of instrument, namely electrometers. (Incidentally, as Whipple notes, the first use of the word 'galvanometer', by Bischoff in 1802, seems to have been to designate an electrometer.[3]) Before the invention of the battery by Alessandro Volta, electricity flowing in a circuit did not exist. Electricity mostly existed in the form of static build-up of charge displaying attractions and repulsions, and its usually sudden release resulting in shocks, sparks, and lightning. In that pre-Voltaic situation, the most important quantity to try to measure was the degree to which a body was charged up with electricity. That was the job of the electrometer.

A superficial look at the history of this instrument may indicate an orderly progression from *electroscopes* giving qualitative detection of charge to *electrometers* allowing quantitative measures. For example, in the Adams electrometer of *c.* 1775 (Wh.6648, Figure 7.1, exhibit 1A) there is a carefully etched scale. This was the work of George Adams the Younger (1750–95), who had succeeded his father as the Mathematical Instrument Maker to George III. When electrical charge is communicated to the instrument through the ball at the top, the wooden indicator arm terminating in a small pith ball turns up away from the main column due to electrostatic repulsion. This is the basic principle behind all simple electrometers: two parts of the instrument are given an electrical charge of the same sign, and they thereby repel each other. John Heilbron summarises the long history of such instruments, starting with those with two strings in the 1740s, up to more sensitive instruments used by Abraham Bennett and Alessandro Volta in the 1780s employing thin straws or gold leaves.[4] In the middle of this history came a variant invented by William Henley in 1770, which was basically the design that Adams employed in his instrument we are examining here. Adams gave his own description of this apparatus in a book on the science of electricity. The 'quadrant electrometer', as he called it, he regarded

3 Whipple, 'The Evolution of the Galvanometer', p. 38.
4 J. L. Heilbron, *Elements of Early Modern Physics* (Berkeley and Los Angeles: University of California Press, 1982), p. 218.

Figure 7.1 Exhibit 1A: an electrometer by George Adams, *c.* 1775. Exhibit 1B: a Curie-type gold-leaf electroscope by Matériel Scientifique, 1905. Images © Whipple Museum (Wh.6648; Wh.1353).

1A   1B

as 'the most useful instrument of the kind yet discovered, as well for measuring the degree of electricity of any body, as to ascertain the quantity of a charge before an explosion; and to discover the exact time the electricity of a jar changes, when without making an explosion, it is discharged by giving it a quantity of the contrary electricity'.[5]

This Adams electrometer in the possession of the Whipple Museum still functions after two and a half centuries, which is a wondrous thing to experience. Usually the functional parts of electrometers were made of very light and fragile materials in order to allow sufficient movement with small amounts of electrical charge. It is a common experience to see gold-leaf electrometers in museums with the crucial gold leaves missing. Adams's skill in working wood allowed him to make an unusually robust instrument.[6]

---

5 George Adams, *An Essay on Electricity, Explaining the Theory and Practice of That Useful Science; and the Mode of Applying It to Medical Purposes, with an Essay on Magnetism*, 3rd edn (London: R. Hindmarsh, 1787), p. 49. A picture of the instrument is given as figure 6 in Plate II attached at the end of the book.
6 Examples of absent gold leaves in the Whipple collection include Wh.1399, a gold-leaf electroscope by Harvey and Peak, presented to the museum by the Royal Institution, which is lacking both gold leaves; and Wh.1344, a Thomson patent electroscope with a wooden drawer that contains gold-leaf scraps.

However, much as we can admire Adams's workmanship, what exactly his or others' electrometers were measuring is not clear. How is the scale on an electrometer calibrated? There were no precise theoretical calculations showing how much deflection of the straw, gold leaf, or Adams's wooden arm should result from a given amount of charge imparted to an electrometer. Not only was Coulomb's law of electrostatic attraction and repulsion not yet established (Coulomb's paper was published in 1785, about ten years after Adams made his instrument), but knowing the inverse-square law is nowhere near enough for actual computations without a lot of additional information about the specifics of the parts of the instrument and its settings. It is difficult to imagine that there would have been consistency of measured values across different instruments, and to use the indications of electrometers in electrical science in a coherent and productive way would have required considerable theoretical savvy and intuition.

Aside from all theoretical considerations, trying to use a simple electroscope, whether it be the original Adams electrometer or a modern toy gold-leaf one, can be a humbling experience. The conceptual simplicity of static electricity imparted by elementary textbooks shows its futility in the face of the practical challenges of the operations of charging, earthing, and insulating. The business of static electricity is not as easy as it might seem in the abstract. Electrometers require a lot of skill and knowledge to build and operate (which those of us with a typical modern scientific education most likely lack). All in all, it seems overdetermined that the numerical readings given by instruments such as the Adams electrometer would not have allowed meaningful arithmetic operations on them, which means that to call them electrometers – measuring tools – as opposed to electroscopes – indicating tools – is clearly a misnomer. Even some of the very early instruments also allowed the measurement of the angle of separation, but that does not constitute an *electrical* measurement. Indeed, the two terms were used quite interchangeably in the early days. By the late nineteenth century the electrometer had been re-conceived more strictly as a measuring instrument, and what it was now seen to measure was electrical potential; the most iconic example of it was the quadrant electrometer invented by William Thomson (later Lord Kelvin).[7]

7 J. A. Fleming, 'Electrometer', in *Encyclopaedia Britannica*, 11th edn, vol. 9 (1910), pp. 234–7. The examples in the Whipple collection include Wh.1325, a Thomson-type quadrant electrometer used by Warren De la Rue. This or a

So, is the story of the eighteenth-century electrometer one of *faux*-quantification and tacit knowledge? Yes, but our next 'exhibit' (Wh.1353, Figure 7.1, exhibit 1B) reveals another whole layer to the story. Well over a century after Adams built his instrument, electrometers made an unexpected return to cutting-edge research in physics. This gold-leaf electroscope, dated December 1905 and following a design developed by Pierre and Marie Curie, was transferred to the Whipple Museum from the Cavendish Laboratory. The Curie-type electroscope had two gold leaves attached at its top to a brass plate, both hanging together vertically when the instrument was not charged.[8] As is the case with many gold-leaf electrometers, the delicate gold leaves are missing.

As the association with the Curies suggests, this electroscope was used to measure radioactivity: a radioactive sample placed on the capacitor plate would cause the charge on the electroscope to decrease. Take the description in an educational text in the 1920s by Noah Ernest Dorsey.[9] First, Dorsey explains, the insulated piece of metal and the attached piece of gold leaf are electrically charged, causing the leaf to deflect away from its normal, vertical position. The position of the leaf is observed by the experimenter peering through a microscope with a ruled scale in its eyepiece. The leaf will gradually lose its charge and move back to its vertical position due to imperfect insulation. The experimenter observes the leaf's movement and times how long it takes for the leaf to move over a few divisions of the scale, and from this can determine a rate of drift in the leaf when there is no radioactive source present. This process is then repeated, but with a radioactive source placed in the instrument, resulting in a second rate of drift, different from the first due to the ionising radiation causing the electrical charge to dissipate at an increased rate. The ratio of the two drifts indicates the intensity of the radioactive source.[10] Dorsey follows his basic description with a discussion about further practicalities, including how the instrument

similar object is pictured in a photograph of De la Rue's laboratory, alongside a moving-magnet pointer galvanometer, as described later in this chapter.

8 J. A. Fleming, 'Electroscope', in *Encyclopaedia Britannica*, 11th edn (1910), vol. 9, pp. 239–40, on p. 240.

9 This description occurs in his outlining of the various steps involved in determining the radium content of a small sealed glass tube.

10 N. E. Dorsey, *Physics of Radioactivity: The Text of a Correspondence Course Prepared Especially for the Medical Profession* (Baltimore: Williams & Wilkins, 1921), pp. 153–5.

should be insulated, whether or not specimens are fully aged, and how corrections should be made for specimens of different sizes.

Typical narratives of progress tend to ignore how old technologies linger on and find other uses, and the Curie electrometer and its use in radioactivity research is an excellent reminder of this. Electroscopes were also required to accurately measure very small quantities of electricity for studies such as C. T. R. Wilson's investigations into atmospheric electricity.[11] The gold-leaf electroscope was, then, an instrument that retained its usefulness well over a century after it had first been developed for electrical measurement. Our pairing of electroscopes from 1905 and 1775, both of which work in essentially the same way to indicate the presence of small quantities of electricity, demonstrates that the arrival of new technologies does not necessarily render older technologies redundant.

Galvanometers in the Lab and in the Field

Now let us turn our attention to galvanometers. First of all, there is a gap to be filled in the understanding of most historians of science about how galvanometers work. A very basic education in physics teaches us the fundamental theoretical principle, which goes back to Hans Christian Ørsted's discovery that a nearby electrical current could turn a compass needle. But there was a very long way to go from that basic electromagnetic connection to making a usable and useful instrument for measuring electrical current. In contrast to the case of electroscopes, in the development of galvanometers the focus on true quantification was very strong, and it was achieved to an impressive degree. In this section we want to start by giving a very brief description of the major steps that were involved in the making of galvanometers, and then show how the desirable configurations depended on the contexts of use. A whole century after Ørsted's discovery of electromagnetic action, significant improvements were still being made to meet the needs of various users. In the remainder of this chapter we want to give a sense of the character of these improvements that made galvanometers what they were, exhibiting the different developments that were made to suit different needs.

Ørsted's set-up, as is well known, was a metallic wire placed above (or below) a compass needle (or, more generally speaking, a pivoted bar magnet), initially resting along the direction of the Earth's

11 C. T. R. Wilson, 'Atmospheric Electricity', *Nature*, **68** (1903), 102–4.

magnetic field. When the wire was connected to a battery and current was established in the wire, the magnet was rotated by a certain angle. This is because the current exerted a certain amount of force on the magnet, which would eventually rest where there was equilibrium between the force exerted by the current and the Earth's magnetic field tending to pull the magnet back to its original position. Johann Schweigger, André-Marie Ampère, Claude Pouillet, and others then did the fundamental experimental and theoretical work, establishing a clear theoretical relation between the angle of deflection and the strength of current.

In order to turn this arrangement into a precise and usable instrument, many major practical steps were needed.[12] Almost immediately after Ørsted's work in 1820, researchers noted that wrapping the wire into a loop going around the magnet would double the action on the magnet: a wire placed below the magnet has the opposite direction of effect to a wire placed above the magnet if they both carry current in the same direction, but in a loop of wire the direction of current is opposite above and below, so the effects on the magnet are aligned in the same direction. Now, the description just given would dictate a long rectangular shape of a loop, with negligible effect from the short sides, but in fact the same effect can be achieved with a circular loop, provided that it is much larger than the magnet and the magnet is placed in the middle of the loop, as in the Helmholtz galvanometer we will describe in the next section (exhibit 3A); this is an example of a 'tangent galvanometer', so named because 'the tangent of the angle of its deflection will be nearly proportional to the current passing through the coil'.[13] The main point, in that configuration, is that the magnet will try to align itself with the magnetic field produced by the loop of current, which points perpendicularly to the plane of the loop. Now, if one makes many turns with one long wire (in other words, one makes a *coil*), the effect can be multiplied.

Another major line of innovation in galvanometer design was to move beyond the reliance on the Earth's magnetic field in the regulation of the movement of the magnet. Aside from the obvious inconvenience of needing to place the apparatus in a particular direction in each setting, there was also the problem of interference

---

12 Most of these steps are briefly summarised in Whipple, 'The Evolution of the Galvanometer', pp. 38–9.
13 J. A. Fleming, 'Galvanometer', in *Encyclopaedia Britannica*, 11th edn (1910), vol. 11, p. 430.

by other magnetic fields that may be present. This problem was dealt with in two different ways, which gave rise to the development of two major types of galvanometers, namely the moving-magnet and moving-coil types. The moving-coil type rested on a larger innovation in one clear sense: departing from the original paradigm of the rotating compass-needle, a small current-carrying coil was put in the magnetic field of a permanent magnet (often a horseshoe-shaped one, as shown in the next section: exhibit 3B). By employing a permanent magnet of a sufficient strength, the influence of any given external magnetic fields, including that of the Earth, could be made negligible. Whipple credits William Sturgeon with the invention of the moving-coil arrangement as early as 1824, which he used to demonstrate voltaic and thermo-electric currents, and notes an early application of this design to telegraphy in 1856 by Cromwell F. Varley.[14] Norman Schneider by 1907 thought that the moving-coil galvanometers of the D'Arsonval type 'had been so perfected that they are being almost universally adopted'.[15]

In the moving-magnet galvanometer, the mechanisms for eliminating the influence of external magnetic fields were more complex. Sometimes one employed a 'compensating magnet' that roughly cancelled out the Earth's magnetic field.[16] But a real line of innovation began with the so-called astatic needle, which consisted in two magnetic needles fixed together in parallel, one a little distance over the other, with their polarities in opposite directions.[17] The net force on such a compound needle exerted by a uniform magnetic field would be zero, as the force on one needle would be equal and opposite to the force on the other needle. It would be 'astatic' in the sense of having no preferred position where it would rest.[18] However, in a non-uniform field, the strengths of the forces on the two needles would be different, so there would be a net force. The simplest situation to arrange is as follows: 'The lower of the magnetic needles is inside the coil which carries the current under

14 Whipple, 'The Evolution of the Galvanometer', pp. 39–40.
15 N. H. Schneider, *Electrical Instruments and Testing: How to Use the Voltmeter, Ohmmeter, Ammeter, Potentiometer, Galvanometer, the Wheatstone Bridge, and Standard Portable Testing Sets, with New Chapters on Testing Wires and Cables and Locating Faults by Jesse Hargrave*, 3rd edn (New York: Spon and Chamberlain; London: E. & F. N. Spon, Ltd, 1907), p. 15.
16 Schneider, *Electrical Instruments and Testing*, p. 10.
17 Whipple, 'The Evolution of the Galvanometer', p. 39, notes that this was an innovation made by Nobili at the suggestion of Ampère.
18 Fleming, 'Galvanometer', p. 428, also discusses a more complicated form in which the two needles are not precisely matched in strength.

test, and alone experiences a torque due to the resulting magnetic field.'[19] In an arrangement that became common later, the two magnets in an astatic arrangement were placed very far apart so as to minimise the effect of the current to be measured on the 'outside' magnet.[20]

Both in the moving-coil galvanometers and in the moving-magnet galvanometers, eliminating the role of the Earth's magnetic field meant that something else had to act as a spring that prevented the indicating needle from automatically deflecting all the way to 90°, regardless of the strength of the magnetic force from the electrical circuit being tested. Almost universally it was a torsion balance that played this role, a wire or string with which the moving coil would be hung; when twisted, the suspension wire or string would exert a stronger restorative (un-twisting) force the larger the angle of deflection was. The torsion balance was the key apparatus used by Coulomb for his measurements of electrostatic force in his famous work published in 1785, and ever since then it had been available for use in various forms in physical measurements.

Our pairing of instruments from the Whipple Museum in this section illustrates very well the contrasting requirements for a precision instrument designed for scientific research carried out in the well-protected space of a laboratory, and for a robust and reliable indicator fit for use in a workshop or even in the field. They are but two very divergent examples from a profusion of different galvanometers developed in the late nineteenth and early twentieth centuries to meet a variety of needs.[21] Both of the instruments we have selected are moving-magnet galvanometers, but, as even their external appearances suggest, they were intended for different types of use. A well-situated physics laboratory would ideally be free from vibration and any large moving metal objects that might interfere with the delicate operation of sensitive electrical instruments. In contrast, engineering workshops would often be housed in an

19 T. B. Greenslade, 'Astatic Galvanometer', in http://physics.kenyon.edu/EarlyAppar atus/Electrical_Measurements/Astatic_Galvanometer/Astatic_Galvanometer.html (accessed 28 October 2018).

20 Cambridge Scientific Instrument Company, *Some Electrical Instruments Manufactured and Supplied by the Cambridge Scientific Instrument Company, Ltd* (Cambridge: Cambridge University Press, 1912), p. 26.

21 A good sense of the variety of galvanometers in use can be gained from trade catalogues such as Cambridge Scientific Instrument Company, *Galvanometers and Accessories, Manufactured and Supplied by the Cambridge Scientific Instrument Company, Ltd*, List No. 126 (Cambridge: Cambridge University Press, 1913).

industrial building with lots of activity, including large stretches of
cable being moved around, which would compromise any delicate
readings that were being made.[22]

The first of those instruments, Wh.0939 (Figure 7.2, exhibit 2A), is
an example of the most famous type of precision galvanometer,
invented in the late 1850s by William Thomson (later Lord Kelvin).
It was originally in the context of telegraph signalling that his work
on galvanometers began. Thomson's main objective was to increase
the sensitivity of the apparatus. As Silvanus P. Thompson describes
in his classic biography of Thomson, 'He wanted an instrument that
would work with a smaller fraction of current. So he determined to
lighten the moving part – the suspended magnet – substituting for
the heavy needle a minute bit of steel watch-spring (or two or three
such bits), which he cemented to the back of a light silvered glass
mirror suspended within the wire coil by a single fibre of cocoon
silk.' And then he added an ingenious method of observation, which
was 'directing upon the mirror a beam of light from a lamp, which
beam, reflected on the mirror, fell upon a long white card, marked
with the divisions of a scale'. The beam of light served 'as a weight-
less index of exquisite sensitiveness, magnifying the most minute
movements of the "needle"'. Thomson took out a patent for this
'mirror-galvanometer' in 1858. It is interesting to note that this
instrument, initially designed for field use, later became an indispensable precision instrument for the laboratory. As Thompson puts
it, 'It served not only as a "speaking" instrument for receiving signals,

22 G. Gooday, 'The Morals of Energy Metering: The Precision of the Victorian
   Engineer's Ammeter and Voltmeter', in M. N. Wise (ed.), *The Values of Precision* (Princeton: Princeton University Press, 1995), p. 247.

but as an absolutely invaluable appliance both at sea and in the laboratory for the most delicate operations of electric testing.'[23]

The Thomson-type mirror arrangement, which could in fact be used either for a moving-magnet or for a moving-coil galvanometer, reached a very impressive degree of precision. By 1910 John Ambrose Fleming could boast that, 'In modern mirror galvanometers a deflection of 1 mm of the spot of light upon a scale at 1 metre distance can be produced by a current as small as one hundred millionth (10^{-8}) or even one ten thousand millionth (10^{-10}) of an ampere.'[24] The sensitivity of the mirror galvanometer recommended it as an instrument of *detection*, which is what is most needed for the telegraph. But it could also be rigged up as an instrument for precision *measurement*.

Our instrument (2A), which had been used at the Cavendish Laboratory before being transferred to the Whipple Museum, was made by Elliott Brothers for the British Association Committee on Electrical Standards. Describing it as a 'very convenient form of Thomson's galvanometer', the *Encyclopaedia Britannica* suggested that 'such a galvanometer as this, provided with a high and low resistance coil, would meet all the wants of most laboratories'.[25] The 1864 report of the British Association Committee gives a few details of what it was like to work with this type of galvanometer:

> The instrument was placed in a deal box, blackened inside, with large apertures to observe through. The spot of light could thus be clearly seen, and the divisions of the scale were sufficiently illuminated to enable the observer to see immediately in which direction the spot of light moved. The instrument was sufficiently delicate to show 0.001 per cent difference in the ratio of any two nearly equal conductors compared, corresponding to 1/10 millim. on scale of bridge. An ordinary galvanometer was also at hand to find about the place of reading on the scale.[26]

23 S. P. Thompson, *The Life of William Thomson Baron Kelvin of Largs*, 2 vols. (London: Macmillan, 1910), vol. 2, pp. 347–9. Fleming, 'Galvanometer', p. 429, states that the mirror was about 1/4 inch in diameter. Similarly, Schneider, *Electrical Instruments and Testing*, p. 12, states that the mirror is less than half an inch in diameter.

24 Fleming, 'Galvanometer', p. 430.

25 *Encyclopaedia Britannica*, 9th edn (Edinburgh: A. and C. Black, 1879), vol. 10, p. 51, quoted in K. Lyall, *The Whipple Museum of the History of Science Catalogue 8: Electrical and Magnetic Instruments* (Cambridge: Whipple Museum of the History of Science, 1991), Part 1, Section 2, Number 20.

26 'Third Report – Bath, 1864', in *Reports of the Electrical Standards Committee of the British Association for the Advancement of Science* (Cambridge: Cambridge University Press, 1913), p. 174.

Indeed, this instrument is a delicate tool, beautifully crafted for precision measurement. There are some clues that the instrument could be adjusted in a number of different ways to accommodate the needs of an experiment: extendable brass tubing that made up the frame of the instrument could alter the distance of the scale from the mirror within the galvanometer; levelling feet were incorporated to help the user ensure it was completely horizontal; an adjustable brass slit below the paper scale allowed the user to alter the amount and focus of light being reflected off the mirror; and the resistance coil in the instrument could be swapped for a higher or lower resistance, depending on the measurements being taken. Together with other material features of the instrument, like the 'single fibre of cocoon silk' described by Silvanus Thompson in his description of the mirror galvanometers, moving-magnet reflecting galvanometers were delicate instruments. Although, with care, they could be used at sea or in the telegraph office, their value in the laboratory came from the tiny variations in electrical current they could indicate.

In contrast, our second instrument, the 'Lineman's Detector' (Wh.3090, Figure 7.2, exhibit 2B), was commonly used by linemen working in the electrical telegraph industry. An early portable meter, it was designed to be used outdoors for indicating faults and tracing circuits.[27] In this instrument the galvanometer is housed in a small wooden box, with rounded corners and a metal ring for ease of carrying. The pointer is protected by a small glazed window, though the instrument can be opened up for adjustments, which in this case also reveals a handwritten note of the resistance at 60 °F. It is a simple instrument with an inch-long magnetic needle mounted on a horizontal axis (so that it sits vertically) in the space within two coils that are fixed side-by-side.[28] It lacks all the delicate controls and shieldings that ensure valid quantitative measurement, but it will easily indicate the presence and direction of a current passing through a circuit element connected up to it.

Telegraph engineers needed to be able to work wherever a fault on the line took them, usually outdoors with only the resources they were able to take to the site to make repairs. As more underground and underwater cables were installed and cable technology

27 For descriptions of field-based fault-finding practices, see R. S. Culley, *A Handbook of Practical Telegraphy* (London: Longman, Green, Longman, Roberts, and Green, 1863).
28 W. Slingo and A. Brooker, *Electrical Engineering*, revised edn (London: Longmans, 1900), p. 134.

improved, new and more sensitive methods were needed to test for current leakage and faults. The galvanometer was a crucial instrument in these tests.[29] In advance of the 1850 attempt to lay a transatlantic telegraph cable, cables were hung over the side of the ship and tested by connecting a battery and galvanometer across them.[30] In more everyday situations, telegraph engineers had to tackle a range of faults with cables laid in the ground. Although quantitative measurements became increasingly important as cable technologies became more sophisticated, many of the faults could be discovered with an uncalibrated galvanometer that simply required observation of whether the movement of the needle was more or less 'strong'. As William Slingo, founder of the Telegraphic School of Science at the General Post Office, wrote with his colleague Arthur Brooker, 'The lineman's detector ... is a very handy instrument when used for tracing circuits and localising faults, but it must not be regarded as a measuring instrument.'[31] That is correct, in the same sense that the electrometers and electroscopes discussed above did not really provide quantitative measurements. However, the important point is that laboratory 'meters' do not have to be true measuring instruments in order to perform useful functions, including the enabling of other measurements.

Instruments for Teaching versus Research

In this section we wish to draw a comparison and contrast between the use of galvanometers in research and teaching. One preliminary remark we need to make is that there were great changes happening throughout the nineteenth century in the institutional and physical settings in which research and teaching took place. Both teaching and research took place at universities, of course, but there was much

29 In describing a long series of steps for testing covered wires, R. S. Culley suggested the following procedure which was unusual in not being entirely dependent on using a galvanometer: 'The wire may be *charged statically* by battery and the time noted during which it will retain the charge. Put one pole of the battery to the earth, and with the other pole touch one end of the wire for a moment. After, say 10 seconds, put the wire to earth through a galvanometer, when the charge will pass to earth, moving the needle. A good wire will retain the charge several minutes. If a galvanometer be not at hand, the charge may be taken on the tongue with equal certainty as to the result, and without any inconvenience from lengths under a mile.' Culley, *A Handbook of Practical Telegraphy*, p. 116.
30 R. M. Black, *The History of Electric Wires and Cables* (London: Peter Peregrinus, 1983), p. 16.
31 Slingo and Brooker, *Electrical Engineering*, p. 134.

3A   3B

Figure 7.3 Exhibit 3A: a moving-magnet pointer galvanometer by Elliott Brothers, *c.* 1880. Exhibit 3B: a moving-coil reflecting galvanometer by the Cambridge Scientific Instrument Company, *c.* 1902. Images © Whipple Museum (Wh.1347; Wh.4190).

else besides. The tradition of industrial research was taking shape, and a small number of non-university institutions such as the Royal Institution maintained their crucial importance.[32] 'Amateur' or 'private' researchers working in their 'spare time' became gradually less important as the century wore on, but they remained a non-negligible force. We want to highlight one such private researcher, Warren De la Rue (1815–89), who gave various instruments including a galvanometer to the Royal Institution, which then presented it to the Whipple Museum (Wh.1347, Figure 7.3, exhibit 3A).

De la Rue was born in Guernsey and educated at the Collège Sainte-Barbe in Paris, after which he entered his father's stationery business in London, and carried out scientific research as an avocation. De la Rue is best remembered now for his pioneering work in the photographing of astronomical objects, but he also carried out interesting and useful research in chemistry and electricity. In the area of electricity, De la Rue's major contributions included the invention of a platinum-coil light bulb, the investigation of electrical discharges in vacuum tubes, and the silver-chloride battery, which he employed for his other studies.[33] Starting with the work on the platinum-coil light bulb, which he undertook at a young age in 1840, De la Rue's electrical research involved high-powered

---

32 The development of industrial research is well represented by a number of periodicals that were published in the nineteenth century, sharing industrial research in the growing telegraphic and electrical industries, including *The Electrician* (1861–1952), *Journal of the Society of Telegraph Engineers* (1872–80), and *The Telegraphic Journal and Electrical Review* (1872–present).

33 The silver-chloride battery is described in J. E. H. Gordon, *A Physical Treatise on Electricity and Magnetism* (London: Sampson Low, 1880), vol. 1, pp. 216–19.

operations using large voltages and currents, in stark contrast to the efforts of William Thomson trying to pick up the very faint telegraph signals coming across the ocean. So it was quite appropriate for De la Rue to employ a moving-magnet tangent galvanometer, which was the instrument of choice for the measurement of high currents. The instrument we have included here was a high-quality but standard-issue Helmholtz-type galvanometer, in which the moving magnet sits in the middle of the instrument, between two coils. This instrument is not unique, but has a few interesting features, including levelling feet, implying it was designed for precision use, and a water-cooling pipe that runs around the coil. Although it was a relatively standard instrument, the uses to which De la Rue turned it were unique to his particular areas of research.

Turning now to the very different context of teaching, the immediate thing to note beyond the academic sphere is the context of trade-based training. When electrical engineering first emerged as a profession, most practitioners using galvanometers and other electrical instruments for their trade learnt about them during a practical apprenticeship. Meanwhile, laboratory-based electrical work was the preserve of a few independent researchers or researchers with university teaching positions.[34] Equipment and techniques in both lab and field were still becoming established as understanding of electrical theories grew rapidly. Early on, galvanometers were non-standardised and often challenging instruments to calibrate and use well. As discussed above, in the field they were used as indicators or detectors more often than as carefully calibrated measuring devices. As the century progressed formal education became increasingly available for both engineers and scientists, and the debates around what should be taught to each group reflected the newly negotiated roles of each profession. The differing roles are also expressed in the types of instruments and the language used by the two groups. Throughout the early decades of the new profession, electrical engineers could by and large manage with rule-of-thumb calculations, and using galvanometers as indicators. By the 1870s, trade publications like *The Telegraphic Journal* were advocating that engineering students should use good scientific practice. An 1873 advert for a prize for students stated that

---

34 For an extensive discussion of the changes in electrical education in the late nineteenth century see G. Gooday, 'Teaching Telegraphy and Electrotechnics in the Physics Laboratory', *History of Technology*, **13** (1991), pp. 73–111.

To be awarded to the Author of the best paper on 'The evidence of the theory of correlation of Physical Forces as applied to Electricity and Magnetism' ... each paper submitted for competition must describe original experiments.[35]

The emphasis on original experiments also suggests a requirement for students to have access to instruments that could measure electrical and magnetic effects. The rhetoric in trade journals in the 1870s repeatedly reinforced the idea, promoted by vocal proponents of technical education, that engineers should think scientifically, and eschew the old ways of doing business:

These words [tension, intensity, and quantity] are daily employed by the majority of those engaged in Telegraphy, and to their use may be attributed, I think, the great want of accuracy in electrical matters displayed by many of this class of men. Nothing is more likely to foster unscientific habits of thought than the constant employment of ill-defined terms.[36]

The education available for telegraph engineers changed dramatically throughout the nineteenth century, reflecting national trends towards awarding qualifications and extending the length of formal study in many professions. The mechanics' institute movement had begun in the late eighteenth century, and by 1850 there were about 700 mechanics' institutes teaching a range of subjects, sometimes including electricity, to workmen.[37] Telegraphy was a large commercial venture, and ensuring that engineering staff were properly trained was an essential part of maintaining the network. Equally, it was important for the telegraph companies to understand the most recent advances in electrical theory in order to improve the network and keep it functioning as well as possible. The availability of instruction manuals and measuring instruments aimed at students grew to meet the demands of the growing numbers of trainee telegraphers and electricians.

By the 1880s, fifty years after William Fothergill Cooke and Charles Wheatstone's landmark demonstration of a practicable telegraph system, non-university technical education in electricity or telegraphy had become a reality. Training centres such as the

---

35 'Prizes to Students', *The Telegraphic Journal*, 1.12 (1 September 1873), p. 214.
36 W. E. Ayrton, 'On the Advantages of a Scientific Education: A Lecture Addressed to the Telegraph Staff', *Journal of the Society of Telegraph Engineers*, **1** (1872), pp. 266–7.
37 R. Bourne, 'The History of Non-university Electrical Engineering Education with Special Reference to South-East London', in *Papers Presented at the 19th IEE Weekend Meeting on the History of Electrical Engineering* (1991), pp. 8.1–8.2.

Finsbury Technical College in North London and the Central Institution in South Kensington were established. This raised a practical question of what could reasonably be taught in a classroom, especially as many engineers felt that the demands of a full-scale telegraph line could not be adequately replicated on a small scale. A compromise was met when William Ayrton – the same Ayrton whose designs and name are associated with many of the electrical instruments in the Whipple Museum's collection – was appointed City and Guilds Professor of Physics in 1879. His inaugural address contained an important message: his lectures and laboratory teaching 'would not compromise the integrity of the workshop as the definitive place for the apprentice to learn his trade'.[38] The classes, he argued, were not meant to take the place of practical experience, but to supplement it. Students would spend laboratory time with teaching instruments as well as field instruments, learning the principles behind the practical work they would do to keep the telegraph lines running. The classes could serve to shorten the duration of an apprenticeship from the then typical seven years to just four years. Instruction in Ayrton's classes met with approval from engineers, and covered topics such as the construction of resistance coils, the construction and use of artificial cables, and the use of various kinds of equipment. To better simulate field conditions, 'artificial lines' were constructed out of several hundred yards of telegraph cable. Bench equipment was used in some instances to create a proxy for field work and on other occasions to enable individual enquiry and theoretical understanding. Universities also expanded their offer for electrical education. Professor George Carey Foster succeeded in establishing a students' physical laboratory at University College London as early as 1866, and was instrumental in teaching practical courses on a variety of physical subjects, including the technical training of telegraph engineers.[39] The arrival of dedicated technical institutes provided specialist practical training in a formal setting that was distinct from natural science teaching. Appropriate scientific instruments were essential for making this work, and equipment like the 'Outfit for teaching Resistance Thermometry' (Figure 7.4) demonstrates the market in instruments for teaching that had emerged by the end of the nineteenth century.

38 Gooday, 'Teaching Telegraphy and Electrotechnics in the Physics Laboratory', p. 95.
39 Gooday, 'Teaching Telegraphy and Electrotechnics in the Physics Laboratory', pp. 87, 92–3.

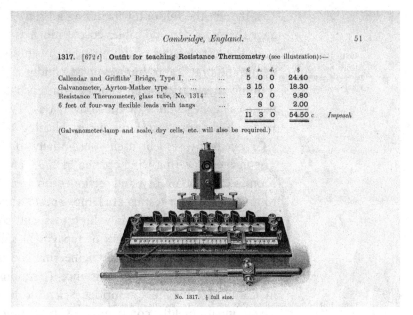

Figure 7.4 An outfit for teaching resistance technology, as advertised on p. 51 of the 1906 Cambridge Scientific Instrument Company catalogue, *Physical Instruments*. Image © Whipple Museum (csi.1).

In this context, our second instrument in this section, the moving-coil reflecting galvanometer (Wh.4190, Figure 7.3, exhibit 3B, the same type of galvanometer as the one illustrated in Figure 7.4), shows the new market that developed specifically around the education of engineers. This is a student version of the 'Ayrton–Mather' galvanometer, described in their 1890 paper on galvanometers for the *Philosophical Magazine*.[40] The Ayrton–Mather type of galvanometer featured a metal coil enclosed in a very long narrow tube between the poles of a permanent magnet. The shape of the coil and its position in the magnet made it very sensitive. In this particular example, a glazed window allows the users, perhaps students, to see the coil's movements. The large horseshoe magnet is not hidden in a case and can be easily viewed.

As in the Thomson-type galvanometer, in the Ayrton–Mather type light was reflected from a mirror to indicate the movement within the instrument, in this case of the long thin coil. The Cambridge Scientific Instrument Company sold this particular model of instrument as a student version bundled with a resistance bridge and resistance thermometer as part of an 'Outfit for teaching Resistance Thermometry'. In this set-up, the galvanometer itself cost only £3,

40 W. E. Ayrton, T. Mather, and W. E. Sumpner, 'Galvanometers', *Philosophical Magazine*, **30** (1890), pp. 58–95.

15s, though the full outfit including a bridge, galvanometer, resistance thermometer, and cables cost as much as £11, 3s (Figure 7.4).[41]

In the last decades of the nineteenth century, the nature of academic and practical training evolved rapidly as the need for practical training in electrical technologies continued to grow both in the technical colleges and in the universities. Twenty-five years after Foster had been successful in establishing a physics laboratory, UCL decided to indebt itself substantially to build an extension.[42] The new engineering wing opened in 1893, and contained a lecture room, a large laboratory, and sectioned-off areas including a dynamo room and areas for testing arc lamps and incandescent lamps.[43] The fact that the university was willing to spend so much money and commit itself to several years of repayment shows how important practical teaching of electrical engineering had become.

The 'Outfit for teaching Resistance Thermometry' was produced in Cambridge by the Cambridge Scientific Instrument Company. The foundation of the company in 1881 was timely, following closely behind the opening of the Cavendish Laboratory at the University of Cambridge in 1874. Its arrival, just as technical training for electrical engineers was gaining traction, placed the company in a strong position as supplier to a new and ambitious physical laboratory at a time when electrical engineering was professionalising and growing rapidly.[44] This goes some way towards explaining the wide range of instruments across the two realms of electrical training held by the Whipple Museum. Apparatus suitable for laboratory research and instruments designed explicitly for teaching both have a place.

Unique versus Standardised Instruments

Our last pairing of instruments illustrates the difference between a standardised instrument and a bespoke piece designed by the user himself. It is not so much that there is an inherent difference

41 See Wh.4191 and Wh.4189, respectively, for examples of the other parts of this equipment.
42 University College London Archives: UCL Annual Reports, 1891, presented on 25 February 1891, pp. 14–15.
43 J. A. Fleming, 1906, 'A Brief History of the Engineering Department of University College, London', in The Times Engineering Supplement, 25 April and 20 June 1906, p. 4.
44 For a general history of the company, see M. J. G. Cattermole and A. F. Wolfe, Horace Darwin's Shop: A History of the Cambridge Scientific Instrument Company 1878 to 1968 (Bristol: Adam Hilger, 1987).

Figure 7.5 Exhibit
4A: a thermal
reflecting
galvanometer by the
Cambridge Scientific
Instrument
Company, 1907.
Exhibit 4B: a
moving-coil pointer
multimeter by
Siemens, *c.* 1930.
Images © Whipple
Museum (Wh.4045;
Wh.2440).

between the two types, but more that they represent different phases
in the invention of an instrument. And, of course, only some
bespoke instruments ever become standardised items, and there will
be interesting reasons for why and how that happens.

The first instrument in this pair (Wh.4045, Figure 7.5, exhibit 4A)
is a thermal reflecting galvanometer, made in 1907 by the Cambridge
Scientific Instrument Company. This type of instrument was
invented by 1904 by William Du Bois Duddell (1872–1917), who
was a student of William Ayrton at the London Central Technical
College and would eventually become the President of the Institute
of Electrical Engineers in 1912. This instrument was an instance of a
thermo-galvanometer, measuring current by the amount of heat it
produces in passing through a resistor. Duddell himself gave a
detailed account of how he came to design the instrument, which
he describes as 'essentially a very delicate Ayrton–Perry twisted
strip-ammeter which has been improved by the addition of a
temperature-compensation device to minimize the zero-creep when
the temperature of the whole instrument changes'.[45] Taking up
Duddell's invention, the Cambridge Scientific Instrument Company
marketed this instrument, along with a whole range of other instruments designed by him.[46] The Company's 1912 catalogue of electrical instruments notes as a special feature of this instrument that it

45 W. Duddell, 'Some Instruments for the Measurement of Large and Small
    Alternating Currents', *Proceedings of the Physical Society of London*, **19**
    (1903), pp. 233–53, on p. 237. Note that the paper is recorded as having been
    'Read May 6, 1904', although the official publication year is 1903.
46 Cambridge Scientific Instrument Company, *Some Electrical Instruments*,
    pp. 40–1. Other instruments by Duddell are described on pp. 34–9 of the same

is 'robust and can be carried about in the pocket' and advertises that 'it is easily set up and requires no levelling'.

The cliché that necessity is the mother of invention comes into force clearly in Duddell's account. First of all, he notes that his concern was to improve instruments for the measurements of alternating currents (AC), which is a completely different game, conceptually and practically, from measuring direct currents (DC). The most common mechanism for AC measurement involved 'suspending within a coil of insulated wire a small needle of soft iron placed with its axis at an angle of 45° to the axis of the coil'. Or a silver or copper disc could be used instead of the soft iron needle. 'When an alternating current is passed through the coil, induced currents are set up in the disc and the mutual action causes the disc to endeavour to set itself so that these currents are at a minimum.' Avoiding the complex electromagnetic interactions involved in such mechanisms, Duddell sought to exploit the heat-generating effects of currents for his AC galvanometers.[47] After hearing Duddell's explanations, it is difficult to dispute that the thermal type of galvanometer, based on a fundamentally different principle from the instruments we have discussed above, was the correct choice for AC measurements.

The central element (literally and figuratively) in Duddell's instrument was a twisted platinum-alloy wire, which was boxed in to have a fixed length.[48] When current passed and heated this wire, it expanded; yet as it was boxed in and not able to stretch out, it twisted itself more tightly to keep its length constant. When Duddell got down to the business of detailed instrument-design for the twisted-strip instrument, his dominant concerns were range and precision. A small mirror attached to one place on this twisted wire allowed a precise monitoring of the amount of extra twisting caused by the heating, and from this angle measurement the amount of current could be inferred. One can expect that this clever arrangement required a lot of skill and knowledge to operate successfully. An important part of the difficulty to be overcome was the fact that it provided a very indirect measurement involving several steps of inference, each of which opened up room for error and uncertainty. As Duddell explains, what is directly measured in this instrument is

catalogue, and also in the Company's other catalogues, including *Galvanometers and Accessories*.

47 Fleming, 'Galvanometer', p. 430.
48 Duddell also invented a 'thermal ammeter' using the heat generated by the current to run a bismuth–antimony thermocouple.

the angle of twisting, from which one infers the change of temperature, from which one infers the rate of heat-production in the wire, from which one infers the amount of current passing through the wire. Duddell conceived his main task as obviating the difficulties in thermal galvanometers arising from the 'fact that what is really measured is a difference of temperature, and not the rate of production of heat'.[49] He invented two different instruments to meet this challenge, which were consciously designed to serve different purposes. Our exhibit 4A was the first of these.[50]

The second object in this final pair (Wh.2440, Figure 7.5, exhibit 4B) is a portable combined voltmeter and ammeter, manufactured by Siemens. Unlike the Duddell instrument, the working parts of the objects are completely obscured literally by a black box. The user can see a needle that points somewhere on a scale marked up with 150 divisions. A second glazed window reveals a small part of the workings, a pivoted coil surrounding a fixed iron core. There are also terminals to connect the instrument up to a circuit, and accessories to modify the range of the instrument. However, the instrument itself gives away little about how it works. The sturdy and carefully padded wooden box it is packed in, however, suggests a delicate lab instrument rather than something that might be used in the field. The lack of intricate knowledge of the inner workings of the box is an indicator of an important shift that took place in electrical measurement as 'direct-reading' instruments were developed. The transition from quantitative instruments that required the user to calculate voltage or current in a manner depending on the specific set-up of the device to an instrument in which the voltage or current was read directly from a dial paved the way for easy-to-use black-box technologies. Direct-reading instruments were initially distrusted by many, with some physicists objecting to 'ammeters masquerading as measuring devices'. Over time, the utility of the instruments won over many, and black-box technologies became commonplace in labs and the field alike.[51]

49 Duddell, 'Some Instruments for the Measurement of Large and Small Alternating Currents', p. 237.

50 This type is described in Duddell, 'Some Instruments for the Measurement of Large and Small Alternating Currents', pp. 237–40. The second type (described on pp. 240–6), to which he gave the name of 'Thermo-galvanometer', though that name might have served just as well for both types, used the measurement of radiant heat coming from the wire.

51 G. Gooday, *The Morals of Measurement: Accuracy, Irony, and Trust in Late Victorian Electrical Practice* (Cambridge: Cambridge University Press, 2004),

As well as varying from exhibit 4A in that this instrument's workings are obscured, another difference between these two instruments is that we know a great deal about the development of the Duddell instrument, but there is very little to be found in the literature about the development of the Siemens device. What we do know is that this instrument is likely to date from the mid 1920s, on the basis of advertisements for similar models produced by Siemens and Halske AG that appeared in *Science* magazine between 1923 and 1926. The company produced a few variations on the precision voltmeter in this period for German, British, and American markets, many of which were packaged in the same casing, and with a small subset that also allowed the user to read off current measurements.[52] The adverts in *Science* magazine appeal directly to their readership of researchers and laboratory scientists, emphasising the use of the instrument in research laboratories. A slightly different model to the Whipple's example by Siemens & Halske, a 'precision Volt-Ammeter with seven ranges', was marketed as

> *More than an ordinary* Voltmeter or Ammeter. This unique instrument is a combination of voltmeter and ammeter which is so accurate, so permanent in its calibration and so completely compensated for temperature changes that it is used for the precise measurement of current and potential in laboratory work as well as for checking and calibrating other instruments.[53]

This portable box, then, could apparently satisfy everything a laboratory might need in one neat package. Other manufacturers were also bundling different types of electrical measurement into single boxes. A notable example is the AVOmeter, produced by the Automatic Coil Winder and Electrical Equipment Co. in 1923. The A, V, and O stand for 'amps', 'volts', and 'ohms', respectively. The British patent for the instrument, with a priority date of 1922, described the apparatus as follows:

quote on p. 47. See Chapter 2 for a wider discussion of the introduction of direct-reading instruments.

52 The Kusdas Collection of historical measuring instruments includes a Siemens & Halske precision voltmeter that reads up to 130 V that appears to have been sold in a case with a resistance box (inventory number 170). The collection also holds a similar instrument to the example in the Whipple Museum's collection, described as a 'Zehnohm-Instrument – Präzisions-Volt- u. Amperemeter für Gleichstrom für äußere Nebenanschlüsse', which also came bundled as a measurement case with a second meter and a range of shunts (inventory number 187). Collection online at www.historische-messtechnik.de (accessed 7 October 2018).

53 *Science*, n.s. 61, no. 1579 (3 April 1925), p. xiii; emphasis original.

A combined portable electric measuring apparatus is arranged to
read current, voltage and resistance on a single moving-coil
instrument. Resistances, a battery, and a switch may be arranged
in connection with the instrument so that, by means of the switch
and without altering the testing leads, the circuits for such
measurements may be changed and the sensitiveness of the
instrument when used as an ammeter or voltmeter may be varied
by shunts or series resistances respectively.[54]

Compared with the high-specification lab equipment being produced and marketed by Siemens, the AVO range was relatively
affordable.[55] The convenience and affordability of what later came
to be described as multimeters, the word used in the Whipple
Museum catalogue to describe this instrument, made them an indispensable tool. Today they are ubiquitous, used routinely by researchers, electricians, and technicians across virtually any industry that
uses electrical or electronic components. However, the technical
development of the multimeter, like many of the other rapid developments in electrical technologies in the early twentieth century,
passed by relatively unnoticed amid the exciting arrival of entirely
new types of technologies. The start of British Broadcasting Company transmissions in late 1922 made 'listening-in' a mainstream
activity in Britain, an activity that led millions of people to take up
home electronics as a hobby. In all of this the multimeter was a
useful tool, but one that developed and repackaged existing technologies rather than broke new ground. Consequently, the history of
this particular black box, and many other standardised off-the-shelf
instruments, remains almost as obscure as its contents.

Learning from Collections

In this chapter we set out to illustrate what can be learnt by closely
studying instruments and their contexts. Each of the eight instruments featured embodies some combination of theories, technical
advances, user needs, and local, national, or international technical
standards. By looking at instruments, their designs, and their users, it
is possible to trace the development of new disciplines, techniques,
and theoretical advances. For instance, telegraph engineers did not
need a measuring device for work in the field, and an indicator like

54 D. MacAdie, 'Improvements in or Relating to Electrical Measuring Instruments', patent number GB200977 (A), 1922.
55 See Wh.2499 for a later model, the AVOminor, purchased by R. G. Stansfield as
   a student at the Cavendish Laboratory in 1936.

the Lineman's Detector was sufficient for fault-finding. In the same period, the moving-magnet reflecting galvanometer used in De la Rue's laboratory had a range of calibration and levelling tools built in and could be used as a sensitive precision measuring tool. Later, engineers working with more sophisticated systems required more refined instruments to keep the network operational. These needs drove an increase in theoretical training, and in turn, the increase in training drove a market in electrical instruments designed for students. As engineers became more adept at applying mathematical rules to their work, portable instruments transitioned from merely indicating electrical current and its direction to providing the ability to directly read measures of current or voltage.

Taking a step back from the individual object and looking at the collection held by the Whipple Museum and other museums can also prove fruitful. There is a tendency in museum collections of scientific instruments towards the pristine and apparently (or actually) unused. This is despite the argument made by Simon Schaffer, among others, that instruments are prone to faults and failure by nature: 'Faults are defaults, yet instruments perform'; however, 'states of disrepair are often not deemed worthy of display, even though – perhaps because – they show signs of use.'[56] The ambition to collect the pristine is made plain in catalogue descriptions. Many of the instruments in the Whipple Museum collection have been catalogued with notes drawing attention to their incompleteness – 'lacking zero-adjustment screw' (Wh.4269); 'glass broken when packing' (Wh.4316); 'lacking wooden case' (Wh.4240); or 'lacking post and controlling magnet' (Wh.1333). The need to include absent parts of an instrument in the new identities of instruments when they are added to a museum catalogue shifts the focus away from the objects' biographies – the means by which they were used, stored, and ultimately added to the museum collection – and instead highlights the partial nature of the object. These nods to wished-for perfection are not reflective of real-world usage of instruments.

There is an asymmetry in catalogue descriptions of instruments that have reached the collection unscathed. Wh.4292 is a moving-coil reflecting galvanometer, given to the Museum by the Cambridge Scientific Instrument Company in 1974. The instrument's appearance is brassy and polished, suggesting that this particular object was a 'masterpiece', produced by a craftsman for the company as the final

---

56 S. Schaffer, 'Easily Cracked: Scientific Instruments in States of Disrepair', *Isis*, **102** (2011), pp. 706–17, on p. 707.

assessed piece of work at the end of their apprenticeship. Unlike many of the other reflecting galvanometers in the collection, this example still has its delicate and easily lost mirror, quite probably because it was never used. The object catalogue describes the technical details of the instrument but says nothing of its pristine appearance; the cataloguer seems to have assumed by default that this is the 'normal' state of the object. If 'mint condition' is the preferred state of an instrument collected for a museum's collection, what reasons might induce a curator to add an incomplete specimen to the collection? Wh.0939 (our exhibit 2A) is a moving-magnet reflecting galvanometer, an instrument 'lacking [its] suspended system'.[57] The instrument features in our story, perhaps, for the same reason it features in the Whipple Museum collection. It was constructed by Elliott Brothers for the British Association Committee on Electrical Standards. Aside from its technical qualities, which would meet the requirements of most laboratories in the 1870s, this instrument was an important witness to the establishment of electrical standards.

All objects and collections offer, like historical texts, partial accounts and insights with inherent biases built into what has been collected, how the objects have been described, and the types of categories they have been placed in. Understanding the contexts of collection is vital in order to identify gaps and value-laden qualities in the collection and its supporting body of data. The contexts and biases of collections can be quite different from those of the written texts that, in principle, would be expected to cover similar ground. That means that they can challenge the narratives that are found in textual sources. For example, the presence of the Siemens volt- and ammeter in the collection provides evidence, hard to find in the written record, of the emergence of convenient instruments that performed multiple types of electrical measuring. Because of the role of the Cambridge Instrument Company in helping to build the collections of the Whipple Museum, and the Company's arrival just as the electrical engineering industry was coming of age and a brand new, ambitious physical science laboratory was becoming established in Cambridge, the range of objects in the Museum's collection is diverse and captures everyday instruments that were not written about extensively at the time they were developed. In choosing the four pairings in this chapter we sought to present narratives that

---

57 Lyall, *Whipple Museum of the History of Science Catalogue 8*, Part 1, Section 2, Number 20.

revealed particular technical and social aspects of the development of galvanometers. However, they also represent a greater whole – the hundreds of electrical and magnetic instruments cared for by the Whipple Museum. Each of those objects has its own story, but by stepping back and looking at the whole collection we begin to see that those detailed investigations of the developments inside individual black boxes also allow us to explore the ways in which, over the course of more than a century, the galvanometer became an embedded and almost invisible part of scientific and technical practice.

8 ❧ Buying Antique Scientific Instruments at the Turn of the Twentieth Century: A Data-Driven Analysis of Lewis Evans's and Robert Stewart Whipple's Collecting Habits

TABITHA THOMAS

Fakes exist only because there is a market for the genuine. Knowledge of fake antique scientific instruments – including their manufacture and identification – would be furthered by consideration of the trade in which these forgeries were bought and sold. In Chapter 9, Boris Jardine explores how the presence of fake scientific instruments at the Whipple Museum was first unmasked by Derek Price in the 1950s. By using a data-driven analysis of the buying and selling of antique scientific instruments in the early years of the trade, I have been able to build up a general picture of the preferences exhibited by different buyers and the features that added value to antique scientific instruments. In this chapter I analyse these factors and how they may have influenced the types of forgery that emerged. This approach has not only made possible these general insights into the trade, but also enabled me to bring to light more specific information regarding fake scientific instruments: they were being sold at public auction as early as the 1890s, and at least one collector actively took measures to spot them and avoid buying them.

Puttick & Simpson's Auction Gallery, at 47 Leicester Square, London, held an impressive 11,000 sales in its 125-year business life from 1846 to 1971.[1] Amongst a diverse array of specialisms, Puttick & Simpson's was notable for having been an early venue for sales of antique scientific instruments. At least seven of these sales from 1894 to 1896 were attended by the instrument collector Lewis Evans (1853–1930), the brother of the eminent archaeologist Sir Arthur Evans (1851–1941). Lewis Evans, a wealthy businessman, specialised

---

1 J. Coover, 'Puttick's Auctions: Windows on the Retail Music Trade', *Journal of the Royal Musical Association*, 114 (1989), pp. 56–68.

Figure 8.1 An example page from one of the six Puttick & Simpson's sales catalogues that Lewis Evans annotated with sale prices, notes, and names of purchasers between 1894 and 1896. Image © Lewis Evans Collection, History of Science Museum, University of Oxford.

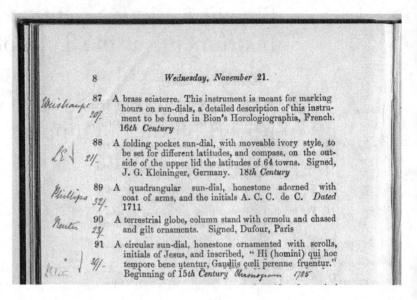

in collecting sundials and astrolabes,[2] and we are fortunate that he annotated six of the seven Puttick & Simpson's sales catalogues of his that survive in the Museum of the History of Science, Oxford.[3] His notes record the sale price of each lot, and often the name of each buyer as well (Figure 8.1). Using the printed information in these catalogues, and Evans's annotations, it has been possible to build up a dataset which made detailed analysis of those sales possible.[4]

Evans's collection (Figure 8.2) would become the founding collection (along with other scientific instruments collected by Robert T. Gunther from Oxford colleges) of the Museum of the History of Science in Oxford in 1924.[5] Similarly, in 1944, the private collection of Robert Stewart Whipple, amassed between 1913 and the time of its donation, was the starting point of the Whipple Museum of the History of Science, Cambridge.[6] As this volume attests, a

2 P. De Clercq, 'Lewis Evans and the White City Exhibitions', *Sphaera*, 11 (2000), www.mhs.ox.ac.uk/about/sphaera/sphaera-issue-no-11/lewis-evans-and-the-white-city-exhibitions/ (accessed 18 November 2017).

3 Puttick & Simpson's sales catalogues annotated by Lewis Evans: 3 April 1894; 18 June 1894; 21 November 1894; 8 March 1895; 20 May 1895; 28 February 1896; 20 March 1896. Lewis Evans Collection, Museum of the History of Science, Oxford. We thank Tony Simcock for bringing these sources to our attention.

4 A copy of the full dataset and a more extensive breakdown of my analysis of it has been lodged with the Whipple Museum.

5 A. V. Simcock, *Robert T. Gunther and the Old Ashmolean* (Oxford: Museum of the History of Science, Oxford, 1985), p. xi.

6 A. J. Turner, 'From Mathematical Practice to the History of Science', *Journal of the History of Collections*, 7 (1995), pp. 135–50.

Figure 8.2 The back of this photograph carries the following note by Evans: 'Photograph of my collection of instruments, taken in my house, "Belswains", Hemel Hempstead in 1890. / Lewis Evans / This part of the collection then was contained in a case over the fireplace in the library / My first purchase was a French dial, when I was about 17.' Image © History of Science Museum, University of Oxford (MS Evans 39).

Department of History and Philosophy of Science would grow up around the Whipple Museum. The preservation of the material culture of science preceded and shaped the study of the history of science, the origins of which have typically been analysed from the perspective of texts, academic journals, and disciplines.[7] Knowledge of how these collections were compiled, from the perspective of auction rooms and individual collectors, has the potential to tell us much about the beginnings of the history of science as a discipline, complementing the text- and teaching-based accounts we already have.[8] These collections were very much the product of a market that was still in its infancy when Lewis Evans was collecting, and which

7 J. A. Bennett, 'The Cambridge Legacy of Robert T. Gunther', in W. D. Hackmann and A. J. Turner (eds.), *Learning, Language and Invention: Essays Presented to Francis Maddison* (Aldershot and Paris: Variorum and the Société Internationale de l'Astrolabe, 1994), pp. 78–83; and J. A. Bennett, 'Museums and the Establishment of the History of Science at Oxford and Cambridge', *British Journal for the History of Science*, 30 (1997), pp. 29–46.

8 A.-K. Mayer, 'Setting Up a Discipline: Conflicting Agendas of the Cambridge History of Science Committee, 1936-1950', *Studies in History and Philosophy of Science*, 31 (2000), pp. 665–89; and A.-K. Mayer, 'Setting Up a Discipline, II: British History of Science and "the End of Ideology", 1931-1948', *Studies in History and Philosophy of Science*, 35 (2004), pp. 41–72. Bennett, 'Museums and the Establishment of the History of Science at Oxford and Cambridge' is the obvious exception to this trend.

was approaching maturity when Whipple was an active buyer. This chapter uses Evans's annotated Puttick & Simpson's catalogues, and a comparable analysis of the surviving records of Whipple's purchasing habits, to build up a picture of the market in antique scientific instruments between the 1890s and the 1940s. The datasets built up from the catalogues, and the Whipple Museum's own accessions database, have been central to the methodology of this project and are important sources that were previously unavailable in this form. The principal findings from this analysis will be briefly presented here; the complete datasets and a more thorough breakdown of the data have been deposited with the Whipple Museum.

This project also contributes to knowledge concerning detection of fakes in the antique scientific instrument trade. What was thought to have begun with Derek Price's work at the Whipple Museum in the 1950s can be pushed back at least sixty years and pinned to Lewis Evans's annotations in the Puttick & Simpson's catalogues. Further, these annotations contribute to the study of forgery itself, which can be understood only in the light of knowledge about the specifics of supply and demand: perceptions of value effectively produce forgeries, and the evidence presented here gives a preliminary account of taste in instrument collecting in its formative years.

Collectors, Dealers, and Museums

Although scientific instruments have been collected in a variety of settings ever since the Renaissance, historians have shown that such instruments began to take on significant value as objects of historic importance in the nineteenth century.[9] By the middle of the nineteenth century public museums such as the South Kensington Museum had been founded, and major institutions such as the British Museum had expanded their collections to include antique scientific instruments.[10] In 1876, the South Kensington Special Loan

9 Turner, 'From Mathematical Practice to the History of Science'; G. Strano, S. Johnston, M. Miniati, and A. Morrison-Low (eds.), *European Collections of Scientific Instruments, 1550–1750* (Leiden and Boston: Brill, 2009); A. Filippoupoliti, '"What a Scene It Was, That Labyrinth of Strange Relics of Science": Attitudes towards Collecting and Circulating Scientific Instruments in Nineteenth-Century England', *Cultural History*, 2 (2013), pp. 16–37.

10 R. G. W. Anderson, 'Connoisseurship, Pedagogy or Antiquarianism?', *Journal of the History of Collections*, 7 (1995), pp. 211–35; A. Macgregor, 'Collectors, Connoisseurs and Curators in the Victorian Age', in M. Caygill and J. Cherry (eds.), *A. W. Franks: Nineteenth-Century Collecting and the British Museum* (London: British Museum Press, 1997), pp. 6–33.

Collection of Scientific Apparatus heralded a new approach to the valorisation of old scientific instruments in Britain. The Loan Collection included instruments of historical interest due to associations with past users, or the important research for which they had been used, as well as more typical teaching and investigatory apparatus.[11] Scientific instruments were not viewed simply as antiques, but were presented as belonging to their own class of objects that have, for example, mathematical functions built into them, and as embodying the progress of precision measurement.[12] Alongside these trends, emergent interest in past heroes of science made instruments with connections to famous scientists of particular interest.[13]

Collecting as a hobby, meanwhile, grew in fashion through the nineteenth century, with specialised collections becoming especially popular towards the end of the century. A small number of collectors, such as Evans and Sir John Findlay, specialised in antique scientific instruments.[14] By the second half of the nineteenth century, the formation of public museums had increased the desire that educated people had for collectables, and such collectables became increasingly available as members of the gentry fell on hard times and sold their possessions to the rising middle classes.[15] The 1876 Special Loan Collection exhibition was well attended and exceptionally well publicised, and contributed to the culture of collecting scientific instruments for their own sake, rather than for the sake of their aesthetics or culture of origin, as with other antiques.[16] Where there is a market, there will be businesses

11 Anderson, 'Connoisseurship, Pedagogy or Antiquarianism?', p. 219.

12 S. Schaffer, 'Metrology, Metrication, and Victorian Values', in B. Lightman (ed.), *Victorian Science in Context* (Chicago: University of Chicago Press, 1997) p. 438.

13 R. Bud, 'Responding to Stories: The 1876 Loan Collection of Scientific Apparatus and the Science Museum', *Science Museum Group Journal*, no. 1 (Spring 2014), http://dx.doi.org/10.15180/140104.

14 A. D. Morrison- Low, 'Sold at Sotheby's: Sir John Findlay's Cabinet and the Scottish Antiquarian Tradition', *Journal of the History of Collections*, 7 (1995), pp. 197–209.

15 M. W. Westgarth, *A Biographical Dictionary of Nineteenth Century Antique and Curiosity Dealers* (Glasgow: Regional Furniture Society, 2009), p. 10.

16 P. De Clercq, 'The Special Loan Collection of Scientific Apparatus, South Kensington, 1876. Part 1: The "Historical Treasures" in the *Illustrated London News*', *Bulletin of the Scientific Instrument Society*, 72 (2002), pp. 11–19; P. De Clercq, 'The Special Loan Collection of Scientific Apparatus, South Kensington, 1876. Part 2: The Historical Instruments', *Bulletin of the Scientific Instrument Society*, 73 (2002), pp. 8–16; P. De Clercq, 'The Special Loan Collection of Scientific Apparatus, South Kensington, 1876. Part 3: Contemporary Publications', *Bulletin of the Scientific Instrument Society*, 74 (2002), pp. 16–21; and P. De Clercq, 'The Special Loan Collection of Scientific Apparatus, South

capitalising on that market. Puttick & Simpson's Auction Galleries was one business that moved to exploit this relatively new fashion for collecting antique scientific instruments, beginning to hold special-ised sales at least as early as 1894.

Puttick & Simpson's and Its Buyers

Puttick & Simpson's was a large and dominant auctioneer, and up until the turn of the twentieth century was as significant in terms of size and value as Sotheby's or Christie's.[17] The sales that Evans attended between 1894 and 1896 advertised 'antique astronomical and scientific instruments' and 'antique sundials from the 16th, 17th and 18th Centuries'.[18] Evans noted down the names of buyers for a majority of the objects sold. There are at least forty-one different buyers that Evans recorded across the seven annotated sales cata-logues that survive. The proportion of objects with the buyer recorded next to them increased from one sale to the next (presum-ably in part due to Evans's increasing familiarity with fellow buyers).[19] Ten of these buyers either spent or bought significantly more than the others and so stand out as either 'serious' collectors or dealers. It is certain that 'LE' was Lewis Evans himself and that 'BM' was the British Museum.[20] Two further significant names recorded in the catalogues were 'Weishaupt' and 'Harding'. George Harding is recorded in the British Museum's online database as having supplied a number of instruments to them in this period. He appears to have been one of the more significant buyers that Evans recorded, and was buying things in his own right apart from those objects he was acquiring specifically for the British Museum. Weishaupt and Co. was also a dealership that sold to the British Museum, supplying at least three instruments (a quadrant and two sundials, none of which

Kensington, 1876. Part 4: Photographs and Copies', *Bulletin of the Scientific Instrument Society*, 76 (2003), pp. 10–15.

17 Coover, 'Puttick's Auctions', 58.
18 Puttick & Simpson's, 1894–6.
19 The catalogue for 8 March 1895 has only very light annotations, and so was not included in the compiled database.
20 The link between 'BM' and the British Museum is less obvious than it might seem. It was not known before this project that the British Museum had been actively acquiring antique scientific instruments in this period. However, a search for one of the distinctive objects purchased by 'BM' – a sundial signed and dated 'Joannes Antonius Ostravsky, 1719' – matched a record in the British Museum's online database. The British Museum's provenance field indicates that this sundial was acquired in 1894 from dealer George Harding. (On Harding, see below.)

appears to have been bought at Puttick & Simpson's) in 1895–7. Aside from 'LE', 'BM', 'Harding', and 'Weishaupt' it has been very difficult to trace the other significant buyers (with the exception of Percy Webster, see below), which could mean that they were simply building their own private collections, or that they were dealers who sold to private clients rather than to museums.

With even this limited set of actors, it is possible to build up a basic picture of their buying preferences. In terms of a desire to acquire instruments carrying an inscription denoting the city or date of manufacture, or the maker, the British Museum, Evans, 'Phillips', and Weishaupt all seemed to express no preference, with about half of their purchases carrying inscriptions (a ratio that reflects the proportion of inscribed instruments sold across all of the Puttick & Simpson's sales). Even though Evans had a reputation for strongly preferring signed and dated instruments, he clearly did not express his preference in his purchases as much as he could have done.[21] In contrast, the purchases of instruments attributed to 'Reuter', 'Thomson', 'Tregaskis', and 'Waters' encompassed only a minority of inscribed instruments. 'Webster' was the only significant buyer who expressed a strong preference that his items be signed, dated, or located to a city. Only four out of seventeen of his purchases had no inscriptions, and for one of those the maker and city were still known. This 'Webster' is almost certainly the clock- and instrument-dealer Percy Webster, a somewhat notorious character with a reputation for 'conjuring unique rabbits out of his hat' or, put more bluntly, dealing at times in '"imaginative" restorations and fakes'.[22] It is notable that both Evans and Whipple purchased objects from Webster that have subsequently been identified as carrying fake inscriptions (Figure 8.3).[23] Had Webster himself been the perpetrator of these fakes, we might expect him to have been in the market

---

21 De Clercq, 'Lewis Evans and the White City Exhibitions'.

22 A. V. Simcock, 'Percy Webster's Stock', *Bulletin of the Scientific Instrument Society*, 40 (1994), p. 28; and J. Betts, *Time Restored: The Harrison Timekeepers and R. T. Gould, the Man Who Knew (Almost) Everything* (Oxford: Oxford University Press, 2006), p. 169.

23 Whipple's purchase (Wh.0226, Figure 8.3) was identified as carrying a fake inscription in David Bryden, *The Whipple Museum of the History of Science Catalogue 6: Sundials and Related Instruments* (Cambridge: Whipple Museum of the History of Science, 1988), no. 377. Webster sold Lewis Evans an instrument carrying a fake Culpeper signature, now in the Oxford Museum of the History of Science (inv. No. 60019). See also B. Jardine, J. Nall, and J. Hyslop, 'More Than Mensing? Revisiting the Question of Fake Scientific Instruments', *Bulletin of the Scientific Instrument Society*, 132 (2017), pp. 22–9.

Figure 8.3 Robert Stewart Whipple paid Percy Webster £3 in 1925 for this inclining dial carrying the signature of George Adams Snr (shown enlarged at the bottom). Whipple Museum curator David Bryden later identified this as a fake inscription added to a cheap nineteenth-century instrument. Image © Whipple Museum (Wh.0226).

for cheaper instruments *without* inscriptions, to which he could add a famous maker's name later – but Evans's annotations, at least, do not bear this out.

The Puttick and Simpson's catalogue descriptions also reveal something about the various factors for which these objects might have been valued. Sometimes the maker would be labelled 'the celebrated', or an object might be dubbed 'very rare', 'very early', or 'exceedingly beautiful'.[24] One obvious feature of these descriptions that sets the instruments apart from non-scientific collectables is that in some cases there are extended instructions on how to use the instrument. Some auction lots have more extended instructions underneath the descriptions of the objects, indicating that it was expected that there would be some interest in using them, or at least understanding their function. Clearly, the working order of an instrument was considered a selling-point. For example, the description of lot 24 offered on Monday 18 June 1894 – an armillary dial on a stand signed by Nairne, London – included the following:

> The horizontal ring, representing the equator, serves as the hour circle, when the vertical ring is set in the meridian of the place of observation. The pin-hole sight through which the sun's rays pass to the hour circle can be adjusted for change of declination by means of the graduated plate on which it slides. The suspending

24 Puttick & Simpson's, 28 February 1896, pp. 3, 7, and 9.

ring with its spring dip can be set for different latitudes with the aid of graduations on the meridian ring; on the dial are marked the latitudes of important places.[25]

The inclusion of such instructions strongly suggests that some technical knowledge of the instruments was presupposed by Puttick and Simpson's, and was considered both interesting and valuable to customers. This might indicate the extent to which collecting instruments had become specialised, and the concomitant desire amongst collectors to acquire items that had retained their functionality – not necessarily so that they could still be used (though sundial enthusiasts like Evans may well have done), but because utility itself was now considered important.

A Question of Value

The price an object realises at auction will not always be a fair reflection of its value. Nonetheless, considered *en masse* as a dataset, the sale prices that Evans recorded in his auction catalogues can be analysed to reveal several general trends and insights, even if they can only be taken as tentative. Evans's annotations vary in detail from catalogue to catalogue, but for six out of the seven catalogues the sale price is written down next to most lots and, more often than not, the name of the buyer too. With this information, it was possible to compile a dataset showing as many key details as were known for each lot.[26] The details included were the sale date, lot number, type of object, any inscriptions or signatures present, city, country, and date of manufacture (to the year, if known), maker, price, the approximate equivalent price in 2017, and any other annotations that Evans added. With this small database compiled, it was possible to extract emerging trends and make cautious projections onto the rest of the trade in antique scientific instruments regarding what features added value to an object.

First, and perhaps surprisingly, the data indicate that whether or not an instrument was dated to a certain year did not seem to correlate with increasing prices. However, and as we might expect, there *is* a reasonably strong positive correlation between knowledge of an instrument's maker and its price. This positive correlation is also seen – though less strongly – between knowledge of an

25 Puttick & Simpson's, 18 June 1894, p. 3.
26 A copy of this full dataset is held by the Whipple Museum and can be supplied to any researcher interested in viewing it.

instrument's place of manufacture and its price. The specific place
of manufacture also correlated with price to a limited extent, as
instruments from continental Europe tended to sell for more than
English instruments, although, as this was an auction taking place in
London, most of the more common instruments were English, which
lowered the average price of English instruments sold. Even so, the
most expensive 10 per cent of objects that sold were heavily dominated by German and French instruments and, perhaps surprisingly,
instruments which had an unknown country of origin too, despite
the positive correlation between known place of manufacture and
price. However, this kind of information clearly adds a kind of
authenticity to the instrument. It anchors it to a place and person,
helping substantiate the genuineness of an object – or providing a
means of (potentially) spotting a fake.

One striking feature of Evans's annotations is that they demonstrate very clearly that whether or not an instrument was what it
appeared to be was of importance at the time. Indeed, one firm
conclusion we can draw from what Evans recorded is that fakes
had already started to enter the market. Evans has developed a
reputation for being able to spot a fake, due to the supposed authenticity of his collection.[27] His annotated sales catalogues do bear out
this reputation. They show that Evans looked for and detected fakes
in the 1890s. The detection of fake scientific instruments has until
now been presumed to have begun in the 1950s when Derek J. Price
started working at the Whipple Museum (as discussed in detail in
Chapter 9 by Boris Jardine).[28] Evans's annotations show that the
date for the first detection of fakes in this niche market can be
pushed back at least sixty years. Table 8.1 summarises the items that
Evans deemed suspicious and includes his annotations. As we see,
there are eight objects that Evans judged suspect, the most common
reason being the practice already linked to some of Percy Webster's
stock: the addition of a fake signature to what had presumably
previously been an unsigned instrument.

Distinct from this form of forgery, and of particular note, is the
sundial that Evans annotated with 'Chronogram 1785' (see lot 91 in
Figure 8.1). This brief note suggests both that the instrument itself is

27 Museum of the History of Science, Oxford, www.mhs.ox.ac.uk/collections/
   library/lewis/lewis-evans-founder-of-the-museum-of-the-history-of-science/
   (accessed 9 November 2017).
28 See also Jardine, Nall, and Hyslop, 'More Than Mensing?'; and G. A. C. Veene-
   man, Scientific Instruments, Originals and Imitations: The Mensing Collection
   (Leiden: Museum Boerhaave, 2000), p. 7.

TABLE 8.1 Lewis Evans's annotations of suspicious objects sold at Puttick & Simpson's

Sale date	Lot	Puttick & Simpson's sale description	Price	Buyer	Annotation
03/04/1894	22	AN ASTROLABE, gilt copper, engraved, 1 moveable plate, unusual size, diameter 15½ inches, French. *16th Century.*	£8	–	*An electrotype*
03/04/1894	24	[An astrolabe], 5 moveable plates (copper), diameter 7 inches, Armenian. *16th Century.*	£3/10	–	*Plates electro*
21/11/1894	91	A circular sun-dial, honestone ornamented with scrolls, initials of Jesus, and inscribed 'Hi [homini] qui hoc tempore bene utentur Gaudiis coeli perenne fruentur.' Beginning of *15th Century.*	£39	Reuter	*(Latin underlined) Chronogram 1785*
08/03/1895	39	A [universal armillary dial] of unusual shape, supplied with alidades, moveable style, lunar calendar and set in a square plate in which it slides, the plate serving as spring dip. Inscribed 'Martin Frey, Regenspure, 1590'. *A very rare instrument.*	–	–	*LE later / from Harding / name false*
20/05/1895	200	A portable gilt brass horizontal sun-dial, reversible style and compass. Signed Adam Perner, Norimbergae, 1596.	£1	Weishaupt	*('Adam Perner' underlined) False*
20/05/1895	226	A brass circular horizontal sun-dial with reversible style and compass, in brass box. Signed Matthias Loebl, Weissenburg.	£12	LE	*False name. Euphic dial*
20/05/1895	238	A folding ivory dial of peculiar construction, composed of 3 instead of 2 plaques. The instrument contains lunar calendar of gilt and engraved copper, scale for ascertaining dial of gilt brass, 1 horizontal with 3-hour-circles for 42, 48 and 54, etc., etc. Germany. *16th Century.*	£4/14	Weishaupt	*Made up*
28/02/1896	73	A quadrangular horizontal brass sun-dial, gilt and engraved, moveable style. Signed Adam Perner, Noribergae 1596 Faciebat.	£20	Webster	*Name false*

almost certainly not from the fifteenth century, and that Evans had a very discerning eye when it came to sniffing out suspect instruments. A chronogram is a sentence in which a date is encoded and can be deciphered. In this case we can reverse-engineer Evans's discovery, arriving at the sum total of 1,785 if we take every letter from the Latin inscription that is also a Roman numeral and add them together. Evans underlined these letters in the Latin inscription (seen in capitals here): hI [homini] qUI hoC teMpore bene UtentUr, gaUDIIs

CoeLI perenne frUentUr. There are five 'I's, six 'U's (which are interchangeable with 'V's), two 'C's, one 'M', one 'D' and one 'L'. If taken to be Roman numerals and converted into their numeric values, they sum as $5 \times 1 + 6 \times 5 + 2 \times 100 + 1 \times 1{,}000 + 1 \times 500 + 1 \times 50 = 1{,}785$. This is not only a bravura piece of detective work on Evans's part, but also a fascinating insight into the history of scientific instruments. It appears that this was a sundial manufactured in 1785, but made to look as if it were much older. Whether or not the maker intended it to deceive is an open-ended question. However, given that the 'true' date is hidden in the inscription and that it takes a keen eye to spot a chronogram, it appears as if the maker set up a hoax to dupe those who did not have sharp eyes. This would, then, be one of a few very early faked scientific instruments, manufactured long before scientific instruments had much of a place in collections or were traded for large sums of money.[29]

As for the other instruments that Evans deemed suspicious, we can compare their sale prices with the average prices of other instruments of the same type, to gauge whether the dubious features were noticeable to the wider salesroom. Puttick & Simpson's sold at least fifteen astrolabes across these sales for prices between £2 and £42, with the mean average being about £12. The two astrolabes noted in Table 8.1 were both sold at below average price, one significantly so. As for the pedometer, we know the prices of just two others, one from the seventeenth century selling for £23 and another signed 'C. H. Opp, Berlin' which went for £14, so the suspect example sold for a lot less than it might have done.[30] Sundials sold within a very wide range of prices, from £1 up to £63, with the average price around £18. The chronogram sundial supposedly from the fifteenth century and the sundial (purportedly) made by Perner clearly sold for very healthy prices, while the other sundials listed in Table 8.1 sold for less than average. For several of these objects, Evans recorded the name of the maker as false. As noted above, there was a correlation between an instrument carrying a signature and a higher sale price. If signatures were being falsified and inscribed on genuine antique scientific instruments to increase their value, then this speaks to the co-production of a marketplace for collectable

---

29 D. J. Price, 'Fake Antique Scientific Instruments', in *Actes du VIIIe Congrès International d'Histoire des Sciences, Florence–Milan 3–9 Septembre 1956* (Vinci: Gruppo Italiano di Storia delle Scienze, 1958), pp. 380–94.

30 Puttick & Simpson's, 3 April 1894, p. 4; 18 June 1894, p. 7.

instruments and the production of forgeries to exploit that market, from the very beginning of this trade.

Finally, we should note what the data say about the significance of sundials at this time. In total, the number of sundials bought vastly outnumbered sales of any other kind of instrument. It is well known that sundials were very popular objects in Victorian Britain.[31] Indeed, in 1872 Margaret Gatty opened her popular *The Book of Sun-dials* with the declaration that 'there is no human invention more ancient, or more interesting than that of the sun-dial'.[32] By 1900 this book was in its fourth edition, with Evans himself contributing an essay on portable dials. The popularity of sundials at this time is an important and largely unaddressed aspect of this formative phase in instrument collecting. It is not surprising that in terms of sales they vastly outnumbered astrolabes, but it is perhaps striking that they also on average sold for higher prices. As for the history of science, astrolabes have long played an important role in the history of astronomy, yet, as Jim Bennett has argued, dialling was a serious and technical discipline in the seventeenth and eighteenth centuries that has perhaps yet to be fully appreciated.[33]

Robert Stewart Whipple's Collection

A brief comparison with Robert Whipple's collecting habits confirms that the insights gained regarding Lewis Evans can be carried forward to the period of the instrument trade's maturity. There were, however, differences between Lewis Evans and Whipple. For one, Whipple was more closely linked to the formation of the discipline of the history of science than was Evans.[34] Whipple also had a much more eclectic approach to collecting scientific instruments than Evans. He collected sundials and astrolabes, just like many of his collecting predecessors, as well as lots of cheaper, more common

31 H. Higton, *Sundials at Greenwich: A Catalogue of the Sundials, Nocturnals and Horary Quadrants in the National Maritime Museum, Greenwich* (Oxford: Oxford University Press, 2002).

32 M. Gatty (Mrs Alfred Gatty), *The Book of Sun-dials* (London: G. Bell, 1900), p. 1.

33 J. Bennett, 'Annual Invitation Lecture: Sundials and the Rise and Decline of Cosmography in the Long Sixteenth Century', *Bulletin of the Scientific Instrument Society*, 101 (2009), pp. 4–9.

34 S. De Renzi, 'Between the Market and the Academy: Robert S Whipple (1871–1953) as a Collector of Science Books', in R. Myers and M. Harris (eds.), *Medicine, Mortality and the Book Trade* (Folkestone: St Paul's Bibliographies, 1998); and Bennett, 'Museums and the Establishment of the History of Science at Oxford and Cambridge'.

items; but he also pioneered the collection of antique optical instruments such as opera glasses, spectacles, microscopes, and telescopes.

Using a similar approach to the Whipple Museum's accession database to that which was taken with the Puttick & Simpson's sales catalogues, it was possible to build up a comparable dataset.[35] Once we extract the optical instruments, we can see that as a collector he displayed the same traits we began to see emerging at sales at Puttick & Simpson's. Notably, there is an even stronger positive correlation between both instruments made by known makers and instruments from known locations, and price.

As historians have already noted, Whipple's collection also reveals the unfortunate continuation – and perhaps growth – of the deliberate manufacture and sale of forgeries to collectors. A number of dealers sold multiple fakes to Whipple, including Gertrude Hamilton in Paris – trading as 'Mercator' – and Antique Art Galleries, London.[36] Whether these dealers were complicit in the selling on of fakes we will probably never know for sure, but we do know that fakes were already circulating in the 1890s and that they appear to have been even more abundant when Whipple was collecting. A tentative contrast that does emerge from a direct comparison between Evans's annotations and the much-studied forgeries in Whipple's collection is that, whilst the majority of suspect objects spotted by the former were genuine antiques embellished with a fake maker's name, most of the forgeries Whipple purchased were fabricated from scratch by a skilled forger and then sold as genuine antiques. However, it will require considerable analysis across larger datasets drawn from many more collections before we can draw firm conclusions about the general trends highlighted here.

---

35 As with the datasets described above, a copy of the full Whipple dataset has been deposited with the Whipple Museum and can be provided to researchers upon request.

36 For more on Hamilton, see W. F. J. M. Bryuns and A. Turner, 'Gertrude Hamilton, An American Instrument-Dealer in Paris', *Bulletin of the Scientific Instrument Society*, 73 (2002), pp. 23–6. See also Jardine, Nall, and Hyslop, 'More Than Mensing?' on Antique Art Galleries and their sales to Whipple.

9   Like a Bos: The Discovery of Fake
Antique Scientific Instruments
at the Whipple Museum*

BORIS JARDINE

It is a curious fact that the first doctoral project in the history of
science at Cambridge ended up with its student, Derek J. Price,
announcing that a number of his sources were in fact forgeries.
And it is a telling detail that in 1956, when Price presented this
finding in a paper entitled 'Fake Antique Scientific Instruments', he
compared his discovery to the recent unmasking of the Piltdown
Skull as a fraud.[1] To be sure, the Piltdown controversy – which
concerned supposedly ancient human remains unearthed in Sussex
circa 1912 – caused more of a stir than the fake astronomical
instruments that Price had found. But there are also strong parallels
between the two cases, and both for Price and for his successors in
the world of scientific instruments the 1956 paper remains a landmark. Underlying both exposés were changes in the nature of collections, the organisation and representation of specimens, and the
ways in which scholars approached their material sources: objects
that had previously been scrutinised one by one were, in the years
after the Second World War, considered *en masse*, and this provided
novel conditions for the detection of forgery.

Since Price's work, a number of studies and an international
working group have uncovered more fakes in collections of scientific
instruments, in particular those with a provenance going back to the

* For their help in the preparation of this chapter I would like to thank Jenny
Bangham, Mirjam Brusius, Richard Dunn, Seb Falk, James Hyslop, Joshua Nall,
David Singerman, Liba Taub, Anthony Turner, and the delegates at the XXXIV
Scientific Instrument Commission Symposium, Turin, 7–11 September 2015,
where this research was first presented. Some of the material here was published
in my essay with Joshua Nall and James Hyslop, 'More Than Mensing? Fake
Scientific Instruments Reconsidered', *Bulletin of the Scientific Instrument Society*
131 (2017), pp. 12–19, and I am grateful to the editor, Willem Hackmann, for his
assistance with that version.
1 D. J. Price, 'Fake Antique Scientific Instruments', in *Actes du VIIIᵉ Congrès
International d'Histoire des Sciences: Florence–Milan 3–9 Septembre 1956* (Vinci:
Gruppo Italiano di Storia delle Scienze, 1958), 380–94.

collector/dealer Anton Mensing.[2] Recent investigations have suggested that the problem of forgery was and is far greater than has been supposed.[3] But Price's own research and its context remain obscure.[4] He had begun work at the Whipple Museum in 1951 under its first director, Rupert Hall, on a project entitled 'The History of Scientific Instrument Making'.[5] Very early on, and apparently at Hall's prompting, he began to question the authenticity of certain instruments, beginning with a fine and apparently early astrolabe signed 'Johannes Bos'. Eventually even the identity and existence of Bos would come into question, and doubt would be thrown on a large number of instruments held in collections around the world.

In this chapter I present an account of Price's methods and motivations, and the context in which he was working.[6] Price was able to uncover forgeries, I argue, owing to new kinds of information that had become available as collections of antique instruments moved from the hands of individuals to institutions. He was working in a post-war age of international cooperation, new techniques of analysis, and a renewed positivism exemplified in the 'science of science' movement. Price's discovery of a group of fakes at the Whipple can be directly related to these developments via his international surveys of scientific instruments and his concept of 'scientometrics'. The general trend is illustrated by contemporary findings in other fields that yielded astonishing findings, specifically regarding deception and fraud – the most famous of these being the unmasking of the Piltdown forgery over the period 1953–5 – just prior to Price's announcement.

2 On Anton Mensing see W. F. J. Mörzer Bruyns, 'The Amsterdam Scheepvaart-museum and Anton Mensing: The Scientific Instruments', *Journal of the History of Collections*, 7 (1995), pp. 235–41; W. F. J. Mörzer Bruyns, 'Frederik Muller & Co and Anton Mensing', *Quaerendo*, 34.3 (2004), pp. 211–39; S. Johnston, W. F. J. Mörzer Bruyns, J. C. Deiman, and H. Hooijmaijers, 'The Anton Mensing Scientific Instrument Project: Final Report', *Bulletin of the Scientific Instrument Society*, 79 (2003), pp. 28–32.

3 B. Jardine, J. Nall, and J. Hyslop, 'More Than Mensing? Fake Scientific Instruments Reconsidered', *Bulletin of the Scientific Instrument Society*, 131 (2017), pp. 12–19.

4 However, for an account of many other aspects of Price's career see S. Falk, 'The Scholar as Craftsman: Derek de Solla Price and the Reconstruction of a Medieval Instrument', *Notes and Records of the Royal Society*, 68 (2014), pp. 111–34.

5 Falk, 'The Scholar as Craftsman', p. 115.

6 For more on Price's methods in particular see J. Nall, 'Finding the Fakes: How to Spot Forgeries Lurking in Collections of Historic Scientific Instruments', *Chemistry World*, 15.2 (February 2018), p. 71. Leads that Price opened up but did not follow are pursued in Jardine, Nall and Hyslop, 'More Than Mensing?'.

Over the course of the twentieth century many disciplines saw a transition of their working collections from private to public hands, and after the Second World War international cataloguing projects produced a new kind of relationship between individual objects and an aggregated way of knowing. Clues to authenticity, which had once been sought out by connoisseurs, were now the possession of the cataloguer, who could marshal and arrange large amounts of information. Material transformations in the collation of data brought about new understandings of material relics, and, in this brave new world, prized specimens became dubious antiques.

'Hall Says a fake!'

A first point to make about collections of early scientific instruments is that, unlike coins, statues, paintings, furniture etc., they are a relatively recent phenomenon – dating back only as far as the late nineteenth century.[7] In Britain, the first sustained attempt to form a collection was undertaken by Augustus Wollaston Franks at the British Museum.[8] Franks purchased from private individuals, took donations, and acquired instruments at the Bernal sale in 1855, building up a small but significant holding, mainly of sundials and astrolabes. By the end of the century a number of museums and private collectors were acquiring scientific instruments – mainly continental – in large numbers. First amongst the private collectors was the paper magnate Lewis Evans, whose collection formed the basis of the Museum of the History of Science, Oxford. Although Price was the first to announce the presence of fake instruments publicly, Lewis Evans had in fact privately expressed suspicions about instruments he had seen at auction more than half a century earlier.[9]

The distinction between Evans's detective work in the 1890s and Price's in the 1950s is instructive. Evans's suspicions operated at the level of the clue (or trace) of forgery: false signatures, not fake instruments. There is a direct parallel between Lewis Evans's minute observations and the clues found by Sherlock Holmes, Sigmund

---

7 See the articles in the special issue of the *Journal of the History of Collections* on historical collections of scientific instruments; in particular A. J. Turner, 'From Mathematical Practice to the History of Science: The Pattern of Collecting Scientific Instruments', *Journal of the History of Collections*, 7 (1995), pp. 135–50.

8 See R. Anderson, 'Connoisseurship, Pedagogy or Antiquarianism? What Were Instruments Doing in the Nineteenth-Century National Collections in Great Britain?', *Journal of the History of Collections*, 7 (1995), pp. 211–25.

9 See Chapter 8 by Tabitha Thomas.

Freud, and the art historian Giovanni Morelli as described in Carlo
Ginzburg's classic analysis of the late-nineteenth-century 'semiotic
paradigm'.[10] Evans used his connoisseurial eye to discern problems
with instruments that he could eliminate from his list of potential
purchases, but there was no scholarly context for his work – it is
unlikely that he considered publishing his findings, which were
recorded only in the margins of auction catalogues. He was working
with his own notes on instruments he had seen, and these were
limited to sales and collections that he could personally visit.

Price operated in entirely different conditions. His (re)discovery
that antique instruments had been faked was based on his work at
the Whipple Museum of the History of Science, founded in 1951 –
the year Price joined the staff. The Whipple was at that point home
to 1,000 or so historical instruments that had been donated to the
University of Cambridge in 1944 by the collector and businessman
Robert Stewart Whipple, augmented by pieces from donations, a
few purchases, and acquisitions from the Cavendish Laboratory of
experimental physics.[11]

It appears that it was Hall who put Price on to the question of
authenticity: in the Whipple's Accession Ledger, alongside the entry
for object Wh.0305 – an astrolabe by the little-known maker
Johannes Bos – there is a note in Price's hand that reads 'Hall says
a fake!' But it was Price who ran with the idea, eventually discovering
that amongst Whipple's founding collection five instruments were
fake, and moreover that these were of a piece with similar forgeries
in collections across Europe and in the United States. Most striking
of all was that these could all be traced back to a single source, the
dealership Frederik Muller & Co. (under the direction of the
collector and dealer Anton Mensing), two of whose sales, in
1911 and 1924, seemed to be linked to all of the forgeries Price
found.[12]

Although Price was tentative in his conclusions, he was effectively
opening up all collections and sales of historical scientific instruments to a scrutiny entirely unknown before. It seems that for

10 C. Ginzburg, 'Clues: Roots of a Semiotic Paradigm', *Theory and Society*, 7
    (1979), pp. 273–88.
11 See L. Taub and F. Willmoth (eds.), *The Whipple Museum of the History of
    Science: Instruments and Interpretations, to Celebrate the 60th Anniversary of
    R. S. Whipple's Gift to the University of Cambridge* (Cambridge: Whipple
    Museum of the History of Science, 2006), in particular the Introduction and
    Part I.
12 See Mörzer Bruyns, 'Frederik Muller & Co and Anton Mensing'.

scientific instruments barely anyone – with Lewis Evans a notable exception – had even *suspected* forgery. As Price pointed out, his revelations could be hugely damaging, not just for the pride of the various collectors of early instruments, but for historians working in the relatively young field of history of science.

For Price and others, the exact role of scientific instruments in the development of science was an important and open question: in a field still dominated by the 'Great Man' style of history – which dealt mainly with ideas and discoveries rather than practices and tools – to work on scientific instruments was unfashionable.[13] For a young scholar like Price, seeking to legitimise his interest in instruments and distance himself from antiquarianism, forgeries had the potential at least to upset the relationship between research and collections, and potentially to alter the historical record itself.

Price had begun his career during the Second World War, working as a lab assistant at South West Essex Technical College, where he subsequently enrolled as a student (in metallurgy) before moving to the University of Cambridge. There, in Easter 1951, he began his researches into 'The History of Scientific Instrument Making', working under Rupert Hall, the first director of the Whipple Museum.[14] From the very beginning of his historical research, Price was interested in the manufacture of instruments. And at this point in the development of history of science as a discipline, he was working almost entirely without precedent. Hall recalled the situation that confronted the two scholars as they attempted to make sense of Whipple's collection:

> How to proceed? Like the whole population of Britain, save a few
> score of individuals, I began with a total ignorance concerning
> the scientific instruments of the period from the sixteenth
> to the early nineteenth centuries. [...] Among other aids,
> at first, I had a copy of Mr Whipple's own *Guide* to the 1944
> exhibition at Cambridge, as well as his numbered acquisitions
> list (in chronological order). [...] Later, I also studied Disney's
> catalogue of the Royal Microscopical Society collection and a very

13 For an assessment of the (lack of) interest in instruments in this period see A. Van Helden and T. L. Hankins, 'Introduction: Instruments in the History of Science', *Osiris* (2nd Series), 9, 'Instruments' (1994), pp. 1–6. On the politics of historiography and scientific instruments in the middle of the twentieth century, see V. Enebakk, 'Lilley Revisited: Or Science and Society in the Twentieth Century', *British Journal for the History of Science*, 42 (2009), pp. 563–93.

14 Falk, 'The Scholar as Craftsman'.

different, more scholarly work by Alfred Rohde, *Die Geschichte der wissenschaftlichen Instrumente*[.][15]

This murky state of affairs was reflected in the circumstances of the Whipple Museum's founding. The 1944 exhibition mentioned by Hall was in fact a rare outing for Robert Stewart Whipple's collection: after he had donated his collection to the University of Cambridge in that year, it was moved a number of times and the Museum did not open until May 1951.[16] Housing a large collection of scientific instruments in the straitened conditions of the late 1940s had simply not been a priority for the University – so, although in 1951 a display was mounted, much of the collection remained packed up in cases. And the practical difficulty of building up a museum was matched by scholarly uncertainty. As Hall himself put it, 'instruments follow lines that may diverge considerably from those pursued by historians of science'.[17] In the context of a relatively new discipline it was unclear how to use the collection in terms of both research and display.

For Hall, who had responsibility for the development of the Museum, this was all becoming daunting. But his understudy Price seems to have found the situation merely enticing: here was a fine collection of early instruments, more or less unstudied and providing the basic material for a subject almost entirely absent from the secondary literature. To judge from his later work, in particular an essay on the 'philosophy of scientific instruments', what motived Price was the search for a lost history of craft know-how, a 'continuous thread' of 'understanding the world through tangible devices'.[18] In line with Hall's comment about the relative independence of the history of science and the history of scientific instruments, Price's project required first and foremost the establishment of sound data, i.e. an accurate record of the material culture of science. This, in combination with Price's background in metallurgy, the links he was

---

15 A. R. Hall, 'The First Decade of the Whipple Museum', in L. Taub and F. Willmoth (eds,), *The Whipple Museum of the History of Science: Instruments and Interpretations, to Celebrate the 60th Anniversary of R. S. Whipple's Gift to the University of Cambridge* (Cambridge: Whipple Museum of the History of Science, 2006), pp. 57–68, quotation on p. 59.

16 See A. R. Hall, 'Whipple Museum of the History of Science, Cambridge', *Nature*, 167 (1951), pp. 878–9.

17 Hall, 'The First Decade of the Whipple Museum', p. 60.

18 D. J. de Solla Price, 'Philosophical Mechanism and Mechanical Philosophy: Some Notes toward a Philosophy of Scientific Instruments', *Annali dell'Istituto e Museo di Storia della Scienza di Firenze*, 5 (1980), pp. 75–85. Price added his mother's maiden name 'de Solla' upon moving to the United States in 1957; here I observe that chronology.

forging with the Department of Physics under Lawrence Bragg,[19] and his more general predilection for applying scientific techniques to historical problems, partially explains his early decision to submit instruments from the Whipple Museum to metallurgical analysis. Using spark spectroscopy Price found that several Whipple instruments were made of modern electrolytically manufactured copper sheet instead of ancient open-hearth metal, and that all five 'Mensing' fakes lacked the tell-tale signatures that indicated appreciable levels of zinc, silicon, aluminium, and silver impurities. This marks the beginning of a line of inquiry that has cast doubt over hundreds of objects in collections around the world, and which continues to this day.[20]

From Connoisseurship to Catalogues

There is, however, another way of thinking about what Price had done: rather than look at his motivations and methods, we can take a step back to look instead at how collections are organised, displayed, and represented – through images, catalogues and checklists – and the ways in which these have changed over time. During the twentieth century collections of antique instruments had shifted from being objects of the connoisseurial gaze to being the subjects of systematic ordering and analysis, and it was in this move, I argue, that the possibilities of detecting forgeries emerged. Another way to put this is that instruments had shifted from being *visible* to being *legible*: earlier they were inspected, coveted, displayed, and traded, but they were not systematically classified and analysed; by the middle of the century this was becoming possible owing to new kinds of institution, as well as to international cataloguing projects.

As I have said, Price began work with the Whipple collection at Easter 1951, and we know from a letter to a colleague at the Cavendish Laboratory that he was already conducting metallurgical analysis – in order to authenticate instruments – by August of that year.[21] The instrument that Hall had fingered as a fake and that Price was now pursuing was a small astrolabe, signed 'Ioannes Bos I / 1597 / Die 24 Martii' (Figure 9.1).[22]

---

19 Falk, 'The Scholar as Craftsman', pp. 114ff.
20 See Nall, 'Finding the Fakes'.
21 Letter from A. A. Moss to D. J. Price, 15 August 1951, Whipple Museum Archives, D 076.
22 Hall, 'The First Decade of the Whipple Museum', p. 66.

Figure 9.1 Astrolabe, signed 'Ioannes Bos I / 1597 / Die 24 Martii', acquired by R. S. Whipple from a dealer in Paris in 1928. Image © Whipple Museum (Wh.0305).

This instrument, noted Price, was listed as item 33a in the 1924 auction catalogue *Collection Ant. W. M. Mensing*, sold by Frederik Muller & Co. But note the very specific date it carries: 24 March 1597. In addition to the 1924 astrolabe and the Whipple astrolabe – which may or may not be the same – Price was able to identify two more Bos astrolabes with the very same date. Hence there were three or possibly four astrolabes made by Joannes Bos *on the very same day* (the uncertainty over the total number stemmed from the fact that Price couldn't be sure whether the astrolabe pictured in the catalogue was one of the ones he had identified).[23] This was the first clue, and from here on Price was hot on the trail (Figure 9.2):

> We started with a very few suspect instruments, found where these had been purchased, and investigated instruments which had been bought from the same source at the same time. We then sought the cooperation of the dealers concerned and traced the collections back, all the time discovering that associated instruments fell into the same category of Strozzi–Mensing copies.[24]

23 Recording of the 1981 conference 'Fakes and Facsimiles (Scientific Instruments)', held at the National Maritime Museum, Greenwich, discussion with D. J. de Solla Price on tape 5. Many thanks are due to Richard Dunn for his help in gaining access to these recordings. Recent research has shown that the only genuine Bos astrolabe is held at the Adler Planetarium, Chicago; see G. B. Stephenson, B. Stephenson, and D. R. Haeffner, 'Investigations of Astrolabe Metallurgy Using Synchrotron Radiation', *Materials Research Bulletin*, 26 (2001), pp. 19–23.

24 Price, 'Fake Antique Scientific Instruments', p. 391.

Figure 9.2 Letter from Henry Nyburg to Derek Price, 16 February 1955, answering questions about the origins of certain instruments. This shows Price's method: from initial suspicions he worked back via provenance to find other instruments that could be examined. Antique Art Galleries sold over eighty instruments to Whipple, between the mid 1920s and the early 1950s. Amongst these about twenty are suspicious, are composites or heavy restorations, or are known forgeries (see Jardine, Nall and Hyslop, 'More Than Mensing?'). Image © Whipple Museum (Wh.0365, Object History File).

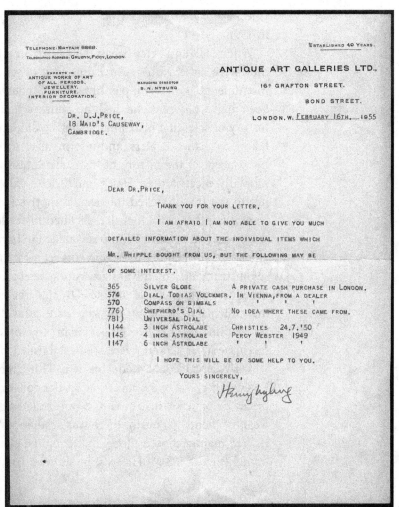

There is no reason to doubt Price's account – but it is not complete. In addition to tracing provenances, from the very beginning Price's method was comparative: while for Whipple, purchasing the astrolabe from a Paris-based dealer in 1928,[25] the existence of other copies could not possibly have been known, by the middle of the century Price was able to consult listings of instruments in numerous institutional collections. Indeed, one of his main preoccupations in this period was the compilation of a large 'International

25 Whipple bought the astrolabe from a dealer called Gertrude Hamilton, who operated a business called 'Mercator', see W. F. J. Mörzer Bruyns and A. J. Turner, 'Gertrude Hamilton, an American Instrument-Dealer in Paris', *Bulletin of the Scientific Instrument Society*, 73 (2002), pp. 23–6.

Checklist of Astrolabes', which was eventually published in two parts in 1955.[26]

When Hall and Price began their work on the Whipple collection, they were faced with a worrisome lack of scholarship on scientific instruments. For Whipple himself, of course, the problem had been quite different: he had no more scholarly expertise in the history of scientific instruments, but he did have a large network of fellow collectors, dealers, and intermediaries to call on. In addition, his concept of the history of instrumentation was less burdened by scholarly niceties than Hall's and Price's would be, though that is not to say that it lacked intellectual motivation. For Whipple there were a number of factors that dictated his collecting habits.[27] First, there was enthusiasm. Whipple's collection began, as he tells it, with an eighteenth-century telescope; this set a pattern for collecting optical instruments, mainly microscopes, telescopes, and spectacles.[28] Second, there was the market. By far the most common early instruments to come up for sale were sundials; these had been in circulation *as antiques* longer than any other kind of scientific instrument, and there were well-established private collections on which Whipple could model his own. Third, there was cost. Whipple was wealthy, but by no means a top-tier collector, as is shown by the prices he paid for instruments from dealers who also sold to more wealthy clients.[29] Fourth, there was aesthetics. Whipple, as Hall put it, 'had mainly chosen pieces that could be placed in cabinets in his home' (Figure 9.3).[30] He was, in short, a connoisseur, with just as

26 D. J. Price, 'International Checklist of Astrolabes', *Archives Internationales d'Histoire des Sciences*, 32 (1956), pp. 243–63 and 33 (1956), pp. 363–81. Further information on Price's method and judgments can be found in J. Holland, 'The David H. H. Felix Collection and the Beginnings of the Smithsonian's Museum of History and Technology', *eRittenhouse*, 26 (2015), pp. 1–18 (available online at erittenhouse.org/articles/vol-26-contents-and-authors/david-h-h-felix-collection, accessed 22 April 2018, via WayBackMachine, owing to the failure of the original website).

27 On Whipple as a collector see S. De Renzi, 'Between the Market and the Academy: Robert S. Whipple (1871–1953) as a Collector of Science Books', in R. Myers and M. Harris (eds.), *Medicine, Mortality and the Book Trade* (Folkestone: St Paul's Bibliographies, 1998), pp. 87–108; and R. Horry, 'Materials for a History of Science in Cambridge: Meanings of Collections and the 1944 Scientific Instrument Exhibition at the University of Cambridge' (unpublished MPhil Dissertation, University of Cambridge, 2008–9).

28 Quoted in 'Robert Stewart Whipple: Founder of the Museum', www.hps.cam.ac.uk/whipple/aboutthemuseum/robertwhipple/ (accessed 22 April 2018).

29 Mörzer Bruyns and Turner, 'Gertrude Hamilton, an American Instrument-Dealer in Paris'.

30 Hall, 'The First Decade of the Whipple Museum', p. 65.

Figure 9.3
Astrolabes on display in the 1920s. This cabinet was set up to display the 'Mensing Collection' in 1924, when it was offered for sale. See M. Engelman, *Collection Ant. W. Mensing, Amsterdam: Old Scientific Instruments (1479–1800), Volume II. Plates* (Amsterdam, 1924), Plate 1.

much interest in the visual appeal of an instrument as in its historical significance. Fifth and finally, there was Whipple's historicist agenda, in which instruments were to serve as examples – arranged in evolutionary sequence – of the progress of manufacture, leading in the end to a modern instrument firm like Cambridge Scientific Instruments, of which he was Chairman.[31]

---

31 A good example of Whipple's commitment to an evolutionary approach is R. S. Whipple, 'The Evolution of the Galvanometer', *Journal of Scientific Instruments*, 11 (1934), pp. 37–43 (see Chapter 7 by Charlotte Connelly and Hasok Chang). The link to modern methods is explicit in his 1939 presidential address to

To Whipple, an instrument was interesting insofar as it satisfied his demands in each of these areas. The fourth and fifth, aesthetics and historicism, are of particular relevance to the question of forgery, and also the much broader question of authenticity as it affects questions of display. We have no evidence of whether Whipple was able to detect forgeries, although the low prices he paid for some of the more obvious imitation instruments in the collection strongly suggest that he was aware of (while still being interested in) such instruments. Yet, akin to the completism of stamp collecting, the compilation of an evolutionary series requires that gaps be filled, and the need for attractive instruments that can be displayed in drawing-room cabinets places demands on the aesthetic appearance of an instrument that would not be imposed by the scholarly historian. In meeting these demands without necessarily knowing the full provenance or authenticity of objects Whipple was by no means alone, as the collections of, for instance, Henry Wellcome and Lt-General Augustus Pitt Rivers attest.[32]

Just as the criteria for a collection like Whipple's made space for forgery, so the inability to cross-reference collections made detecting forgery all but impossible. While instruments were largely in private hands, knowledge of their scarcity and distribution was unobtainable. Price could confidently assert that Joannes Bos was the maker of only one authentic instrument – the astrolabe shown in the Mensing catalogue,[33] but for Whipple such an assertion would be

Section A (Mathematical and Physical Sciences) of the British Association for the Advancement of Science, R. S. Whipple, 'Instruments in Science and Industry', *Nature*, 144 (1939), pp. 461–5.

32 On these see, respectively, F. Larson, 'The Things about Henry Wellcome', *Journal of Material Culture*, 15 (2010), pp. 83–104; and A. Petch, 'Collecting Immortality: The Field Collectors Who Contributed to the Pitt Rivers Museum, Oxford', *Journal of Museum Ethnography*, 16 (2004), pp. 127–39. On the notion of a 'series' in collections of antiquities see N. Schlanger, 'Series in Progress: Antiquities of Nature, Numismatics and Stone Implements in the Emergence of Prehistoric Archaeology', *History of Science*, 48 (2010), pp. 344–69.

33 This issue (along with many others of interest to the present chapter) is discussed in the recordings of the 1981 conference 'Fakes and Facsimiles (Scientific Instruments)', held at the National Maritime Museum, Greenwich (see n.23). It seems to have been Price's working assumption that the catalogues showed authentic instruments, which were then copied by craftsmen who were collaborating with Anton Mensing in restoration work. That the situation was considerably more complicated than this seems certain, yet Price was far more interested in establishing authenticity than he was in apportioning blame. In addition, insofar as Price was interested in forgery *per se*, he seems to have considered it as much a matter of exuberant and experimental craftsmanship as of fraud. Forgers, to Price, and in line with his views about the tradition of craft experimentation, were simply copying things for fun or to test their prowess.

meaningless: for him the astrolabe was a beautiful object, showing the development of astronomical instrumentation on the Continent in the sixteenth century.[34] Whipple's choices were based on his historicism and on aesthetic critera, and were dictated by the market and his wallet.

Price, meanwhile, without even necessarily studying the object,[35] could draw on a wholly different set of resources and was working with a different set of assumptions. Gone were aesthetics, finance, and historicism in the earlier evolutionary mould, and in their place came card catalogues, an international network of scholars and curators, and an interest in the instrument-making trade as an end of historical research in itself. Although Price was no less progressivist in his attitude than Whipple, his method was to amass data first and make inferences about the nature of craftsmanship and handing-on of techniques on this basis. As Price put it in his 1956 paper on fakes, the instruments with which he was dealing were 'of such exceptional workmanship that they could not be detected as spurious *except by comparison with the rest of the series*'.[36] This, moreover, was also Price's general historical method, as he explained in the introductory notes to the 'International Checklist of Astrolabes', where he states that 'the full significance of any one instrument cannot properly be realised except by comparison with the corpus of all such instruments extant'.[37] The 'Checklist' boasts a huge volume of data – some 700 instruments in around 200 collections. At the end of the 'Checklist' a graph reveals the chronology of astrolabe production in the East and West,

34 It is interesting to note that Whipple did in fact have a copy of the catalogue in which the Bos astrolabe is first shown (i.e. the 1924 catalogue of the Mensing collection itself). Yet he seems not to have made the connection between the instrument advertised there and the one he bought in Paris, let alone to have suspected forgery (the catalogue is in the Whipple collection, with the inventory number Wh.6494). Indeed, two of the forgeries in the Whipple collection identified by Price were purchased by Whipple in 1952 – long after he had donated his collection to the University of Cambridge. One of these latter instruments was signed Bos, but again Whipple's suspicions seem not to have been raised. After Price had reached his verdict of forgery an embarrassed Hall discussed the case (specifically of the fake Bos astrolabe) with Whipple, 'who took it very well, understanding (I believe) the various dubious points in our particular instrument very clearly when his critical attention was drawn to them'; see Hall, 'The First Decade of the Whipple Museum', p. 66.
35 It is telling that, although Price carried out metallurgical analysis on a number of objects in the Whipple collection, for the Bos astrolabe he in fact presented no evidence for forgery beyond the coincidence of the various dates.
36 Price, 'Fake Antique Scientific Instruments', p. 382 (my emphasis).
37 Price, 'International Checklist of Astrolabes', p. 243.

with peaks around 1700 and 1600, respectively. Historical conclusions were to be drawn from aggregating information about instruments in collections around the world, and individual instruments acquired meaning only through this process. In the case of Bos this was a particularly pertinent point, as there existed no biographical information beyond the instruments, and no Bos instruments beyond the group of identical astrolabes Price identified. As Price put it, Joannes Bos 'becomes incomprehensible as an historic person' unless the fakes are identified and discounted from the record.[38]

It was this groundwork on which Price based his short 1956 paper on fake scientific instruments. Gone was the connoisseurship of a collector journeying to an antique shop or sale-room, gone too the visual arrangement of instruments in a private display case. Although some of Whipple's collection was, by the early 1950s, on display in the new Museum, much of it remained in packing cases, a situation both typical and unavoidable in museums then and since. Yet in spite of its relative invisibility, through Price's tireless data-gathering the collection took on a new legibility, even as the instruments themselves were transformed from personally appraised objects to lines in a printed table. It was precisely this move that permitted Price to make his claim that there existed numerous forgeries, not only in the Whipple, but in collections by then already dispersed around the world.

Post-war Internationalism and the Changing Nature of Collections

Collections arranged in a developmental series created gaps – gaps that could be filled by unscrupulous dealers and eager collectors. And where anomalies arose, the collector's imperative – to have something no one else did – could play its part. Collections systematically catalogued, on the other hand, created anomalies – objects that didn't fit and suddenly looked suspicious. Price's 'big data' approach to the history of science was not favourable to highly anomalous instruments – for him, comparison was key. The exception that proves this rule was, of course, the Antikythera mechanism, on which Price spent a large portion of his career, and about which he crafted elaborate arguments.[39] For historians of instruments and technology of Price's generation, the progressivist model of the development of instruments still held, but the

38 Price, 'Fake Antique Scientific Instruments', p. 393.
39 Price, 'Philosophical Mechanism and Mechanical Philosophy'.

pressure it placed on outliers was all the greater because of the new historiographic (rather than dilettante) criteria. The attitude of 'filling the gaps' was replaced by attempts to amass data, sort out anomalies, and place historical arguments on a sure footing.

But this was not merely the historical pursuit of an enthusiast: in fact Price was participating in a range of post-Second World War developments that affected a number of fields. One way to illustrate this is to consider the ideology of Price's method. Statistical techniques could be used, he argued, to shed light on technical aspects of history, such as the development of the astrolabe, and could be applied to the growth of science itself. This involved the development of what Price called 'scientometrics' – the 'scientific' use of statistics to represent and assess the development of scientific training, publishing, and institutional support.[40] This had historiographic and technocratic implications: scientometrics used the analysis of large numbers of papers and citation indices to establish the ways in which scientific networks developed, thus downplaying the role of individual papers and scientists. For Price, 'great men' were outliers – far less significant than the scientific structures that supported them. This had political implications, and implications for the role of sociologists in crafting policy. It is clear from Price's increasingly active contribution to science policy in the late 1960s and 1970s that he saw his role as both analytic and normative.[41] For example, only careful management, he contended, would overcome the tendency to overproduction in the sciences – itself a result of the rapid transformation of scientific education after the Second World War.

Price's work on instruments, then, was not merely an intriguing side-line, but a central part of an ambitious project that applied large quantities of data to historical problems. Instruments were the materials of history in a strong sense, because science was a matter of experimental tinkering, invention, and craft know-how. For instruments as well as people, large numbers showed smooth trends and steady development that could be shifted by government policy. Nor were these concerns limited to historiography and the politics of science. If we look elsewhere in the sciences in this period we can see a similar pattern, of large quantities of data, the interpenetration of

40 See D. J. de Solla Price, *Little Science, Big Science ... and Beyond* (New York: Columbia University Press, 1986 [1963]).
41 See, for example, D. J. de Solla Price, 'Principles for Projecting Funding of Academic Science in the 1970s', *Science Studies*, 1 (1971), pp. 85–94.

different disciplines and experimental techniques, and the commitment to modernist technocracy and internationalism.

As mentioned at the outset, Price gave a direct comparison for his uncovering of fake antique scientific instruments: the Piltdown forgery. This was no mere passing allusion. As Price knew, the Piltdown story very closely mirrored that of fake instruments, both in the specifics of the argument and in its political context. The so-called 'Piltdown skull' in fact consists of only a few fragments of bone, discovered in the years prior to 1912 by the amateur archaeologist and antiquarian Charles Dawson, in Pleistocene gravel beds in Sussex. Dawson's collaborator Arthur Smith Woodward christened it *Eoanthropus*: 'The Dawn Man', and to many it came to be known as the 'missing link' between humans and their primate ancestors.[42] Initial doubts about the legitimacy of the skull never quite disappeared, however, and in the mid 1950s it was conclusively shown to be a composite, though the identity of the forger has never been absolutely settled.[43] In one sense, the story is quite straightforward: a controversial scientific breakthrough came to be increasingly doubted and was eventually discredited. Of greater relevance for Price and fake instruments is the complex way in which the *visibility* of the skull was bound up with its authenticity: at first an internationally important find was made into a piece of scientific theatre – in the end it was undone by someone who boasted that he hadn't even needed to see it in person.

Owing to the fragmentary nature of the skull, the very first Piltdown controversy hinged on its correct reconstruction. At the first presentation of the remains, anatomist-anthropologists Grafton Elliot Smith and Arthur Keith began a short-lived but fierce confrontation over the cranial capacity of the reconstructed skull, with Keith's estimate giving a larger and therefore more human brain-case than Elliot Smith's.[44]

---

42 See M. Goulden, 'Boundary-Work and the Human–Animal Binary: Piltdown Man, Science and the Media', *Public Understanding of Science*, 18 (2009), pp. 275–91. On the scientific context of Piltdown see F. Spencer, 'Prologue to a Scientific Forgery: The British Eolithic Movement from Abbeville to Piltdown', in G. W. Stocking, Jr (ed.), *Bones, Bodies, Behavior: Essays on Biological Anthropology* (Madison: University of Wisconsin Press, 1988), pp. 84–116.

43 Anon., 'J. S. Weiner and the Exposure of the Piltdown Forgery', *Antiquity*, 57 (1983), pp. 46–8. Charles Dawson is considered by most to have been the forger; see M. Russell, *Piltdown Man: The Secret Life of Charles Dawson and the World's Greatest Archaeological Hoax* (Stroud: Tempus, 2003).

44 For details of this episode see Keith's Royal Society obituary: W. E. Le Gros Clark, 'Arthur Keith. 1866–1955', *Biographical Memoirs of Fellows of the Royal Society*, 1 (1955), pp. 144–61, especially pp. 150ff.

Figure 9.4 The examination of the Piltdown skull, by John Cooke, 1915. Arthur Keith, whose interpretative reconstruction of the skull carried the day, is seated and wearing a white coat. Note the portrait of Charles Darwin hanging behind the gathered scientists, conferring not just authority but also an evolutionary justification for the existence of 'Piltdown Man'. (Public domain image from https://commons.wiki media.org/wiki/File:Piltdown_gang_(dark).jpg.)

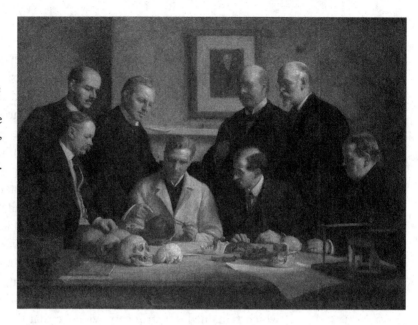

Keith's finding, and his estimate of an age of about 500,000 years, were the reasons the skull was considered so important.[45]

Just as remarkable as the Piltdown skull's antiquity, however, was the method by which the scientific community was convinced of the fact. So confident was Keith in his ability to reconstruct a skull – *any* skull – from the merest fragments, he performed a demonstration to the Royal Anthropological Institute in 1914. Keith had two colleagues select a modern skull, prepare a cast and isolate fragments of similar scale to those of the Piltdown specimen. Keith's reconstruction, done in complete ignorance of the nature of the skull from which these new fragments came, was then unveiled alongside the cast of the original skull.[46] Showmanship and the anatomist's expert way of knowing established the authenticity of the artefact: the nature of Piltdown Man was demonstrated primarily by close examination of, and direct working with, the fragments themselves (Figure 9.4).

In stark contrast, the revelation of forgery some forty years later was not a continuation of close visual analysis. Rather it was the product of distant appraisal. J. S. Weiner, the physical anthropologist

45 A. Keith, *The Antiquity of Man* (London: Williams and Norgate, 1915), Chapter 19.
46 Le Gros Clark, 'Arthur Keith. 1866–1955', p. 151.

at the University of Oxford who is credited with exposing the Piltdown forgery, first became interested in the topic after attending a
talk by the palaeontologist Kenneth Oakley, at which Oakley
announced that fluorine-absorption tests had shown a possible discrepancy between the Piltdown mandible and skull cap.[47] Weiner
was less impressed by the discrepancy (fluorine tests having a large
margin of error) than by Oakley's relatively recent date for *all* of the
fragments. Since 1912 many more early hominids had been discovered and the Piltdown skull was looking increasingly – and
concerningly – like an outlier.[48] As the high status of the Piltdown
skull had caused Oakley's fluorine tests to be called into question, the
immediate issue was whether another means of accurately dating the
specimen could be found. Weiner's first idea was that X-ray crystallography might reveal differences in fossil bones from different
geological eras, but this did not prove immediately feasible. His next
approach was to study the published images of the Piltdown teeth,
which remained the strongest evidence that the skull combined ape
and human features – the very fact that had become especially
incongruous in the light of Oakley's tests. Weiner found that he
was able to fabricate similar teeth artificially by filing and staining
ape teeth, and from here it was only a short leap to calling the
authenticity of Piltdown into question, a move that negated the need
for accurate dating. But note that, even as Weiner was attempting to
convince others that the Piltdown skull was a fake, he had not yet
examined the skull in person. As Wiener's research assistant put it
later, only after revealing the forgery did Wiener 'need access to the
original fossils, which [he] had never seen'.[49] It was only once
Weiner had convinced himself and others that there was a high
likelihood of forgery that he considered it necessary to look at the
fragments themselves. While the original validity of Keith's interpretation of the skull was bound up with the performance of his own
expertise in reconstruction, the ultimate unmasking of the Piltdown
forgery was based on Weiner's distance from the artefact itself.

In addition to this move – from *visibility* to *legibility* – there are
two other features of the Piltdown story that are common to most

---

47 Anon., 'J. S. Weiner and the Exposure of the Piltdown Forgery'. On the use of
    fluorine tests see M. R. Goodrum and C. Olson, 'The Quest for an Absolute
    Chronology in Human Prehistory: Anthropologists, Chemists and the Fluorine
    Dating Method in Palaeoanthropology', *The British Journal for the History of
    Science*, 42 (2009), pp. 95–114.
48 Anon., 'J. S. Weiner and the Exposure of the Piltdown Forgery', p. 47.
49 Anon., 'J. S. Weiner and the Exposure of the Piltdown Forgery', p. 47.

exposés: changes in the techniques used to analyse the suspicious artefact, and close links between the unmasking and an attempted disciplinary reform. In this case, in addition to Wiener's replication of the teeth, the new techniques were the fluorine test and X-ray crystallography (which was eventually used by Wiener, who published the results in his 1955 book on the Piltdown forgery); the discipline was the 'New Physical Anthropology' – whose main advocate in Britain was J. S. Weiner, and which emphasised large-scale survey work, international collaboration, and the study of populations over individual specimens.[50]

Over the course of the first half of the twentieth century the Piltdown skull went from being an object appraised in its own right, in particular by Keith, to being just one part – and a highly anomalous one at that – of an international catalogue of early human remains. Keith had only a handful of remains to refer to and to fit within his theory of human evolution; Weiner, with more material to examine in collections around the world, was primarily concerned with establishing the exact ages of the specimens themselves, in order to provide a solid foundation for a reformed physical anthropology. It is this move, from consideration of the particular artefact to its relocation in a catalogued collection, that provides a direct parallel to Price's work. This is as much about disciplinary reform as it is about the nature of collections as they move between private display and museum accession. Forgery is typically seen as an accusation levelled by connoisseurs, able to determine authenticity by the 'eyeball test' – but sometimes the pattern is reversed, and it is not proximity but distance that enables detective work.

Conclusion

In his classic essay 'Clues', Carlo Ginzburg suggests that in the years around 1900 a 'semiotic paradigm' took hold of a range of disciplines: psychoanalysis (via Freud), art history (via Giovanni Morelli), and criminal detection (via Conan Doyle). The last of these shows that the search for clues stretched into the realm of human imagination. But Ginzburg also found concrete links between his

50 On Weiner and physical anthropology see M. A. Little and K. J. Collins, 'Joseph S. Weiner and the Foundation of Post-WW II Human Biology in the United Kingdom', *American Journal of Physical Anthropology*, 149, suppl. 55 (2012), pp. 114–31; on the New Physical Anthropology see J. Mikels-Carrasco, 'Sherwood Washburn's New Physical Anthropology: Rejecting the "Religion of Taxonomy"', *History and Philosophy of the Life Sciences*, 34 (2012), pp. 79–102.

protagonists: this was the microhistory technique applied to an ambitious thesis about the relationship between parts and wholes, seeing and knowing, deception and truth. My intention has been to provide a 'version 2.0' for Ginzburg's argument, by documenting changes brought about by large quantities of data in the middle of the twentieth century.

In this chapter I have argued that the question of forgery eventually became a question of data: where earlier detection had rested on close appreciation of individual objects, by the middle of the twentieth century large collections of instruments could be compared internationally and appraised *en masse*. As antique instruments moved from the collector's cabinet to the museum catalogue, they entered into new kinds of relationships with each other, with systems of classification and recording, and with historians and curators. In the 1950s the first international databases of scientific instruments were being put together, in the context of UNESCO's scheme for systematic international cultural cooperation. In these conditions and precisely as a result of the new kinds of data being generated, the question of forgery became acute. The Whipple Museum in its early years was host to the discovery of fake antique scientific instruments, and therefore played a special role in the history of instrument studies.

The irony of this situation is that earlier in the century instruments were subject to the classic scrutiny of the connoisseur: precisely the conditions in which the detection of forgery is traditionally thought to occur, as it had done in the art world for centuries. In those earlier circumstances, instruments were highly (if selectively) *visible* – but they were not *legible*. Later, in the age of the card catalogue and the international survey, instruments achieved a legibility that made unusual individual instruments into anomalies that had to be dealt with using special techniques: metallurgical analysis, examination with precision tools of measurement, and complex historical reconstruction. The general trend was towards a history of the diffusion and role of craft techniques and expert manufacture tending towards standardisation: outliers could be significant, but they were also more suspicious, didn't necessarily add to the historical narrative, and might be fake. Ideas of the public record and the public interest were invoked in an age very different from that of the private drawing-room museum.

Forgeries, as many others have pointed out, are peculiarly revealing sources for the history of scholarship, the history of aesthetics, and the history of commerce.[51] One reason for this is obvious, though not often stated: forgers can respond to the market in a much more systematic and coherent way than the historical record itself. In extreme cases, scholars themselves have fabricated their material: here the historical record appears to become identical with scholarly interest, at least until the deception is uncovered. Possibilities and failures of detection can also provide clues to the ways in which attitudes towards authenticity, connoisseurship, commodification, and tradition have shifted. In cases when apparently obvious forgery goes undetected or objects are mistakenly attributed, we get a glimpse of how recent and selective our positivist mentality really is.[52] Authenticity and its opposite are not conditions of objects out there waiting to be discovered: they are processes involving networks of objects, scholars, publics, spaces, and techniques, and as such they are subject to the forces of historical change. As we move into an age of greater reflexivity within museums concerning all aspects of provenance, curatorial voice, participation, and representation, the question of authenticity can be raised again – not as a means to get the historical record straight, but as a means of understanding the relationship between the kinds of structures that have governed ownership and interpretation of objects and the conclusions that are drawn from and about them.

51 For fakes and scholarship the exemplary works are A. Grafton, *Forgers and Critics: Creativity and Duplicity in Western Scholarship* (Princeton: Princeton University Press, 1990); and C. S. Wood, *Forgery, Replica, Fiction: Temporalities of German Renaissance Art* (Chicago: University of Chicago Press, 2008). For fakes and aesthetics see A. Nagel and C. S. Wood, *Anachronic Renaissance* (Cambridge: MIT Press, 2010). For fakes and commerce see, for example, M. Jones (ed.), *Fake? The Art of Deception* (Berkeley: University of California Press, 1990); and C. Helstosky, 'Giovanni Bastianini, Art Forgery, and the Market in Nineteenth-Century Italy', *The Journal of Modern History*, 81 (2009), pp. 793–823.
52 See Nagel and Wood, *Anachronic Renaissance*.

10 ❧ Wanted Weeds: Environmental History in the Whipple Museum*

HELEN ANNE CURRY

'A plant where you don't want it'. 'A noxious or useless plant'. 'A plant out of place'. 'A troublesome plant'. 'A plant not edible, so far as known, nor medicinal, or otherwise serviceable to man, and which always thrives where not wanted'. 'A plant for which we have no use so far as we know'. 'Any plant from which its situation or inherent properties is harmful to human interests; a vegetable malefactor.'[1] These definitions of 'weed', gathered via an American botanist's informal survey in 1892, might just as easily have been collected today. We all know that weeds are unwanted pests. Or do we? In the late nineteenth century, some agronomists and botanists came to see the very same plants as desirable, useful, and well-placed for solving a particularly tricky question. Their thinking transformed vegetable malefactors into benefactors.

This re-imagining of the useless weed is evident in a small seed herbarium owned by the Whipple Museum, which contains seeds of plants originating across Europe, the Middle East, and the Americas: *Crepis biennis*, or rough hawksbeard, a lanky biennial herb with bright yellow flowers that was originally native to Europe; *Rudbeckia hirta*, the black-eyed susan, a showy little sunflower hailing from North America; and *Delphinium consolida*, also known as forking larkspur or royal knight's-spur, a purple-flowering annual common to Eastern Europe.[2] The Whipple's herbarium (Figure 10.1) comprises small samples of seed from these and ninety-seven other species, labelled and arranged in a commercially manufactured microscope slide box. Little was known about this seed collection when the Whipple Museum acquired it. There was no place or date

---

* I am grateful to Josh Nall and Dominic Berry for their help in tracking down primary source materials and references for this chapter, and to Dominic for his incisive comments on an early draft. I extend thanks also to Josh and Liba Taub for their editorial advice.
1 G. McCarthy, 'American Weeds', *Science*, 20.493 (1892), p. 38.
2 These are the Latin names given on the instrument label; in 2018, *D. consolida* is classified as *Consolida regalis*.

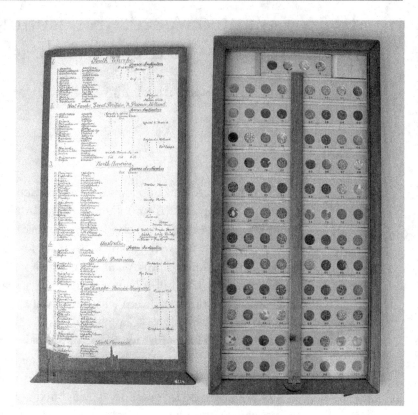

Figure 10.1 The Whipple Museum's seed herbarium. The manuscript nameplate on its lid reads 'The Origin of Seeds Source Indicators'. Image © Whipple Museum (Wh.6624).

of creation, no record of its ownership, and, most pressingly, no knowledge of the uses for which it had been intended. The hand-written outer label, 'Origin of Seeds Source Indicators', and contents list within, which provides information about each sample's status as something called a 'source indicator' via the listing of miscellaneous forage crops, provided no immediate answers.

In pursuing the history of this object for the Whipple Museum, I discovered it was itself an unusual specimen. Like other seed herbaria, such as those kept by individual botanical researchers or by institutions such as botanical gardens and arboreta as reference tools for identification and classification, this small collection enabled its handler to identify the species or genus of a seed via its visible characteristics.[3] The ultimate aim of this initial identification,

3 Most seed herbaria were and are assembled through the labour of botanical researchers, either individually or collectively at a particular institution. For a description of an institutional seed herbarium of the twentieth century, see C. G. Gunn, 'Seeds: U.S. National Seed Herbarium', in *Systematic Collections of the Agricultural Research Service*, Miscellaneous Publication No. 1343 (Washington: US Department of Agriculture, 1977), pp. 79–82; see also the (now digitised) seed

however, was not to learn more about the seed being compared with the herbarium specimens, as one might typically expect. Instead, it was to correctly categorise still other seeds, ones with potentially far more value. The Whipple's herbarium was not an all-purpose reference tool but one designed to facilitate a specific agricultural task. To the extent it was ever used, it enabled its user to deploy the unwanted weed seeds inherent in commercial seed stocks as much-needed evidence of the geographical origins of those stocks – that is, as 'source indicators'. To that end, it contains only the seeds of weeds commonly found among forage or fodder crops and of circumscribed geographical distribution.[4]

This simple explanation of the contents and intended uses of the Whipple's enigmatic herbarium belies the challenges that are likely to have confronted its users. As I describe here, the need to assess the geographical origin of commercial seeds had arisen as a consequence of an increasingly international seed market. Although seed-testing stations adopted the use of weed seeds as indicators of provenance, this mobilisation of agricultural pests (weeds) in the service of agricultural improvement (accurately labelled seed) proved troublesome. By highlighting the folly of premising a static laboratory instrument, intended to manage the tumult of international agricultural exchange, on the assumption of a stable global plant biogeography, my effort to shed some light on the Whipple Museum's seed herbarium ends with a lesson that is as much environmental history as object history.

Noxious Plants

It is rarely straightforward to label a plant as a weed. Botanists have been at pains to point this out for decades, and environmental historians have recently joined them. Consider that the opportunistic Tree of Heaven is vilified as a noxious invasive in Australia, the United States, and parts of Europe but celebrated as a medicinal

---

herbarium of the Arnold Arboretum, at www.arboretum.harvard.edu/plants/herbaria/seed-herbarium. An example of how an individual reference collection might be assembled is given in J. W. Harshberger, *Text-book of Pastoral and Agricultural Botany, for the Study of the Injurious and Useful Plants of Country and Farm* (Philadelphia: Blakiston's Son & Co., 1920), pp. 153–4, 270–1.

4 Elsewhere in this chapter, I use 'forage' as a catch-all term for both forage and fodder crops. Forage typically picks out plants that are grazed directly by livestock in the field, while fodder refers to plants that are cut and delivered to animals as feed.

plant in its native China. Or that one person's troublesome lawn pest is another's dandelion salad. Or that an oat seedling is a weed when found in a field of wheat. What counts as a weed depends not so much on the plant as on the person looking at it and the place they happen to be standing.[5] Nonetheless, even if the label is constrained by time, place, and culture, a plant 'growing where it is not desired' is typically thought of as a weed.[6] Perhaps it is more accurate to say that such a plant is typically *reviled* as a weed. Weeds have a truly bad rap, so much so that the menace of weedy plants to agriculture is often sufficient to mobilise military language, and military tools, in response.[7] Equally tellingly, the metaphors of weeds and weeding have appeared wherever certain kinds of people have been designated undesirable, whether criminals, immigrants, or perceived eugenic threats.[8]

Many of the plants commonly designated as weeds share a suite of characteristics that help them thrive in recently disturbed habitats – places like agricultural fields, home gardens, construction sites, and roadsides – and to disperse themselves quickly and widely.[9] Humans

5 On challenges of defining the term 'weed', see C. L. Evans, *The War on Weeds in the Prairie West: An Environmental History* (Calgary: University of Calgary Press, 2002), Chapter 1; and J. Wegner, 'A Weed by Any Other Name: Problems with Defining Weeds in Tropical Queensland', *Environment and History*, 23.4 (2017), pp. 523–44. Other extended reflections on the cultural and environmental history of weeds include D. W. Gade, 'Weeds in Vermont as Tokens of Socioeconomic Change', *Geographical Review*, 81.2 (1991), pp. 153–69; F. Knobloch, *The Culture of Wilderness: Agriculture as Colonization in the American West* (Chapel Hill: University of North Carolina Press, 1996); and Z. J. S. Falck, *Weeds: An Environmental History of Metropolitan America* (Pittsburgh: University of Pittsburgh Press, 2010). For illustrative accounts by botanists, see McCarthy, 'American Weeds', p. 38; E. Anderson, *Plants, Man, and Life* (Boston: Little, Brown and Co., 1952), esp. Chapters 1 and 2; and J. R. Harlan and J. M. J. deWet, 'Some Thoughts about Weeds', *Economic Botany*, 19.1 (1965), pp. 16–24.
6 T. J. Monaco, S. C. Weller, and F. M. Ashton, *Weed Science: Principles and Practices*, 4th edn (New York: John Wiley & Sons, 2002), p. 3.
7 B. M. H. Larson, 'The War of the Roses: Demilitarizing Invasion Biology', *Frontiers in Ecology and the Environment*, 3.9 (2005), pp. 495–500. See also E. Russell, *War and Nature: Fighting Humans and Insects with Chemicals from World War I to 'Silent Spring'* (New York: Cambridge University Press, 2001).
8 See, for example, T. Cresswell, 'Weeds, Plagues, and Bodily Secretions: A Geographical Interpretation of Metaphors of Displacement', *Annals of the Association of American Geographers*, 87.2 (1997), pp. 330–45; O. Santa Ana, '"Like an Animal I Was Treated": Anti-Immigrant Metaphor in US Public Discourse', *Discourse & Society*, 10.2 (1999), pp. 191–224; and G. O'Brien, 'Anchors on the Ship of Progress and Weeds in the Human Garden: Objectivist Rhetoric in American Eugenic Writings', *Disability Studies Quarterly*, 31.3 (2011), http://dx.doi.org/10.18061/dsq.v31i3.1668. Thanks are due to Dominic Berry for suggesting that I include this point.
9 Monaco, Weller, and Ashton, *Weed Science*, pp. 14–15.

are incredibly talented at creating spaces in which these peregrinating plants thrive.[10] As a result, they follow us everywhere. In the words of the botanist Edgar Anderson, 'The history of weeds is the history of man.'[11] A handful of environmental historians have made hay from this insight, casting weeds as decisive actors in human history. Among other achievements, weeds are said to have enabled European colonisation in temperate climates, attuned displaced settlers to new environments, and united prickly farmers to common purpose in the American West.[12] It is the mobility of weeds, their movement with us and despite us, that has made them so unexpectedly influential. This was true even in the laboratory, where the movement of weeds underlies the creation, and very probably the abandonment, too, of the Whipple's weed seed herbarium.

To understand the intended functions of this herbarium, it is essential to first understand some dysfunctions of the international seed market of the late nineteenth century. Expanding markets for grain and other agricultural products had created new demand for commercial seed, and many eager producers, vendors, and middlemen clamoured to meet it.[13] If contemporary accounts are to be believed, their number included all sorts of unscrupulous individuals. Reports of deceptive sales boomeranged across Europe and North America, tales of miscellaneous seeds sold as pure strains, old stock coloured to look fresh, seeds of worthless plants used to

10 The phrase 'peregrinating plants' is from Gade, 'Weeds in Vermont as Tokens of Socioeconomic Change'.

11 Anderson, *Plants, Man, and Life*, p. 21.

12 A. Crosby, *Ecological Imperialism: The Biological Expansion of Europe, 900-1900* (New York: Cambridge University Press, 1986); J. E. McWilliams, 'Worshipping Weeds: The Parable of the Tares, the Rhetoric of Ecology, and the Origins of Agrarian Exceptionalism in Early America', *Environmental History*, 16.2 (2011), pp. 290-311; and M. Fiege, 'The Weedy West: Mobile Nature, Boundaries, and Common Space in the Montana Landscape', *Western Historical Quarterly*, 36.1 (2005), pp. 22-47.

13 On the emergence of the American seed industry (amidst this international market), see J. R. Kloppenburg, Jr, *First the Seed: The Political Economy of Plant Biotechnology*, 2nd edn (Madison: University of Wisconsin Press, 2004); and D. J. Kevles, 'A Primer of A, B, Seeds: Advertising, Branding, and Intellectual Property in an Emerging Industry', *University of California, Davis Law Review*, 47.2 (2013), pp. 657-78. On Britain, see P. Palladino, 'The Political Economy of Applied Research: Plant Breeding in Great Britain, 1910-1940', *Minerva*, 28.4 (1990), pp. 446-68; and J. R. Walton, 'Varietal Innovation and the Competitiveness of the British Cereals Sector, 1760-1930', *Agricultural History Review*, 47.1 (1999), pp. 29-57. See also C. Fullilove, *The Profit of the Earth: The Global Seeds of American Agriculture* (Chicago: University of Chicago Press, 2017).

bulk up prized varieties, and uncleaned seed sold rife with weeds.[14] One widely circulating story held that a German firm was discovered offering seed dealers quartz stone 'so agreeing in size and colour with red or white clover [seed] that the farmer could not distinguish them'.[15]

These abuses were particularly rampant in the sale of seed for forage crops such as clover, alfalfa, timothy, and rye grass. 'Red clover is usually the foulest seed sold on the market,' advised an 1894 US agricultural bulletin.[16] In comparison with grain crops, forage tended to be poorly weeded, if at all, which meant that weed seeds ripened alongside crop seeds and were most often caught up in the harvest.[17] And whereas farmers tended to select with care the seed they used for corn, wheat, and other grains, choosing either from their own harvest or from some nearby producer, far less deliberation went into obtaining seeds for forage. This led to more instances of adulteration and badly cleaned seed than seen in other seed crops. 'It may almost be said that the average farmer buys the cheapest seed in the market and trusts entirely to luck for it to produce the desired crop,' lamented one botanist of forage seed sales. He rued in particular the crop of weeds that tended to spring up from such thoughtless plantings, as their subsequent eradication inevitably cost the farmer far more than the premium on a bag of good, clean seed would have cost.[18]

Seed Testing

The notoriety of bad seed, and the real and imagined havoc that it wreaked for individual farmers as well as for regional and national productivity, led private firms and national governments to varied methods for ensuring the circulation of 'pure' seed. These ran the

14 For some representative claims about the problems with commercial seed, see G. Hicks, 'Pure Seed Investigation', in *USDA Yearbook of Agriculture 1894* (Washington: US Government Printing Office, 1895), pp. 389–408; 'Seed Testing: Its Uses and Methods', *Bulletin of the North Carolina Agricultural Experiment Station*, no. 108 (Raleigh: North Carolina Agricultural Experiment Station, 1894); and T. Johnson, 'The Principles of Seed-Testing', *Science Progress in the Twentieth Century*, 1.3 (1907), pp. 483–95.

15 For example A. J. Pieters, 'Seed Selling, Seed Growing, and Seed Testing', *USDA Yearbook of Agriculture 1899* (Washington: US Government Printing Office, 1900), pp. 549–74, on p. 571; Johnson, 'The Principles of Seed-Testing', p. 486; and C. V. Piper, *Forage Plants and Their Culture* (New York: MacMillan, 1916), p. 72.

16 'Seed Testing: Its Uses and Methods', p. 353.

17 Piper, *Forage Plants and Their Culture*, p. 68.

18 Hicks, 'Pure Seed Investigation', pp. 389–90.

gamut, from instructional materials that taught farmers how to assess the quality of their seed purchases to laboratories where seed analysts (typically women) evaluated seed, to regulations allowing, or even compelling, state evaluation of commercial seed. Central to all of these was the seed test, a set of methods and tools for judging the quality of a seed stock: its genuineness, purity, and capacity for germination. Many accounts credit the botanist Friedrich Nobbe of Saxony with formalising seed testing. In 1869 he opened what is considered to have been the first seed-testing station at Tharandt, simultaneously setting out some basic principles for running such a station. His ideas proved immensely popular among farmers and governments, and seed dealers, too, who benefited from being able to sell guaranteed pure seed. By the end of the century, there were reportedly 119 seed-testing stations operating along similar lines in nineteen different countries.[19]

Although procedures varied from site to site, especially with regard to the extent of involvement of the state and the nature of an institution's relationship with commercial seed dealers, the actual process of testing followed a set pattern. After preparing a representative sample of a given stock, a seed analyst determined the genuineness of the sample – that it was indeed seed of the indicated crop species, originating from the indicated country or region – and its purity. The latter involved her examining a subset of seeds from the sample, say 1,000, and separating whole healthy seeds of the desired crop from dirt, straw, and seeds of other species to arrive at a percentage of pure seed. She would then find the average weight of the seed lot, possibly assess its moisture content, and make a final assessment of quality by testing its germination rate. The results of these assessments were then compiled, often on standardised forms, and provided to whomever had requested the test.[20]

19 For an early-twentieth-century account of the operation of several of the earliest European stations, see 'Seed Control Stations on the Continent', *Journal of the Board of Agriculture*, suppl. no. 13 (August 1914). For recent chronicles of the early international history of seed testing, see A. M. Steiner and M. Kruse, 'Centennial – The 1st International Conference for Seed Testing 1906 in Hamburg, Germany', *Seed Testing International: ISTA News Bulletin*, no. 132 (October 2006), pp. 19–21; M. Muschick, 'The Evolution of Seed Testing', *Seed Testing International: ISTA News Bulletin*, no. 139 (April 2010), pp. 3–7; and A. M. Steiner, M. Kruse, and N. Leist, 'The 1st Meeting of the Directors of Seed Testing Stations in Graz, 1875', *Seed Testing International: ISTA News Bulletin*, no. 142 (October 2011), pp. 29–32.
20 A comparative account of the seed-testing procedures at several European seed-testing stations can be found in 'Seed Control Stations on the Continent'.

Figure 10.2 A seed analyst weeds out the impurities from a sample of red clover. From B. O. Longyear, 'Seed Testing for Farmers', *Michigan State Agricultural College Experiment Station Bulletin*, no. 212 (April 1904), p. 4. Widener Library, Harvard University, HD Sci 1635.15.3.

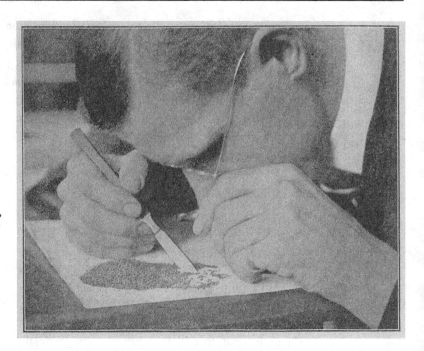

Some aspects of seed testing were demanding of an analyst's patience and knowledge, not least the sorting of hundreds of minute seeds and their subsequent identification (Figure 10.2). Everyone agreed that a seed herbarium – that is, a collection of identified seed specimens – was an essential aid to the latter, even for the most knowledgeable botanists. These reference collections were often assembled over time, through aggregation and exchange among individual researchers or collectively at botanical institutions. And there was a fair amount of equipment beyond a herbarium involved in seed testing. As an 1895 textbook specified, the 'very simple' necessary equipment included 'a small magnifying glass, some sieves of various grades, bellows, forceps, delicate scales, thermometers, jars, test-plates, chemical tests, and a good knowledge of botany'.[21] At the large seed-testing stations, equipment grew significantly more complex, especially over time, and microscopes, mechanical separators, incubators, germinators, and other devices increasingly crowded laboratories' spaces.[22] Still, the processes of testing were

21 W. J. Beal, *Grasses of North America* (Lansing: Thorpe & Godfrey, 1887), pp. 208–9.
22 See, for example, images in 'Seed Control Stations on the Continent'; see also the seed-testing equipment described in J. S. Remington, *Seed Testing* (London: Sir Isaac Pitman & Sons, 1928) or offered for sale in scientific catalogues such as

straightforward enough – save one. Even with all the best equipment, determining the place of origin of a seed stock could be a real pain.

The chief obstacle to identifying a seed's origin was that seeds of the same species from different countries or regions or even continents did not necessarily look any different from one another. Yet they were decidedly not the same. When cultivated in a particular area over a long period of time, crops become locally adapted, responding to the climate and soils of the place where they are grown. As a result, seed harvested from one location might not perform as well in another, and in some cases might not grow at all. For example, seeds from a southern latitude might fare poorly at a northern one, as a result of shorter seasons or colder winters. A Canadian agronomist summarised in 1925 what was by then a commonplace: 'The superiority of home grown seeds over imported seeds has been demonstrated in nearly all countries and for so many kinds of crops that enumeration of the experiments would lead too far.'[23]

Unfortunately, the vagaries of harvests from year to year and the inevitable roller-coaster of agricultural prices led to uneven seed supply and demand, which in turn meant that seeds often came from far afield. This was especially true in the case of forage and fodder crops, as farmers were less likely to produce seed for these on-farm.[24] The general lack of attentiveness to the quality of forage crop seeds compounded the problems created by a shortage of locally produced seed. The result? An anything-goes international market in which the origin of forage crop seeds was often misrepresented. The Swiss agronomist Friedrich Stebler characterised the problem in his

A. Gallenkamp & Co., Ltd, *Catalogue of Chemical and Industrial Apparatus*, 9th edition (1931), Whipple Museum (Gall.11) or Chas. Hearson & Co., Ltd., *Hearson's Apparatus* (1930–1), Whipple Museum (Hea.3). According to Dominic Berry, excellent resources for tracing the history of seed testing, including methods and instrumentation, can be found at the National Institute of Agricultural Botany (NIAB), Cambridge, England. See his work on the history of seed testing and the management of synonymy at NIAB: D. Berry, 'The Plant Breeding Industry after Pure Line Theory: Lessons from the National Institute of Agricultural Botany', *Studies in History and Philosophy of Biological and Biomedical Sciences*, 46 (June 2014), pp. 25–37.

23 F. T. Wahlen, 'The Determination of the Origin of Agricultural Seeds with Special Reference to Red Clover', *Scientific Agriculture*, 5.12 (1925), pp. 369–74, on p. 369.

24 On the forage seed market in England in the nineteenth century, see M. Ambrosoli, *The Wild and the Sown: Botany and Agriculture in Western Europe: 1350–1850*, translated by M. M. Salvatorelli (Cambridge: University of Cambridge Press, 1997), Chapter 7.

country with respect to red clover in the 1880s: 'American seed, for example, is often sold as English or German seed ... American is inferior to European seed; as it is cheaper, it is often advantageous for the seedsman to substitute the former for the latter.'[25] Across the Atlantic, the perspective was the reverse. Canadians needed seed testing as protection against 'imported southern grown seeds' coming from Europe and the United States, meanwhile US farmers were advised to use American-grown seed rather than European.[26]

The problem of origins resulted in a kind of seed nationalism, one exacerbated by knowledge of further harms arising from imports. A lacklustre harvest might be only the start. Because poor-quality forage seeds were often badly cleaned or even deliberately bulked with the detritus left after cleaning other crops, they were a chief source of weeds. Americans knew well that the most aggressive weeds tended to have arrived from abroad. 'Nearly all of our worst weeds are of European origin, and by far the greater part of them have been introduced into American soil through impure seed', claimed a typical rant of a US agronomist.[27] This particular researcher thought the problem had been made worse by seed regulations – specifically, by the imbalance in these between the United States and Europe. 'While seed-control agitation in Europe has resulted in a marked improvement of home stocks, it does not prevent the shipment of poorly cleaned seed to other countries,' he explained. '[A]s a result a large proportion of our inferior seed comes from abroad.'[28] Though perhaps more exercised on the issue, Americans were not alone in their concerns about the introduction of new weeds through imported seeds. The transfer of dodder seeds from one part of the globe to another was a near-universal concern, in part because everyone already knew the local kind of these parasitic plants to be a real pain.[29]

25 F. G. Stebler and C. Schröter, *The Best Forage Plants, Fully Described and Figured*, translated by A. N. McAlpine (London: David Nutt, 1889), p. 132.
26 Wahlen, 'The Determination of the Origin of Agricultural Seeds with Special Reference to Red Clover', p. 369; A. J. Pieters, 'Red Clover Seed's Origin Is Important', in *USDA Yearbook of Agriculture 1927* (Washington: US Government Printing Office, 1928), pp. 627–9; and A. J. Pieters and R. L. Morgan, 'Field Tests of Imported Red-Clover Seed', *USDA Circular*, no. 210 (February 1932).
27 Hicks, 'Pure Seed Investigation', p. 390.
28 Hicks, 'Pure Seed Investigation', p. 390.
29 This is evident in almost any discussion of seed testing or forage crop cultivation dating from this period. See, for example, A. D. Selby and J. F. Hicks, 'Clover and Alfalfa Seeds: Their Purity, Vitality and Manner of Testing', *Bulletin of the Ohio Agricultural Experiment Station*, no. 142 (June 1903); E. Brown and F. H. Hillman, 'Seed of Red Clover and Its Impurities', *Farmers' Bulletin*, no. 260

Foreign seeds, in short, portended various farm disasters. But how could a seed analyst discover potential immigrant stocks circulating amidst those native-born, when the seeds themselves looked nearly identical?

Here is where the worthless, and sometimes worse-than-worthless, weeds proved their merit. Although crop species had globalised, and some weeds, too, many common weed species remained geographically circumscribed. Where this was the case, the presence of their seeds amidst a stock of crop seeds could be used to identify the region of the world, or in some cases the part of a country, in which that stock had been produced. Friedrich Stebler in Switzerland expanded this general insight into a system for origin identification in the 1880s and 1890s using his meticulous observations of the weed seeds that accompanied stocks to his seed testing station in Zurich. 'Source indicators' were weed seeds that he felt faithfully linked a tested seed to some world region, such as Southern Europe or North America. What he called 'companion seeds' gave some insight into origins, though their presence was not sufficient to confirm it.[30]

Useful Weeds?

Stebler's method (though not necessarily his terminology), soon became standard practice for identifying place of origin in laboratory evaluations of seed.[31] The Whipple Museum's herbarium was created to facilitate such identification, either by its use in direct comparisons or in training seed analysts to recognise different species. It contains only those weed seeds considered to be so-called source indicators or companion seeds, and not the whole gamut of weed species that would be expected to emerge from sacks and samples amidst routine testing. More typical seed herbaria kept for

(Washington: US Department of Agriculture, 1906); Johnson, 'The Principles of Seed-Testing'; and 'Seed Control Stations on the Continent'.

30 See the explanation in Harshberger, *Text-book of Pastoral and Agricultural Botany*, pp. 266–7. Stebler opened the first-ever international conference on seed testing with a discussion of this work. See the report in H. Th. Güssow, 'International Seed-Testing Conference at Hamburg, 1906', *Journal of the Royal Agricultural Society of England*, 67 (1906), pp. 265–7.

31 For a discussion that refers to 'leading' and 'accessory' species (rather than source indicators and companion species), and mention of the uptake of weed surveys to discover seed provenance, see F. T. Wahlen, 'A Survey of Weed Seed Impurities of Agricultural Seed Produced in Canada, with Special Reference to the Determination of Origin', *Proceedings of the International Seed Testing Association*, no. 3 (1928), pp. 19–60.

Figure 10.3 This 1906 reference collection of 'economic plants' includes both crop and weed seeds of Canada. Reproduced courtesy of the Nova Scotia Museum Botany Collection, Nova Scotia Archives (Harry Piers accession number 3058).

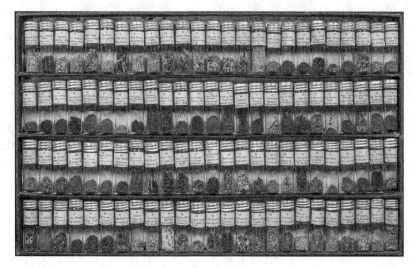

use at agricultural research institutions, including seed-testing stations, would contain examples of agricultural crops as well as weeds, the latter regardless of whether they could be used as source indicators (Figure 10.3).[32] By virtue of including a wide range of plant material, local and global, these herbaria were (and are) suited to dealing with varied needs of researchers and farmers. In contrast, the Whipple's seed herbarium was tailored to a single task: the identification of a seed's place of origin.

The deployment of weeds as instruments of seed testing turned some of the most common definitions of these notoriously pesky plants on their head. The noxious, useless plants out of place were now wanted, beneficial, and perfectly positioned – at least some of the time. The contradictions inherent in appreciating weeds as contributors to agricultural efficiency surfaced almost immediately. Seed testing prized purity. The best stocks were free of seeds from other species, whether crop or weed, and of other debris that might add to their weight and therefore their price without adding to the value of the future crop. But a well-cleaned bag of seed, earning high marks for purity, contained few weeds, and therefore its origins were more difficult to certify. By the 1920s, very clean seed could be construed as an obstacle for seed analysts, as much as a goal, as it necessitated the

32 On herbaria as essential tools of seed testing, see, e.g., Hicks, 'Pure Seed Investigation', p. 408; Remington, *Seed Testing*, p. 12; S. P. Mercer, *Farm and Garden Seeds* (London: Crosby Lockwood & Son, 1938), p. 110; and US Department of Agriculture, *Manual for Testing Agricultural and Vegetable Seeds*, Agricultural Handbook no. 30 (Washington: US Government Printing Office, 1952), p. 20. For a representative guide to weed seeds, see Remington, *Seed Testing*.

development of new methods for determining origins. 'The combined result of modern improvements in cleaning machinery and the introduction of the Testing of Seeds Order [mandating assessment of commercial seeds prior to sale] is ... for cleaner and cleaner samples to be put on the market as time goes on. It was, therefore, decided to start investigations with a view to establishing, if possible, a "country of origin test" other than "impurity",' explained a British botanist who was leading investigations into what he called a 'Nationality Test' for red clover.[33] Where weeds were wanted, purity was a problem.

Even if improvements in seed cleaning had not been a concern, the method of determining the origin of crop seeds by virtue of the weeds that travelled with them was far from foolproof. The problem was movement. An important feature of the species frequently identified as weeds is that they turn up in new places unbidden, often much to the chagrin of cultivators. Some of the travelling of weeds that caused a problem for seed analysts was small-scale and human-engineered. Already classed as an impurity in the seed-testing world, seeds of reliable source indicator species were soon also being understood as potential adulterants. Savvy but unsavoury seed merchants or wholesalers could deliberately add 'certain weed seeds suggestive of an origin heavily in demand' with the purpose of misleading buyers or testers.[34]

The more critical undermining of Stebler's source indicator method did not result from these premeditated movements, however, but rather from those more inherent to weed species. As I have already described, these plants out of place are notorious globetrotters, hardy travellers whose cross-country and cross-continent journeys are typically and often unknowingly facilitated by human companions.[35] It was utterly reasonable to assume, therefore, that weeds that were still comparatively provincial would prove themselves more cosmopolitan in time. Their transport was already arranged: easy transit from a Russian farm to a Canadian or Swiss

---

33 R. G. Stapledon, 'Seed Studies: Red Clover with Special Reference to the Country of Origin of the Seed', *Journal of Agricultural Science*, 10.1 (1920), pp. 90–120, on p. 91. The Testing of Seeds Order was a 1917 British mandate that seeds of many agricultural crops be tested prior to sale. For a history of seed testing in Britain, see D. Berry, 'Agricultural Modernity as a Product of the Great War: The Founding of the Official Seed Testing Station for England and Wales, 1917–1921', *War & Society*, 34.2 (2015), pp. 121–39.

34 Wahlen, 'The Determination of the Origin of Agricultural Seeds with Special Reference to Red Clover', p. 370.

35 Perhaps the most famous world-travelling weeds are those that accompanied European colonists; see Crosby, *Ecological Imperialism*.

or British one via the boats and trains and carts and sacks of the international seed market. 'The opinion has often been expressed that the leading species will in time become worthless as clues [to origin] because of the international trade in seeds,' reported a Swiss agronomist in 1925. Curiously, he expressed his scepticism of this 'opinion' but then enumerated several cases in which weeds had been rendered useless precisely because they had become 'too cosmopolitan'.[36] The propensity of weeds to behave as weeds so famously do meant that using these to gauge a seed stock's place of origin had to be done more carefully with each passing year. A British seed manual of 1938 advised that, '[I]n view of the international traffic in seeds, which has been ongoing for generations, weeds have been transported to new areas all over the globe'. As a result, they were only rarely 'diagnostic' of origin. The seed analyst could, however, still 'form a shrewd opinion of origin from the *profusion* of some characteristic species'.[37] Considering the quantity of weed seeds and their particular combinations was advised, rather than simply looking for the presence of any one species. Weed peregrinations also necessitated that the lists relied upon as indicators be thought of as ever-changing, rather than set in stone.[38]

In short, transforming weeds into reliable laboratory instruments proved difficult. Seed testing created a purpose for useless plants, a desire for the undesirable. In spite of this re-categorisation, however, weeds continued to be weeds. The wild and woolly seed trade they were meant to tame instead encouraged their own unruly behaviour. Just like the crops they travelled with, they set down roots in new places – but, unlike those crops, they had not been invited to do so. Although we still know few specifics about the Whipple Museum's 'Origin of Seeds Source Indicators' herbarium, it seems safe to speculate that it had a limited lifespan. It ultimately served more as a snapshot of global agricultural history at the time of its creation than as an enduring tool for assessing the products of global agriculture.

36 Wahlen, 'The Determination of the Origin of Agricultural Seeds with Special Reference to Red Clover', pp. 370–1.
37 Mercer, *Farm and Garden Seeds*, pp. 77, 97. Emphasis in the original.
38 US Department of Agriculture, *Manual for Testing Agricultural and Vegetable Seeds*, p. 176. Changing ideas about the reliability of weed seeds as indicators of origins can also be traced through the *Proceedings of the International Seed Testing Association* and its rules for seed testing. A helpful mid-century summary of the precautions to be taken in using weeds as indicators of origin can be found in *Proceedings of the International Seed Testing Association*, 18.1 (1953), pp. 1–69, on p. 38.

11 ✳ What 'Consul, the Educated Monkey' Can Teach Us about Early-Twentieth-Century Mathematics, Learning, and Vaudeville[1]

CAITLIN DONAHUE WYLIE

The values of an era are embedded in objects. They can be difficult for those of different generations to interpret or even recognise; nonetheless, objects offer silent witness to bygone cultural moments. Here I investigate the tacit and explicit meanings of an object in the Whipple Museum's collection that is at once a mechanical calculator and a depiction of a monkey. This unusual amalgamation offers us a window into the world that made and used it, including how people thought about mathematics, education, and childhood.

"Consul", the Educated Monkey' is a 14 cm × 15 cm metal backplate printed with a number chart, to which is attached a moveable monkey figure, wearing a suit and bow tie and made of thin enamelled steel (Figure 11.1).[2] If you position the monkey's feet to point at two numbers in a row of 1–12, let's say 4 and 8, then the monkey's metal-pin joints force its arms and torso to change position until its hands cradle the two numbers' product printed on the backplate: 32. The device groans as you gently coax its resistant feet into position, but the rest of Consul's body moves with the precise coordination of a dancer. Its gritted-teeth smile doesn't move. The monkey's pose can be rather gymnastically grotesque for certain calculations. As a fun surprise, as Consul's head approaches its ankles to multiply far-flung numbers such as 1 and 12, the metal

---

1 Aspects of this chapter first appeared in C. Wylie, 'Monkeys and Mathematics in Twentieth-Century America', in Kelley Swain (ed.), *The Rules of Form: Sonnets and Slide Rules* (Cambridge: Whipple Museum of the History of Science, 2012), pp. 55–81.
2 Accession number Wh.5821. The Whipple also owns Consul's instruction list (printed on a paper envelope that packaged the object) and a paper-printed addition table, which temporarily converts Consul into an adding machine instead of a multiplier and includes further instructions on the reverse side (see Figure 11.3 later in this chapter). Missing from the collection is a printed description of a game called Multe, which has been preserved with a few other Consuls.

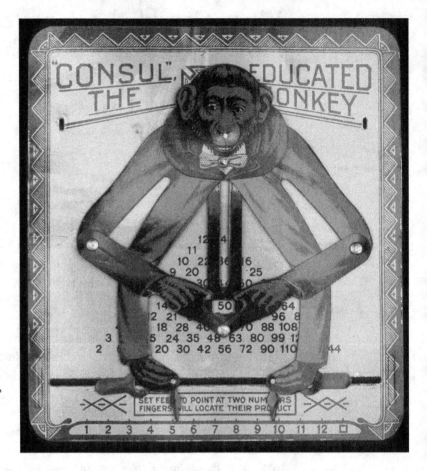

Figure 11.1 Consul, the Educated Monkey. Image © Whipple Museum (Wh.5821).

bar connecting its hands and torso sticks up over the monkey's head, transforming into a strikingly realistic tail painted with brown and black furry stripes. As an additional touch of charm, next to the number 12 is a small empty square, signifying the 'square' of the number chosen for the other foot.

The Whipple has displayed this grinning simian beside mechanical calculating devices, such as Napier's bones and abacuses, and included it in the 'Science Toys Trail' activity, illustrating how this object exemplifies several categories of things. This object has always been a hybrid; even its inventor patented it as both a calculator and a plaything. William Henry Robertson filed for two US patents in 1915: the first patent was for 'certain new and useful Improvements in Calculating Devices' and the second, filed six months later, was for 'certain new and useful Improvements in Toys'.[3] He assigned the

---

3 W. H. Robertson, 'Toy, 1,188,490', US patent, issued 1916; and W. H. Robertson, 'Calculating Device, 1,286,112', US patent, issued 1918.

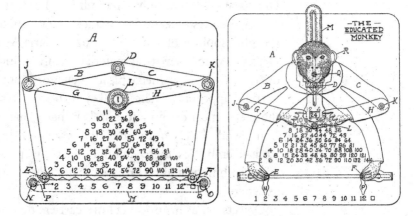

Figure 11.2
Robertson's
'Calculating Device'
patent diagram (left,
granted 1918) and
'Toy' patent diagram
(right,
granted 1916).

rights to the inventions to a company he founded, the Educational Novelty Company, in Dayton, Ohio. Robertson's toy patent cleverly incorporated his earlier calculating device to create an unusual hybrid, as he explains:

> A number chart has never before been combined with any object resembling and representing a living creature which can be adjusted in relation to the chart to perform computations, thereby suggesting the idea of a calculating animal.[4]

The US Patent Office issued the toy patent first, on 27 June 1916, followed by the calculating device patent on 26 November 1918 (Figure 11.2).[5]

In addition to solutions to arithmetic problems, Consul offers a glimpse into twentieth-century views about education. Its invention coincides with the rise of public education and Progressivism in the

---

4 Robertson, 'Toy, 1,188,490'.
5 Consuls were manufactured probably beginning in 1915, because the earliest models have 'patent applied for' printed on the metal behind the monkey's head. A second version was likely to have been manufactured between the two patent dates, because the text reads 'PAT. JUNE 27, 1916/OTHER PATENTS PENDING'. A third version lists both patent dates. (See W. Denz, 'Rechenaffen – Educated Monkeys', *Rechnen ohne Strom* (2018), www.rechnen-ohne-strom.de/ rechner-galerie/tabellen-rechenhilfen/rechenaffen/. Denz wrote this website about his private collection of historic calculators.) The Whipple's Consul is the third version, which means it was made in late 1918 at the earliest. Consuls are all the same in shape and materials, except for the differences in the printed patent dates, the colour of the monkey's suit (the first version is blue), and where the name of the manufacturer is printed (on the metal below the monkey's feet in the first version versus on the separate paper instructions in the second and third versions).

United States.[6] Its use by one individual at a time fits the contemporary theories of learning as hands-on and child-directed, based on the philosophies of Johan Pestalozzi, Maria Montessori, and John Dewey.[7] As mathematical material culture, it includes the manually operated design of mechanical calculators[8] but with a consumer-friendly appearance. The monkey distracts from the complexity of the mechanical design and number charts that achieve the calculations, while bringing to mind the whimsy of performing animals. Consul embodies these cultural trends by providing fun ways to learn arithmetic and by bringing mathematics into the home as well as school.

Objects act as historical culture-carriers that offer unique insights into past lifestyles. Lorraine Daston asserts that 'thinking with things is very different from thinking with words, for the relationship between sign and signified is never arbitrary – nor self-evident'.[9] Everything about objects is significant, but interpreting what exactly is 'signified' requires in-depth understanding of objects' contexts. According to Samuel Alberti, objects must be placed in their world of people and ideas, as productions of a culture and not as isolated items.[10] With these approaches in mind, a calculating monkey becomes a guide to the mathematics, education, and playthings of a century-old culture. Unfortunately, a lack of recorded contemporary information, such as about how the object was designed, manufactured, and used, makes Consul's physical construction and accompanying packaging the best available sources, as is often the case in object studies.[11]

6 J. Kilpatrick, 'Mathematics Education in the United States and Canada', in A. Karp and G. Schubring (eds.), *Handbook on the History of Mathematics Education* (New York: Springer, 2014), pp. 323–34.

7 D. L. Roberts, 'History of Tools and Technologies in Mathematics Education' in A. Karp and G. Schubring (eds.), *Handbook on the History of Mathematics Education* (New York: Springer, 2014), pp. 565–78.

8 P. A. Kidwell, A. Ackerberg-Hastings, and D. L. Roberts, *Tools of American Mathematics Teaching, 1800–2000* (Baltimore: Johns Hopkins University Press, 2008).

9 L. Daston (ed.), *Things That Talk: Object Lessons from Art and Science* (New York: Zone Books, 2004), p. 20.

10 S. J. M. M. Alberti, 'Objects and the Museum', *Isis*, 96 (2005), pp. 559–71, on p. 561.

11 Several Consuls survive, including in the collections of London's Science Museum, Chicago's Adler Planetarium, the Smithsonian National Museum of American History in Washington, the Computer History Museum in Mountain View, California, and the Strong National Museum of Play in Rochester, New York.

On the basis of the object itself, its printed instructions, and the two patents granted for its design, '"Consul", the Educated Monkey' is at once a calculating device, an educational aid, and a toy. How successfully did Consul fulfil these diverse purposes in the United States in the early twentieth century? I begin by exploring the object's technical and mechanical meanings, then I analyse how it fits its context of educational theory and practice, and finally I consider its playfulness and cultural references, such as the evolutionary connotations of monkeys and the widespread popularity of vaudeville. This order matches Consul's own development, on the basis of the chronology of its patents. It also matches typical object-study methodology, by first assessing the object as physical evidence and then investigating its more elusive cultural context. I suggest that Consul bridged the boundaries between school and home, work and play, and adulthood and childhood, making the red-suited calculating monkey a valuable informant about early-twentieth-century American culture.

Consul as Calculator

Consul is an efficient mechanical calculator. Its geometric design and clever mechanism suggest a creative and mathematically skilled inventor. After teaching high-school mathematics in Texas, Robertson moved to Dayton, Ohio, where by 1910 he was working as a draftsman and later as a designer for the National Cash Register Company.[12] He also founded the apparently short-lived Educational Novelty Company (1915–17).[13] Robertson designed his calculating

---

12 P. A. Kidwell, 'Consul the Educated Monkey, or the Inventions of William H. Robertson', O Say Can You See? Stories from the National Museum of American History, 2015: http://americanhistory.si.edu/blog/consul-educated-monkey-or-inventions-william-h-robertson (accessed 28 August 2018).

13 Kidwell, 'Consul the Educated Monkey'. There is evidence of at least three production companies, suggesting that Consul was mass-manufactured. These companies probably licensed Robertson's patents, because the objects are so similar to each other and to Robertson's descriptions and drawings. Unfortunately, the manufacturer's name is recorded only on the paper-printed instruction list for the second and third versions of Consul. Paper components are liable to become separated from objects and lost, and I have found only a few surviving examples in online photographs of Consuls from auction houses, private collections, and museums' databases. Luckily, the Whipple has an almost complete set of affiliated papers for Consul, which identifies TEP Manufacturing in Detroit, Michigan, and its factory in Dayton, Ohio, as the object's manufacturer. I have found no information about this company and no other Consuls with a TEP imprint, suggesting that it was not a major producer.

device to make arithmetical number charts easier to use, as he explained in his 1918 patent: 'To provide a quick and simple method of finding results on the chart by mechanical means thereby relieving the eye of having to follow columns and of making mistakes by locating results at the wrong intersections.'[14] The device makes arithmetic more accessible and reliable by simplifying chart-reading.

Both Robertson's calculating device and toy patents contain terms and conventions from Euclidean geometry, reflecting his expertise as a maths teacher. Mathematical terms describe the object's design: 'The product of any two numbers of the series lies on an imaginary line which is the perpendicular bisector of another such line connecting the two numbers of the series.'[15] This sentence is found in both patents. Robertson uses more mathematical language in his calculating-device patent (perhaps because this patent and device are intended for a more mathematical audience), writing that the device's ability to locate correct answers 'can be proven by geometry'.[16]

Robertson's mathematical language is striking when compared with the descriptive language of a British patent for 'a device or calculator for multiplying, dividing, adding and subtracting integers' that is an exact replica of Consul, granted on 2 December 1918.[17] This patent, taken out by Charles Allaun of Leeds, uses inexact and non-technical language to describe 'a jointed figure' on the device, which in the diagrams is a monkey: 'The legs of the figure are wide apart and disposed at an obtuse angle, the head is almost in a line with the body, the arms are bent at an angle and the forearms rest on the knees.' A more detailed but not more mathematical description follows: 'The above described jointed figure is mounted on the base plate *a* by means of studs *q* secured to the feet *r* of the figure and engaging with a slot *s* formed in the base plate.'[18] Allaun refers to diagram points with lower-case, italicised letters, unlike the upper-case letters used in geometric proofs and in Robertson's patent. Neither does Allaun mention exact spatial dimensions or the geometric relationships of the device's parts. These differences, and the apparent lack of connection between Allaun and Robertson, suggest

14 Robertson, 'Calculating Device, 1,286,112'.
15 Robertson, 'Calculating Device, 1,286,112'; and Robertson, 'Toy, 1,188,490'.
16 Robertson, 'Calculating Device, 1,286,112'.
17 C. Allaun, 'A device or calculator for multiplying, dividing, adding and subtracting integers, 120,985', UK patent, issued 1918.
18 Allaun, 'A device or calculator for multiplying, dividing, adding and subtracting integers, 120,985'.

that Allaun may have claimed a British patent by describing an American-manufactured Consul itself, as he didn't copy the language or diagrams in Robertson's patent. Thus, Robertson's Euclidean language and mathematical descriptions, as well as his original invention of a number-chart-reading device, suggest that he was first a mathematical inventor and later applied these skills to the toy industry, especially since he applied for the calculating-device patent before the toy patent (though the toy patent was granted first).

Detailed instructions on how to multiply, divide, factor, add, and subtract using Consul present it as a do-it-all calculating machine. Conducting these operations using Consul is ostensibly so easy that a child can do it (though Consul's fragility makes this rather unrealistic). Mass manufacture of its simple design and small size would have been cheap and efficient, making Consul accessible to adults who wanted to simplify their use of arithmetic number charts without investing in a newfangled expensive calculating machine.[19] Thus, Consul offered an easier version of number charts and a cheaper, though perhaps less dignified, alternative to mechanical calculating machines designed for adults.

Consul as Teacher

Although Robertson's second patent is for a 'toy', he specifies that 'my invention relates to toys for educational purposes.'[20] Specifically, this invention 'is intended to interest the child and increase his knowledge of numbers and number tables'.[21] Irrespective of whether educational toys were playthings that claimed to teach or learning tools that imitated playthings, there was a growing market for them in early-twentieth-century United States. This market is evident in the name of the company that manufactured most of the surviving Consuls: The Educational Toy Manufacturing Co. of Springfield, Massachusetts. Robertson's company, the Educational Novelty Company of Dayton, Ohio, is the assignee of the two patents. But was Consul a good 'educational' object according to philosophies of learning at the time?

Consul meets the contemporary demand for manipulative, hands-on objects as physical representations of knowledge that children

19 Kidwell, Ackerberg-Hastings, and Roberts, *Tools of American Mathematics Teaching*, p. 246.
20 Robertson, 'Toy, 1,188,490'.
21 Robertson, 'Calculating Device, 1,286,112'.

could interact with directly. The importance of learning through experience, inquiry, and physical objects was stressed by education scholars such as J. H. Pestalozzi in early-nineteenth-century Switzerland, Maria Montessori in early-twentieth-century Italy, and John Dewey in the late-nineteenth- and early-twentieth-century United States. Visual and physical ways of learning arithmetic grew in popularity, exemplified by abacuses in nineteenth-century classrooms. In the late nineteenth century, 'tools for group demonstration such as the teaching abacus gave way to devices for a single child.'[22] By 1900, 'the material used in the schoolroom as objective aids is limited and highly artificial, consisting of tiles, pegs, splints, toothpicks, squares of cardboard, etc.'[23] These simple objects served as counters to help children learn the physical meaning of arithmetic.[24]

Consul matches this trend of single-user educational tools, but it does not serve the accompanying educational philosophy. Pestalozzi, Montessori, and Dewey called for hands-on activities to encourage children to ask their own questions and explore their world directly.[25] Likewise, and in contrast to the previous pedagogy of memorisation and recitation, educator David Smith argued in his 1913 book *The Teaching of Arithmetic* that mathematics learning should be more active for students:

> There has for a century been a tendency away from what is called the direct method of imparting number facts, and toward the rational method. This means that instead of stating to a class that 4 + 5 = 9, and drilling upon this and similar relations, the schools have generally tended to have the pupils discover the fact and then memorize it. The experience of a century shows that this tendency is a healthy one.[26]

Smith believes this 'rational' method is successful because it gives a student the freedom to think: 'A child likes to be a discoverer, to find out for himself how to add and multiply, always under the skillful

22 Kidwell, Ackerberg-Hastings, and Roberts, *Tools of American Mathematics Teaching*, p. 139.
23 D. E. Smith, *The Teaching of Arithmetic* (Boston: Ginn and Company, 1913), p. 46.
24 Kidwell, Ackerberg-Hastings, and Roberts, *Tools of American Mathematics Teaching*, p. 139.
25 H. G. Good and J. D. Teller, *A History of American Education*, 3rd edn (New York: Macmillan, 1973), p. 337; R. L. Church and M. W. Sedlak, *Education in the United States* (New York: Free Press, 1976), p. 261; and L. A. Cremin, *The Transformation of the School: Progressivism in American Education, 1876–1957* (New York: Knopf, 1961), p. 141.
26 Smith, *The Teaching of Arithmetic*, p. 43.

guidance of the teacher.'[27] Educational mathematics objects were tools, not answer-givers. Counters had to be counted accurately, and rulers and protractors only provided measurements to serve as the basis of a student's further calculations, such as of area and volume. A child must understand how these tools work and how to interpret the information they yield to reach a correct answer.

Consul's packaging personifies an ideal teacher because 'it makes no difference to the monkey whether children are bright or stupid. He never loses patience at having to answer their questions,' suggesting that children 'discover' and learn from Consul by asking 'him' questions. But Consul does not allow such 'discovery' of arithmetical principles because it can answer only a very limited type of question, not the open-ended investigative questions that twentieth-century educators believed led to true learning. Specifically, after the user has pointed the monkey's feet at appropriate numbers, the answer appears without further user input or thought-processing. Consul thus yields answers regardless of the user's knowledge of mathematics. The user must only arrange the monkey correctly: 'To multiply, adjust each of the monkey's feet to point directly at a number. The monkey's fingers will then locate the product of the two numbers.' These directions imply that the monkey itself carries out the calculation. Similarly, to subtract the user must move one monkey foot and one monkey hand to the numbers in question, and then 'the other foot will be found pointing at the difference.' The answer is literally pointed to, and a child learning arithmetic would have no idea how that answer was reached. Robertson even writes in his 1916 patent that 'the idea of a calculating animal' would hold a child's attention; he does not mention Consul's innovative mathematical design as appealing for young users.[28]

Nevertheless, the packaging presents Consul as teacher as well as calculator: 'It teaches the complete multiplication table. It teaches the complete addition table. It can add, subtract, multiply, divide, or factor elementary numbers.' The abilities to teach and to do arithmetic are merged, suggesting an underlying pedagogy in which giving answers to arithmetic problems is the same as teaching arithmetic. Thus, it seems the device could not help children understand arithmetic facts or learn mathematics according to contemporary pedagogy, but could only provide correct answers – the opposite of Smith's 'rational method'.

27 Smith, *The Teaching of Arithmetic*, p. 43.
28 Robertson, 'Toy, 1,188,490'.

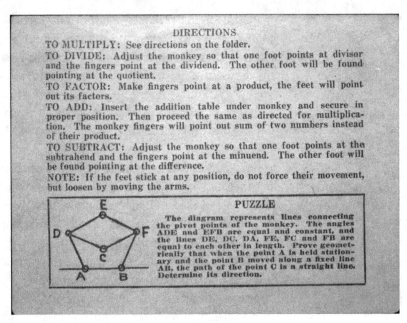

DIRECTIONS

TO MULTIPLY: See directions on the folder.
TO DIVIDE: Adjust the monkey so that one foot points at divisor and the fingers point at the dividend. The other foot will be found pointing at the quotient.
TO FACTOR: Make fingers point at a product, the feet will point out its factors.
TO ADD: Insert the addition table under monkey and secure in proper position. Then proceed the same as directed for multiplication. The monkey fingers will point out sum of two numbers instead of their product.
TO SUBTRACT: Adjust the monkey so that one foot points at the subtrahend and the fingers point at the minuend. The other foot will be found pointing at the difference.
NOTE: If the feet stick at any position, do not force their movement, but loosen by moving the arms.

PUZZLE

The diagram represents lines connecting the pivot points of the monkey. The angles ADE and EFB are equal and constant, and the lines DE, DC, DA, FE, FC and FB are equal to each other in length. Prove geometrically that when the point A is held stationary and the point B moved along a fixed line AB, the path of the point C is a straight line. Determine its direction.

Figure 11.3 The reverse side of the Addition Table included in Consul's packaging, offering 'DIRECTIONS' and the 'PUZZLE' of Consul's geometric design. Image © Whipple Museum (Wh.5821).

However, the magic or perhaps the 'education' behind Consul's ability to calculate is revealed in an included 'mathematical puzzle which has advanced students guessing' (Figure 11.3). This puzzle presents Consul's construction as a geometric diagram of Consul's joints with information about angle congruence and segment length, whose relationships a student must prove. Although the answer is not provided, phrasing Consul's construction as a geometry problem reveals the object's seemingly mysterious calculating ability as merely an application of familiar mathematics, at least for advanced students. I have found no records of early-twentieth-century users' impressions of Consul; however, twenty-first-century mathematics teachers recommend that high-school students build their own Consuls with paper and metal fasteners.[29] This hands-on activity requires students to apply geometry and algebra as well as arithmetic. As an embodied geometric proof, Consul also appeals to today's hobbyist model-builders, who replicate the device using various media.[30] It is possible that teachers taught the same activity

29 S. J. Kolpas and G. R. Massion, 'Consul, the Educated Monkey', *The Mathematics Teacher*, 93 (2000), pp. 276–9.

30 For example, Fischertechnik construction pieces (D. Fox, '"Consul", the Educated Monkey', *ft:pedia*, Heft 1/2015 (2015), pp. 19–24, www.ftcommunity.de/ftpedia_ausgaben/ftpedia-2015-1.pdf) and a virtual Consul ('consul', tan-gram (2018), www.tan-gram.de/consul.pl).

a century ago. Consul thus can serve as a geometric proof as well as a calculator, though the device's workings were most likely to have been a black box to young users.

Calling the geometric proof a 'puzzle' presents it as fun, an important component of learning according to Pestalozzi, Montessori, Dewey, and Smith. Pestalozzi advocated mathematics lessons even for young children, challenging the norm of introducing arithmetic only at age twelve. According to Smith, Pestalozzi 'concealed the drill under the guise of play, but play with a definite purpose' so as to interest young children.[31] Montessori designed 'materials' to guide children to understand arithmetic, such as labelled boards on which to arrange beads, which children found so engaging that they begged to take them home. One youngster reportedly threatened that, "'Unless she gives us the material for the multiplication table we won't come to school any more.'" Montessori took this demand as a sign of success: 'The multiplication table, the bug-bear of all children, had become so attractive and tempting a thing that it had made wolves out of my lambs!'[32] The right materials thus had the power to make dreaded topics enjoyable, even if children acted rudely in their efforts to access them. Smith also valued students' enjoyment. He filled twenty pages of his book with 'number games for children', including competitions and interesting 'tricks' about arithmetic to stimulate students' interest in mathematics.[33]

The inclusion of 'an entertaining and instructive game for children' in Consul's packaging matches this trend.[34] This game, called Multe, an acronym for 'Many Useful Lessons Taught Enjoyably' and perhaps also short for 'multiplication', involves children competing for accuracy in arithmetic facts. After all of the players have completed ten facts, 'each player who is in doubt as to whether the answer is correct is allowed to consult his monkey' (with 'consult' perhaps being a play on Consul's name). Consul is the answer-checker. The instructions seem to expect that Multe will be played in a classroom. These directions reflect the Progressive ideology of

31 Smith, *The Teaching of Arithmetic*, p. 107.
32 M. Montessori, *The Montessori Elementary Method* (New York: Frederick A. Stokes, 1917), pp. 219–20.
33 Smith, *The Teaching of Arithmetic*, pp. 105–25.
34 This game is missing from the Whipple Museum's Consul but survives with other Consuls. See 'Kids' Page', Early Office Museum (2016), www.earlyoffice museum.com/kids.htm; and R. Atzbach, 'Consul the Educated Monkey,' Rechenwerkzeug (2018), www.rechenwerkzeug.de/consul.htm.

education as social reform, and thus teachers as valuable reformers:[35] 'The mechanism of the Educated Monkey device ... offers teachers an opportunity to develop a fine art in teaching children numerical tables and stimulating even the dullest to their best.' Defining teaching as 'stimulating' children portrays teachers as the driving force behind effective educational use of Consul, in line with Pestalozzi's, Montessori's, Dewey's, and Smith's argument that teachers should lead students towards understanding through activities rather than memorising facts.

Multe is intended to make mathematics fun and engaging, the game's instructions claiming it 'can be used to turn certain kinds of work into play'. Like many arithmetic games, including some described by Smith, Multe uses competition to motivate children to memorise arithmetic facts. Thus, it more closely resembles an interactive form of mathematical drill than the child-directed, exploratory learning that early-twentieth-century educators recommended. I have found no evidence of Consuls being purchased or used by school classrooms, so it's possible that Multe was written only in the hope of appealing to the large economic market of school supplies. Furthermore, the presentation of Consul as a plaything suggests that its main target audience was parents, not schools.

Consul as Toy

Consul is a self-professed 'classic in the toy line', as printed on its paper folder packaging, and its charming clothes-wearing monkey and bright colours are assumedly meant to appeal to children. But, upon closer examination, Consul is a poorly designed toy. The thin, lightweight backplate and monkey figure would be damaged by clumsy handling, and the monkey's joints are fragile and could easily be knocked out of alignment, which would ruin the device's calculating ability. Consul may seem fragile today because the surviving examples are up to a century old. However, a warning on the packaging suggests that Consul's finicky operation is original: 'If the feet stick at any position, do not force their movement but loosen by moving the arms.' Consul's small size suggests intended use by small hands, but I suspect most school-age children lack the dexterity and patience necessary to work its calculating function.

35 Church and Sedlak, *Education in the United States*; Cremin, *The Transformation of the School*; and Kilpatrick, 'Mathematics Education in the United States and Canada'.

Furthermore, Consul's use is not intuitive, and the included instructions are not child-friendly. The instructions printed on the object itself (in its post-1916 versions) are brief: 'Set feet to point at two numbers/Fingers will locate their product.' The instructions on the packaging are small, dense text. More instructions are printed on the back of the enclosed addition number-chart, which explain how to divide, factor, add, and subtract using Consul (Figure 11.3). But these are also in small print and use technical terms that were probably alien to children learning basic arithmetic. For example, 'To subtract: Adjust the monkey so that one foot points at the subtrahend and the fingers point at the minuend. The other foot will be found pointing at the difference.' To understand these directions, a child must be a good reader, familiar with mathematical terms, and patient enough to locate the small-printed words in an obscure location on the packaging.

Consul's metal construction resembles contemporary mechanical toys like train models and building sets, such as the popular 1914 Tinkertoys. Animal toys were also common, such as the posable figurines in the Humpty Dumpty Circus of 1910 and the teddy bear craze of 1906.[36] This educated monkey, however, is not cute. It is a garishly awkward creature. Who then is the intended audience for Consul?

One selling point may be that its name and appearance, from red-polka-dot bowtie to comb tracks on its furry pate, differentiate it from just any monkey. These details specifically link the object with a trained chimpanzee called Consul that hit the New York vaudeville stage in 1909, six years before Robertson's patent applications.[37] A star like Consul would have been well known, as vaudeville was a popular and reputable form of entertainment among people of all ages and social classes from the 1890s to the 1920s in the United States. Vaudeville defined itself as more moral than 'common entertainments of the concert saloons, the dime museums, and the circus'. 'Polite vaudeville' in particular promised 'to provide a respectable place and decent entertainment that families could patronize without damage to their reputations'.[38] Animal acts were considered

36 G. Cross, *Kids' Stuff: Toys and the Changing World of American Childhood* (Cambridge: Harvard University Press, 1997), pp. 40, 74, 96; and R. O'Brien, *The Story of American Toys: From the Puritans to the Present* (London: New Cavendish Books, 1990), pp. 75, 78, 81.
37 This connection is suggested by H. Ball and G. Auckland, 'Educated Monkey', Grand Illusions (2017), www.grand-illusions.com/educated-monkey-c2x21140029.
38 F. Cullen, *Vaudeville Old and New: An Encyclopedia of Variety Performers in America* (New York: Routledge, 2007), p. xviii.

respectable and moral, and they especially appealed to children, perhaps drawing on the prominence of anthropomorphised animals in children's literature since the middle of the eighteenth century.[39] Polite vaudeville's popularity meant that animal stars were widely recognisable to children and adults.

Consul's arrival in New York from Europe was big news. News articles claimed that this 'educated monkey' had human abilities – he wore clothes, played shuffleboard, and lit and smoked cigarettes. He could also talk: 'Drowski [Consul's manager] spoke to Consul in French, and the chimpanzee responded with gutteral [sic] sounds that seemed to be understood by his manager.'[40] In June 1909, a *New York Times* article documented this priceless exchange: 'To the next [reporter's] question, "Do you like wearing clothes?" the chimpanzee replied, "garrrrr-egre-grummm-goora-umn." This was translated by Drowski to mean: "Have any of you got a cigarette, I want to smoke."' These decidedly un-monkey-like behaviours made Consul a big hit, and, according to Robertson's device, 'educated'.

What then did it mean for a monkey to be educated? According to psychologists and popular culture, it meant being able to behave like a human. In 1909, Pennsylvania psychologist Dr Lightner Witmer used intelligence tests designed for 'backward' children to test a performing chimpanzee named Peter, 'who has been appearing in vaudeville throughout the country as an example of "a monkey with a mind".' In a 1909 *New York Times* article, Witmer concluded that 'the chimpanzee is educated in a real sense and that he has the power of reasoning ... Peter, under the proper conditions, might be taught to read, write, and speak.'[41] Psychologist William Hornaday, who also studied trained chimpanzees, was not surprised by Witmer's findings. He described the trend of performing monkeys: 'In 1904 the American public saw Esau. Next came Consul, – in about three or four separate editions! In 1909 we had Peter.'[42]

There seem to have been many Consuls, or perhaps, thanks to these vaudeville stars, 'consul' became a generic word for a trained monkey or a chimpanzee. This usage even crossed into scientific

39 T. Cosslett, *Talking Animals in British Children's Fiction, 1786–1914* (Aldershot: Ashgate, 2006).
40 'Consul a Lively Ship Passenger', *New York Times*, 21 June 1909.
41 'Trick Chimpanzee Fulfills Mind Test', *New York Times*, 18 December 1909.
42 W. T. Hornaday, *The Minds and Manners of Wild Animals* (New York: Charles Scribner's Sons, 1922), p. 62.

Figure 11.4 'Consul Peter – smoking', 1909. Library of Congress, Prints & Photographs Division (LC-DIG-ggbain-04090; www.loc.gov/resource/ggbain.04090/).

terminology when, in 1933, the paleontologist Arthur Hopwood named a new genus of ancient primate *Proconsul*, meaning 'before chimpanzee'.[43] This generalisation of the name may also explain the caption of a 1909 photograph of a suit-wearing, cigar-smoking chimpanzee aboard a transatlantic ship: '"Consul Peter" – smoking' (Figure 11.4). Is this smoker Consul or Peter? According to Hornaday's chronology of trained chimps in America, it could be either. Or perhaps the author intended 'consul' to mean 'primate' or perhaps even 'educated monkey', followed by an individual's name, Peter. There is similar confusion about a 1909 film made about Consul's arrival in the United States (Charles Urban's *Consul Crosses the Atlantic*), in that sources disagree whether Consul or Peter was its simian star.[44] Others explain that the movie was about the famous Consul, whose behaviour was then imitated by 'Peter the Great' and other celebrity chimpanzees.[45] Whichever particular primate Robertson's device was named after, 'consul' was a well-known word associated with humanlike monkeys in the early twentieth century.

43 V. Morell, *Ancestral Passions: The Leakey Family and the Quest for Humankind's Beginnings* (New York: Simon and Schuster, 1995), p. 130.
44 For example, 'Consul Peter: 1909' and 'Consul Crosses the Atlantic,' *New York Dramatic Mirror*, 11 December 1909.
45 A. Balducci, *The Funny Parts: A History of Film Comedy Routines and Gags* (Jefferson: McFarland & Company, 2012), p. 36.

These chimpanzees are documented onstage as dressing like humans, eating with table manners, smoking, dancing, riding bicycles, and roller-skating, but no mention is made of any Consuls doing mathematics.[46] Thus, the real Consuls did quite different things from Robertson's calculating Consul. Does this imply that arithmetic is so easy that even monkeys can do it? Or, on the other hand, is it *more* impressive for a monkey to do maths, as a uniquely human ability? The latter makes sense in the light of late-nineteenth-century psychologists' supposedly successful attempts to teach monkeys to count and then perform their 'simian arithmetic' for interested colleagues and journalists as evidence of intelligence.[47] But language was the telltale marker of humanlike animals, not mathematics.

The aesthetics of Consul the calculator speak more to performance than to intelligence. For example, its backplate is muted yellow with an ornately patterned dark green trim (marked with tiny plus signs), suggesting the sepia tones of early photographs and cinema, and the monkey's bright red suit invokes the spectacle of vaudeville. There are no scholarly symbols on the object or instructions, such as books, mathematical instruments, or eyeglasses to represent Consul's 'educated' status. Consul looks more like an entertainer than a mathematician.

The idea of a monkey doing mathematics may have appealed to children for its novelty, but the underlying theme of a humanlike monkey would have been more meaningful to the Darwin-aware adult population. Evolutionists and anti-evolutionists alike believed that children and primates shared many developmental and psychological characteristics, such as cuteness, mischief (especially a love of stealing), and emotional expression.[48] Hornaday described the appeal in 1922: 'During the past twenty years, millions of thinking people have been startled, and not a few shocked, by the amazing and uncanny human-likeness of the performances of trained chimpanzees on the theatrical stage.'[49] As Constance Clark wrote in *God – or Gorilla: Images of Evolution in the Jazz Age*, 'monkeys were everywhere in the 1920s.'[50] Although Clark does not mention

46 Hornaday, *The Minds and Manners of Wild Animals*; 'Consul a Lively Ship Passenger'; and 'Trick Chimpanzee Fulfills Mind Test'.
47 S. Shuttleworth, *The Mind of the Child: Child Development in Literature, Science, and Medicine, 1840–1900* (Oxford: Oxford University Press, 2010), pp. 258–60.
48 Shuttleworth, *The Mind of the Child*, Chapters 13 and 14.
49 Hornaday, *The Minds and Manners of Wild Animals*, p. 62.
50 C. A. Clark, *God – or Gorilla: Images of Evolution in the Jazz Age* (Baltimore: Johns Hopkins University Press, 2008), p. 1.

primate performers or toys in her book, the association between monkeys and evolution was so ubiquitous that the highly publicised trial of 1925 concerning the teaching of evolution in public schools was called the Scopes 'monkey' trial.[51] Both the performing and the calculating Consuls drew on the public fascination with monkeys as symbols of evolution and human development.

Robertson's 1916 patent diagram is labelled 'The Educated Monkey', not 'Consul'. Perhaps 'Consul' was added later for its attractive vaudeville association. But the famous Consuls appeared before 1909, and Peter died in 1910.[52] Hornaday mentioned no famous chimps after Peter. Robertson's calculating toy was probably first produced around 1915, when he applied for the two patents.[53] Would a vaudeville star, even one as famous as the Consuls and Peter, still be recognised by children at least five years later? Both childhood and stardom are short-lived. But it is likely that adults would still remember the famous chimps of the 1900s in 1915, even if children did not. Designing toys that appealed to parents more than children was common before the 1930s, when companies began to target children as consumers in their own right.[54] This focus on adults is also acknowledged on the packaging, which describes Consul as 'a device which interests both young and old'. Furthermore, the Educational Toy Company of Springfield, Massachusetts, advertised Consul in 1920 in *Illustrated World*, a magazine for adult science and technology enthusiasts. The advertisement does not market Consul as a fun novelty or as a calculator for adults, but as an educational device: '"CONSUL", THE EDUCATED MONKEY should be in every home, for he points out The Royal Road to the Multiplication Table ... He Makes Arithmetic Fun.'[55] This angle matches the company's speciality in making education fun, as shown by its name. The advertisement's placement on a page of ads for other gadgets, including a telescope and a newfangled kerosene

---

51 Clark, *God – or Gorilla*.
52 Hornaday, *The Minds and Manners of Wild Animals*.
53 The Whipple's Consul was made later, as shown by the printed date on the object of Robertson's second patent, 26 November 1918. According to the packaging, it was made by the TEP Manufacturing Company of Detroit, Michigan, a different company from Robertson's. But the design had not spread far: the packaging bears the name of a TEP factory located in Dayton, Ohio, as well as a stamp from the Gebhart Folding Box Co. in Dayton.
54 G. Cross, 'Toys and Time: Playthings and Parents' Attitudes toward Change in Early 20th-Century America', *Time and Society*, 7 (1998), pp. 5–24, on p. 7.
55 '"Consul", the Educated Monkey', *Illustrated World*, 33.4 (1920), p. 765.

254 CAITLIN DONAHUE WYLIE

burner for stoves, targets parents rather than kids. The advertisement invites consumers to send the company 35 cents[56] for an educated monkey, 'or a dollar for three'; an order of only one or three monkeys seems more appropriate for families than schools. Consul's marketing seems to appeal to a dual audience of school-age children and their parents, thanks to the adults' nostalgia for vaudeville as well as their aspiration for their children to learn mathematics.

Conclusion

Finding a vaudeville celebrity remade as a mechanical calculator may seem surprising. This uniquely twentieth-century cultural meaning gives this charming metal device an air of sophistication, fame, and fun. The object invites children to practice their sums with not just any suit-wearing monkey bent into acrobatic poses, but with an individual suit-wearing monkey. The star of *Consul Crosses the Atlantic* can participate in something as seemingly mundane as multiplication, thereby lending mathematics an exalted status.

The calculating monkey dissolves some of the boundaries we imagine about historic lifestyles, such as between school and home and between adulthood and childhood. For example, children can now 'perform' calculations, at home or at school, as mathematical drill disguised in objects, games, and subtle hints of vaudeville. Consul brought aspects of school, such as arithmetic facts, to children's homes in the form of an educational toy. Likewise, it may have brought aspects of home, such as fun and games, to school, if Multe's instructions for classroom play were actually followed. Also, this supposed toy was not marketed in ways that would capture children's attention. It is not a cute, sturdy object like the other animal figurines in a child's toy box at the time. Its name probably meant little to children of the late 1910s and early 1920s, though their parents would have well remembered the individual performing chimps of the first decade of the twentieth century. Thus, this object uses nostalgia for past celebrities plus future hopes for children to learn mathematics to convince adults to purchase this mechanical calculator.

56 The equivalent of about US $4.17 today, according to the 'Consumer Price Index Inflation Calculator', Bureau of Labor Statistics (2018), https://data.bls.gov/cgi-bin/cpicalc.pl.

The combination of a calculator, teacher, and toy in one object presents mathematics as important in early-twentieth-century American society. Mathematics was considered an avenue to individual learning and social progress, hence the rise of toys that were claimed to help children learn arithmetic. Robertson's device perhaps functions best as a calculator, with marketing and decoration that reflect contemporary trends in education and popular culture. His sequence of patents supports this interpretation, as his geometry-based mechanical calculator later acquired aspects of education and play. Robertson added a toy-like representation of a monkey, and later the name of a cultural icon, to make his 'calculating animal', and the object was packaged in the language of education. The connection between these issues, as well as the social value placed on educational toys and on learning as an experiential process, reveals the intricate ways in which we create physical forms to meet social functions. Historians then reap the benefits, and face the challenges, of interpreting these forms as rich evidence of the culture they represent.

12 ❧ Robin Hill's Cloud Camera: Meteorological Communication, Cloud Classification

HENRY SCHMIDT

In the summer of 1923, a young Cambridge chemistry student named Robert Hill (nicknamed 'Robin') pasted a newspaper article into his sketchbook and cloud journal. It calls for public contributions to a scientific project of French origin: 'Meteorology is a science that needs international cooperation more than perhaps any other', it exhorts the British reader, '[a]s the number of official [meteorological] stations is limited it is proposed to ask professional and amateur photographers … to cooperate voluntarily.' Participating photographers, professional and amateur alike, are asked to take 'five photographs at each [appointed] hour, one facing each of the cardinal points and one with the camera pointing to the zenith'.[1] The International Survey of the Sky, as the programme was called, sought to compile photographic records of the sky in order to map the entire European cloud sheet. It was organised in Britain by Captain C. J. P. Cave, then ex-President of the Royal Meteorological Society. Hill answered Cave's call for amateur contributions with gusto. Within months his name and several cloud photographs, shot with a camera of his own invention, appeared in a range of prominent weather-related journals and magazines.

Hill's fame derived from the enthusiastic reception of his cloud camera and its novel distinguishing feature: the fish-eye lens. His photographs of cloud formations over Cambridge provided Cave's survey with an unbroken perspective of the sky, from horizon to horizon in all directions. The homemade, wood-bodied cameras that Hill used to take those first photos now reside in the Whipple Museum of the History of Science, alongside later prototypes for a commercial version produced by R. & J. Beck Ltd.[2] The words 'Robin Hill 180° Cloud Camera' are printed on the commercial version's

1 'Study of Clouds', *Times*, 8 September 1923, p. 11.
2 Their Whipple Museum accession numbers are, respectively, Wh.4416 and Wh. 5732.

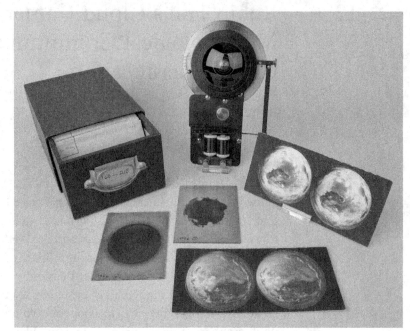

Figure 12.1
A selection of photographs taken by Hill and preserved with his cloud cameras in the Whipple Museum. Image © Whipple Museum (Wh.4416).

brass rim. Small, circular photographs of clouds that survive with these artefacts (Figure 12.1) illustrate the distortive effect of the lens.

The many versions of Hill's cloud camera that now reside in the Whipple Museum survive thanks to its inventor's long career as biochemistry researcher at the University of Cambridge.[3] Later versions of the camera held in the Whipple Museum are no longer marked as 'cloud cameras'[4] – Hill's original was by the middle of the century out of use in cloud study, employed instead primarily by ecologists rather than meteorologists. Most such analogue techniques were phased out of meteorology by the 1950s, with technological innovations – many of military origin – having ushered in methods of quantitative, as opposed to photographic or visual, weather prediction.[5] The style of cloud research underlying both the International Survey of the Sky and Hill's cloud camera became hopelessly antiquated.[6]

---

3 See the obituary in Wh.4416, and D. J. Mabberley, 'Hill, Robert [Robin] (1899–1991)', in *Oxford Dictionary of National Biography* (Oxford: Oxford University Press), www.oxforddnb.com/view/article/49777 (accessed 29 February 2016). Hill was formally employed by the Agricultural Research Council.

4 See Wh.5732.

5 See K. C. Harper, *Weather by the Numbers: The Genesis of Modern Meteorology* (Cambridge: MIT Press, 2008), pp. 96–104.

6 J. Fleming, *Historical Essays on Meteorology, 1919–1995* (Boston: American Meteorological Society, 1996), pp. 25, 59.

This chapter examines the reception of new photographic perspectives enabled by Hill's camera. In doing so, it indirectly reveals the imagined futures of meteorological research on clouds shortly after the First World War. My analysis captures a snapshot of the institutional, social, and technological dynamics within the field of meteorology during that period. The reception of Hill's camera shows how it coincided with attempts to remake cloud study, namely by considering clouds primarily in relation to weather systems at the scale of the 'whole sky' rather than individual specimens. Entwined with those initiatives was an attempt to reform the patterns of communication between centralised meteorological offices and their dispersed contributors, many of whom were amateur volunteers.[7] Before 1923, professional meteorologists circulated exemplary images of each cloud type in order to standardise contributors' records of cloud occurrences, generally communicated by correspondence or telegram. The Hill Cloud Camera earned recognition for its role in a project that instead collected photographic data from peripheral contributors, thereby assigning those observers a purely technical role and reserving challenges of classification for experts in centralised meteorological offices, especially those in Greenwich and Kew. By contextualising the camera's reception within these social and institutional networks, we can relate its technological capacities to the field's broader problems of representation and communication. The scientific and political stakes of cloud photography come into focus. A connection may be drawn between this chapter's approach and synoptic uses of the cloud camera itself: it aims to represent less the early history of the camera and its photographs than the dynamic networks of research, communication, and instrumentation that shaped its reception.

I first provide an outline of events and controversies in cloud study during the forty years preceding the Hill Cloud Camera's invention. I then describe how Hill's new instrument promised to mediate visual representations of different kinds of meteorological data. Finally, I relate the camera's reception to practical problems in meteorological knowledge, namely the use of verbal versus visual communication and the role of amateurs in research. In this way, I show how the Hill Cloud Camera's novel capabilities represented

---

7 The primary sites of British cloud research were the Royal Observatory in Greenwich and the Kew Observatory. Well over 100 other observatories compiled British meteorological data. For a list, see J. Glasspoole, 'The Driest and Wettest Years at Individual Stations in the British Isles, 1868–1924', *Quarterly Journal of the Royal Meteorological Society*, 52, no. 219 (1926), pp. 237–49.

a potential solution to two problems in meteorological practice: one observational and representational, the other pertaining to social organisation and the circulation of knowledge. Cloud photography's impact on central meteorological projects such as the *International Cloud Atlas* (first published in 1891, revised often) and the International Survey of the Sky (1923) is an important thread throughout this story. The designs of these projects reflect conceptions of the role of photography and organisation of cloud study in social and instrumental ensembles. By examining the camera's reception in relation to such projects, answers to an important question emerge: how did Hill's camera represent an ingenious solution to a whole host of problems that scientists would soon cease to recognise and, in doing so, how does it register some conceptual discontinuities of early-twentieth-century science of the atmosphere?

Clouds and 'Synoptic' Meteorology

Why was Hill's cloud camera considered particularly suited to photographing clouds? The history of cloud knowledge and the problems it faced prior to Hill's invention provide insight into the perceived promises of his camera.

When late-nineteenth- and early-twentieth-century meteorologists talked about clouds, they utilised Luke Howard's concise Linnean classification – consisting of *cirrus, stratus, cumulus,* and *nimbus* – which he first developed in 1803.[8] Problems emerged, however, when emphasis shifted from the correctness of Howard's names to more practical concerns regarding communication within an international meteorological community. '[T]he name of a cloud is of far less importance than that the same name should be applied to the same cloud by all observers,' the British cloud expert Ralph Abercromby wrote in 1887.[9] Cloud study was among the last and most stubbornly resistant of the many initiatives to standardise the production of meteorological knowledge.[10] A few key efforts

8 L. Howard, 'On the Modifications of Clouds, and on the Principles of Their Production, Suspension, and Destruction; Being the Substance of an Essay Read before the Askesian Society in the Session', *Philosophical Magazine*, 16, no. 64 (1803), pp. 344–57.

9 R. Abercromby, 'Suggestion for an International Nomenclature of Clouds', *Quarterly Journal of the Royal Meteorological Society*, 13 (1887), pp. 140–55.

10 For related articles on the projects of standardisation and metrology in late-nineteenth- and early-twentieth-century science, see S. Schaffer, 'Late Victorian

inaugurated the international standardisation of Howard's cloud theory. They are central in understanding how the clouds pictured by Hill's cloud camera came to be valued in relation to their meteorological context: the 'whole sky'.

Abercromby, a key figure in *fin-de-siècle* cloud study, presented photographs and a lecture to the Royal Meteorological Society in 1887 in which he detailed his travels throughout the world and attempted to 'illustrate the fact of the identity of the forms of clouds'.[11] By travelling to exotic locales and documenting their cloud forms, Abercromby demonstrated that clouds everywhere could be classified by Howard's vocabulary. His celebrated cloud photographs secured the global legitimacy of terms like *nimbus*, *cirrus*, and *stratus*. Cloud forms' universality posed as many problems as it solved, however – as Abercromby noted, 'shape alone is not sufficient to give a true prognostic value. Clouds always tell a true story, but one which is difficult to read; and the language of England is not the language of Borneo. The form alone is equivocal; the true import must be judged by the surroundings, just as the meaning of many words is only known by the context.'[12] Abercromby thus showed meteorology based on cloud-watching to be a global form of knowledge, but one nonetheless premised on the local adaptation of a universal vernacular. Local context for him and his contemporaries largely consisted of other meteorological phenomena measurable by barometers, thermometers, or anemometers. He forecast by positioning cloud forms on diagrams like that in Figure 12.2, which shows where certain cloud types form within 'cyclonic' systems, represented by isobaric lines derived from barometric readings.[13]

Metrology and Its Instrumentation: A Manufactory of Ohms', in R. Bud and S. E. Cozzens (eds.), *Invisible Connections: Instruments, Institutions, and Science* (Bellingham: SPIE, 1992), pp. 23–54.

11 R. Abercromby, 'On the Identity of Cloud Forms All over the World and on the General Principles by Which Their Indications Must Be Read', *Quarterly Journal of the Meteorological Society*, 13 (1887), pp. 140–6.

12 Abercromby, 'On the Identity of Cloud Forms'.

13 For an example of such analysis, see R. Abercromby, 'On the General Character, and Principal Sources of Variation, in the Weather at Any Part of a Cyclone or Anticyclone', *Quarterly Journal of the Meteorological Society*, 4, no. 25 (1878), pp. 1–2. For a general history of cyclonic theory, see G. Kutzbach, *The Thermal Theory of Cyclones: A History of Meteorological Thought in the Nineteenth Century* (Boston: American Meteorological Society, 1979). Some cloud types in Figure 12.2 reflect Abercromby's early interest in folk-knowledge of clouds, a preoccupation that was largely purged from meteorological research by the 1910s.

Figure 12.2
Abercromby's
Cyclone Diagram,
describing the
relative positions of
certain cloud forms
within a larger
'cyclonic system'.
From N. Shaw,
Forecasting Weather
(London: Constable
and Company,
1911), p. 87.

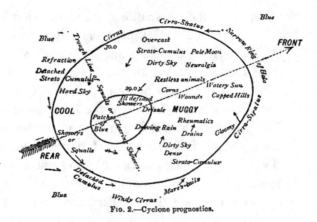

FIG. 2.—Cyclone prognostics.

This was the characteristic late-nineteenth- and early-twentieth-century approach to cloud observation.[14] Abercromby described the state of the art clearly in 1878:

> Two methods have to be combined to obtain a complete idea of
> Cyclone or Anticyclone Weather. The first, or synoptic method,
> consists in drawing an isobaric map, and marking on it the
> position of rain, clouds, &c., by which their position relatively
> to each other, and to the cyclone or anticyclone centres, can be
> ascertained ... The second method consists in recording the
> sequence of phenomena which occur to a single observer, as a
> cyclone or anticyclone passes over him.[15]

Katharine Anderson shows how both these modes reflect a visual sensibility that seeks to distill complex meteorological phenomena into images of immediate, graphic clarity, open to interpretation 'at a glance'. The second mode motivated greater use of photography for cloud analysis.[16] The photographic record of an individual observer's perspective, however, proved difficult to reconcile with Abercromby's other mode of visualisation and analysis, namely synoptic isobaric

---

14 For a crucial text during this period, see N. Shaw, *Forecasting Weather* (London: Constable & Company, 1911). Shaw identifies two approaches to cloud study: one classificatory, discussed here; and the other laboratory-based, recently discussed in P. Galison and A. Assmus, 'Artificial Clouds, Real Particles', in D. Gooding, T. Pinch, and S. Schaffer (eds.), *The Uses of Experiment* (Cambridge: Cambridge University Press, 1989), pp. 225–74; and R. Staley, 'Fog, Dust, and Rising Air', in J. Fleming, V. Janković, and D. Coen (eds.), *Intimate Universality: Local and Global Themes in the History of Weather and Climate* (Sagamore Beach: Science History Publications, 2006), pp. 93–113.
15 Abercromby, 'On the General Character', pp. 1–2.
16 K. Anderson, *Predicting the Weather* (London: University of Chicago Press, 2005), pp. 187–219.

maps. As he noted, 'These two methods, the plan and section as it were, are so different that it is difficult practically to combine them, and to realize how a difference in a cyclone, on a synoptic map, will affect the sequence of weather as it passes over an observer.'[17]

The meteorological projects of Hill's time placed his cloud photographs in direct dialogue with those mapping practices. The camera's importance to the International Survey of the Sky (1923) was tied to that project's intent to revise the *International Cloud Atlas*, which was first published in 1891 as the standard of cloud knowledge and classification.[18] Meteorology's prevailing information order attempted to ensure the reliability of vast research networks of amateur contributors by distributing exemplary cloud photographs. The *International Cloud Atlas* comprised the central focus of this effort. The earliest form of the revised cloud atlas – and a direct inspiration for the International Survey of the Sky – was a treatise published by the Office National Météorologique de France entitled *Les systèmes nuageux*.[19] It defines each cloud type not only by form, but also by its relation to larger pressure systems, or 'cyclones', using barometric pressure gradients, narrative description of formation processes, and photographic representation. A review of the text, published by the Royal Meteorological Society, situates the project clearly:

> With the development of synoptic meteorology, attention was naturally directed to the relation between clouds and weather and travelling systems of isobars, and in our own country the work of Abercromby and Clement Ley was especially prominent ...
> It was recognized that the synoptic charts explained most of the local weather signs and to a large extent superseded them ... In consequence the interest in cloud structure as an aid to forecasting appears to have declined.[20]

The French project aimed to halt that perceived decline in interest in cloud structure by defining the complementary relation between cloud knowledge and the synoptic understanding afforded by isobaric mapping.

Here we see the hybrid approach to cloud analysis growing in complexity, scope, and institutional support. The individual cloud form, identified visually, is situated in relation to unseen phenomena

---

17 Abercromby, 'On the General Character', p. 2.
18 C. J. P. Cave, 'The International Survey of the Sky', *Nature*, 113 (1924), pp. 279–80.
19 P. Schereschewsky and P. Wehrlé, *Les systèmes nuageux* (Paris: L'Office National Météorologique, 1923).
20 C. K. M. Douglas, 'Review of *Les systèmes nuageux*', *Quarterly Journal of the Royal Meteorological Society*, 50, no. 212 (1924), pp. 390–2.

such as wind and pressure. Schematic diagrams mediate that gap. *Les systèmes nuageux* attempts to represent a cloud's role in larger 'cloud systems' by triangulating text, photograph, and diagram: text performs narrative work, diagrams show systematic relationships between measurable weather phenomena, and photographs render such phenomena visible and identifiable. Such divisions of labour reflect distinct valuations of instrumental ensembles, situating cameras amongst barometers and anemometers.

Hill conducted his photographic experiments in this atmosphere of ever more socially, technically, and scientifically complex cloud study. His early cloud cameras, however, were simple and low-tech. The Whipple Museum holds two wood-bodied, pinhole cameras that Hill used for his earliest wide-angle photographs, which depict him before his Cambridge home. In later versions, like that produced by R. & J. Beck, automatic shutters replaced manual, brass replaced clumsy wooden bodies, and the lens was refined for enhanced range and clarity. Nonetheless, Hill used the earlier versions to produce the photos that he submitted to the International Survey of the Sky and that subsequently so impressed the European meteorological community.

Colonel Delcambre, director of the French meteorological office, noted his international survey's ambitions and linked them directly to Hill's new cloud camera. He wrote to Hill in late 1923, remarking that 'The very interesting photographs that you attached will soon be put under consideration in the complete revision of the Cloud Atlas.'[21] Hill's camera represented a new way to depict clouds at the scale afforded by such measurements of pressure, wind, and temperature. The distributed measurement of those phenomena had provided data for producing synoptic weather maps for decades prior, and caused the lapse of cloud research in the years preceding 1914.[22] The outbreak of war, according to commentators, heightened the need for meteorological exactness achievable only by coordination of cloud knowledge with quantitative measurement.[23]

C. J. P. Cave, the coordinator of Britain's contribution to the international survey, praised Hill in his summary of the survey's results by noting that his camera provided 'a far better representation of the cloud distribution than can be obtained with an ordinary camera unless a prohibitive number of plates are exposed'.[24] The

---

21 Cambridge University Library (hereafter CUL) MS Add.9267 [C].
22 Abercromby, 'On the General Character', p. 1.
23 See, for example, Douglas, 'Review of *Les systèmes nuageux*', p. 392.
24 Cave, 'The International Survey of the Sky'.

Quarterly Journal of the Royal Meteorological Society's review of *Les systèmes nuageux* in 1924 noted the usefulness of Hill's camera for its ability to capture rapidly changing phenomena across the whole sky. The author remarks that

> In the first volume [of *Les systèmes nuageux*] it is rightly
> emphasized that the important features from the forecasting
> point of view are the appearance of the whole sky and the changes
> over a period of some hours. The details of cloud structure change
> so quickly that a complete photographic reproduction, even
> for a short period, would require the expenditure of an inordinate
> number of plates, though the use of Mr. R. Hill's lens would
> reduce this difficulty considerably.[25]

The place of individual clouds in cyclonic systems could be seen at a glance. Photographs challenged diagrams' superior ability to represent visible weather phenomena at the scale of the whole sky, rather than the individual specimen. Relations between visual and non-visual phenomena – usually cloud types and pressure gradients – demanded representational compromise. The language of the camera's reception reflects enthusiasm for combining the geographical advantages of, say, isobaric mapping with the naturalism of photography. Hill's camera landed amongst meteorologists keen to expand photographic records to a synoptic scale, and so mediate between Abercromby's 'two modes' of analysis.

If Hill's camera promised to transform the visible record of cloud phenomena, how did he prove the camera's compatibility with existing modes of identifying and defining clouds? After all, his camera produced hugely distorted images, utterly unlike those of the *International Cloud Atlas*. On comparing Hill's fish-eye lens with those of two contemporaries, Wood and Bond, whose designs provoked no reaction from the meteorological establishment, we uncover an important difference in the cameras' public presentation and construction that may explain the enthusiastic reception of Hill's invention. In an article of 1906, and a revised edition of his book, *Physical Optics* (1911), R. W. Wood showed how the 'external world appears to a fish below the surface of smooth water' by way of a pinhole camera filled with water.[26] In 1922, Wilfred Noel Bond published a brief article in the *Philosophical Magazine* describing

25 Douglas, 'Review of *Les systèmes nuageux*', p. 392.
26 See both R. W. Wood, 'Fish-Eye Views, and Vision under Water', *Philosophical Magazine*, 6, no. 12 (1906), pp. 159–62; and R. W. Wood, *Physical Optics*, 2nd edn (New York: The MacMillan Company, 1911), pp. 65–8.

Figure 12.3 Bond's own photograph of the sky, using his 'cloud lens', with grid superimposed. Plate VII in W. N. Bond, 'A wide angle lens for cloud recording', *Philosophical Magazine*, 44, no. 263 (1922), pp. 999–1001.

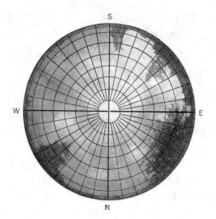

an adapted version of Wood's camera, in which he first suggested its use for photographing clouds.[27] Bond's camera closely resembled that developed by Hill only a year later, but Bond used one hemispherical lens situated below a pinhole, while Hill's version placed another convex lens above the pinhole in order to more clearly capture images near the horizon.

Why did Hill's camera garner acclaim, while Bond's publication went largely unnoticed? We cannot say for sure. Hill may have simply enjoyed superior timing, and more actively communicated his results to coordinators of the International Survey of the Sky. We should note, however, that their designs and their techniques for representing visual distortion differ in one important respect. Bond adequately describes the real nature of the camera's effect, but also misrepresents it visually. Figure 12.3 shows how concentric bands divide the image, but also deceptively frame the picture-plane as a globe with receding lines of latitude and longitude.

Hill, in contrast, applied a huge amount of effort towards correcting for the lens' distortion, with far better results. He wrote to a nearby optical physicist requesting a ray diagram for representing the lens' distortion, but eventually settled for a more practical solution: merely photographing illuminated grids. The camera's fabricator, R. & J. Beck, used those photographs to assure the Royal Air Force of the camera's value for photographic surveys from the

27 W. N. Bond, 'A Wide Angle Lens for Cloud Recording', *Philosophical Magazine*, 44, no. 263 (1922), pp. 999–1001. Cave's note to Hill on 20 September 1924 mentions this article (see Hill Papers, CUL MS Add.9267).

Figure 12.4 Hill's diagram depicting the translation from photographs produced by his cloud camera to a conventional camera format. The outlined portion of the image would, in principle, produce the image on the right when backlit and projected through the lens in reverse direction. Plate VI in R. Hill, 'A Lens for Whole-Sky Photographs', *Quarterly Journal of the Meteorological Society*, 50, no. 211 (1924), pp. 228–9. Reproduced by permission of the Royal Meteorological Society.

underside of planes.[28] In an article for the *Quarterly Journal of the Royal Meteorological Society*, Hill also calculated the precise distortion relative to the image's central point.[29]

Another one of Hill's most important marketing tactics, highlighted in R. & J. Beck catalogues and in his own article for the *Quarterly Journal of the Royal Meteorological Society*, was to stress the camera's ability to translate between the distorted, synoptic image produced on the plate and a direct, conventional, flat image. The primary means of doing so entailed projecting the image back through the lens and onto a flat surface, thereby reversing the distortive effect (Figure 12.4).[30]

Hill's outermost, curved lens, absent in Wood and Bond's designs, enabled this reversal of perspective. The camera's physical construction not only registered information with capabilities beyond those of the human eye, but also allowed it to translate them back. The need to discern crucial information in primarily non-naturalistic

28 See notes in CUL MS Add.9267 [C].
29 R. Hill, 'A Lens for Whole-Sky Photographs', *Quarterly Journal of the Meteorological Society*, 50, no. 211 (1924), pp. 228–9.
30 Hill, 'A Lens for Whole-Sky Photographs', p. 233.

representations (such as isobaric diagrams) 'at a glance' defined the preferred observational practices and units of meteorological analysis. The capacity of Hill's camera to translate between cloud specimens and cloud sheets, a new unit of observation commensurate with isobaric diagrams, endeared it to meteorologists.[31] Hill emphasised that 'each small area of the whole is a faithful representation of the corresponding part of the sky ... When the nature of the distortion is realised, there is no difficulty in interpreting the results.'[32] He repeatedly emphasised the modified lens' ability to secure simultaneous synoptic visibility and local fidelity. Photographic representation could clearly identify familiar cloud types, even while the photographs' unfamiliar visual forms afforded 'at a glance' knowledge of 'whole-sky' phenomena. Representations of that translation, like Figure 12.4, proved the camera's commensurability with existing technologies of photography and visualisation.

Hill's cloud photographs were published in a moment when the future of cloud study and of meteorological prognosis relied on one's ability to envision the interrelations of visual and non-visual components in large-scale weather systems.[33] Hill's camera extended the vividness, visual clarity, and objectivity of photographic depiction to analysis over geographical scales rivalling those of traditional weather mapping.[34] It did so without compromising perceptions of its representational faithfulness of the single specimen. Hill's camera surpassed the eye, but did not betray it.[35]

Beyond Clouds: Photographs and Photographers

The history of cloud knowledge shows how meteorologists struggled to reconcile synoptic charts of non-visual data (derived from barometers and anemometers) with photographs. That struggle entailed the coordination of many different meteorological observers with varying levels of scientific status and technical proficiency. Did the Hill cloud camera, which promised to alter photography's place in

31 Anderson, *Predicting the Weather*, pp. 187–219.
32 CUL MS Add.9267 [C], Letter to the Editor of *The Amateur Photographer*.
33 For example, Abercromby, 'On the General Character'; and E. van Everdingen, 'Clouds and Forecasting Weather', *Quarterly Journal of the Royal Meteorological Society*, 51, no. 215 (1925), pp. 191–204.
34 On the association of photography with the epistemic virtue of 'mechanical objectivity', see L. Daston and P. Galison, *Objectivity* (Boston: Zone Press, 2007), pp. 121–2.
35 See CUL MS Add.9267 [C.18] for newspaper clippings describing Hill's camera as the 'Magic Eye'.

the field's representational ensemble, in turn affect these social aspects of cloud research? One way to answer this question is by examining how the uses of different media – in this case, photographs and lexical data communicable by telegraph – were delegated to or controlled by different members of the meteorological polity. This section relates the reception of Hill's cloud camera by meteorologists keen to reorganise social dynamics in meteorological research and communication. Its place within the International Survey of the Sky is an instructive case. The novelty of that survey lay in its attempt to exploit the camera on this grand, popular scale: it was the first major meteorological effort that requested the contribution of photographs from non-professionals. Amateur contributions had long played a central role in other kinds of meteorological research. *Symons' Meteorological Magazine*, one of the field's most important organs of communication by the late nineteenth century, began as a publication by and for amateur meteorologists.[36] R. H. Hooker, then President of the Royal Meteorological Society, wrote in 1922 that '[i]t has been almost traditional in this country to regard scientific progress as necessarily associated with voluntary effort.'[37] Professional offices made do with whatever volunteer contributions they could attract.[38] With ever greater amateur access to cameras, large-scale meteorological dynamics over the course of a week could thus be photographically reconstructed with unprecedented geographical scope. Looking at the meteorological community's composition alongside its techniques of communication and standardisation – especially how they organise circulation and delegate control of words and images, naming and observing – reveals the professional and political stakes of cloud photography in meteorological practice.

The professional–amateur relation even shaped the language of meteorology itself. Standardising the vocabulary of cloud classification, and thus synchronising a community sufficiently large for synoptic analysis, presented nineteenth-century cloud scientists with

36 R. Gregory, 'Amateurs as Pioneers', *Quarterly Journal of the Royal Meteorological Society*, 55, no. 230 (1929), p. 104.
37 R. H. Hooker, 'The Functions of a Scientific Society, with Special Reference to Meteorology', *Quarterly Journal of the Meteorological Society*, 48, no. 201 (1922), p. 1.
38 For comments on the professional–amateur distinction in Victorian science, see R. Barton, '"Men of Science": Language, Identity, and Professionalization in the Mid-Victorian Scientific Community', *History of Science*, 41 (2003), pp. 73–119. I use those terms in accordance with most meteorological publications between 1880 and 1930.

a major problem. In the 1880s a controversy stimulated by the English don of *fin-de-siècle* cloud knowledge, the Reverend Clement Ley, focused on the complexity of Howard's cloud classifications. Ley situated the problem of classification clearly:

> Before the dawn of synoptic meteorology Luke Howard's system filled a need, though it did little to promote inquiry. Since that era it may safely be made the basis of a carefully discriminating and eclectic system of terminology. But any endeavour to restrict ourselves to its use, cuts off the possibility of obtaining what becomes more and more necessary, viz. the power of either communicating from distant localities the actual aspect of the sky, so that this may be represented synoptically, or of recording such an aspect in a journal so as to call up any vivid idea of the observed phenomena to the reader of the journal.[39]

Ley's contemporaries disliked his more vivid and synoptically accurate system due to its excess of terms. D. W. Barker called for something simpler than Ley's classification 'for the use of ordinary observers', and H. Toynbee echoed the same idea.[40] G. M. Whipple, Superintendent of Kew Observatory (and father of the Whipple Museum's founding donor Robert Stewart Whipple), suggested in response the formation of a committee 'to inquire into the question of cloud classification' and to produce standard photographs of cloud forms that could be distributed to an audience beyond those in regular contact with the Meteorological Office.[41] That committee's work would provide the basis for the *International Cloud Atlas*, which was first published in 1891. The negotiation of a descriptive language for clouds thus emerged in relation to controversies over the role of non-professional observers in meteorological research. It had enormous stakes: Ley and his fellows assumed that only lexical communication could produce 'synoptic' understanding. In this exchange and others, the material character of cloud knowledge was considered in relation to two problems. Photographs legitimised the universal application of Howard's terminology, but their role in the communication of cloud data across the vast distances and durations studied by modern meteorology remained subservient to that of imprecise verbal transmission (due largely to the nature of

39 See the 'Discussion' section in H. Toynbee, 'General Remarks on the Naming of Clouds', *Quarterly Journal of the Royal Meteorological Society*, 12 (1886), pp. 99–101.
40 Toynbee, 'General Remarks on the Naming of Clouds', p. 101.
41 Toynbee, 'General Remarks on the Naming of Clouds', p. 101.

telegraphic communication). Meanwhile, the limits of that verbal lexicon reflect a reliance on amateur contributors.

Professional and amateur meteorologists alike had to be taught how to distinguish, say, *cirro-stratus* from *strato-cirrus* clouds in their own research. The meteorological community circulated drawings, paintings, and photographs that could serve as exemplary renderings of cloud forms. In 1887, Abercromby responded to Ley by calling for the circulation of photographs: 'One difficulty in the way of any accordance of nomenclature arises from the impossibility of expressing the varying forms of clouds in words, and I have long been driven to the conclusion that no international accordance of cloud names can be obtained till typical photographs could be circulated at a moderate price.'[42] Four years later, that idea justified the creation of an *International Cloud Atlas*. The *International Cloud Atlas* embodies Ley's assumption that meteorological observations must be communicated by the circulation of written tables, and Abercromby's conviction that accuracy can be secured only by way of shared, standard images. The accurate circulation of words, in the form of tables that recorded total and proportional counts of different cloud types at particular locations, would require a reverse circulation of standard images. Only then could distant contributors be adequately trained, and their observations considered reliable. Even that strategy of centrally standardising visual references for cloud classification, however, came with problems: 'Some attempts I made to get fleecy clouds engraved were expensive failures; and the photographs were often of unsuitable size,' wrote Abercromby.[43] The definitive visual standard for cloud types, namely the *International Cloud Atlas*, contained some paintings due to difficulties in procuring acceptable photographs of some clouds. The camera's reliability was not absolute, even in the hands of expert technicians.

Hill advertised his camera's central role in attempts to subvert the *Cloud Atlas'* approach. His camera was celebrated for its innovative contribution to the International Survey of the Sky, which attempted a major reorientation of photography's place in the relation between professional and amateur meteorologists. That survey's novelty lay less in its scope and community of contributors than in its technological substructure. Cloud photography had previously been used to prove the homology of cloud forms throughout the world and represent classificatory differences, but never to produce visual data

42 Abercromby, 'Suggestion for an International Nomenclature of Clouds', p. 154.
43 Abercromby, 'Suggestion for an International Nomenclature of Clouds', p. 155.

for synoptic modelling of large-scale systems. There simply weren't enough willing cloud photographers to be enlisted, within or without institutional ranks. The project outstripped the meagre resources of meteorological institutions: 'As the number of official stations is limited it is proposed to ask professional and amateur photographers who are willing to cooperate voluntarily in the work to take photographs at the appointed times.'[44] The need for enthusiastic, volunteer photographers derived, as a bulletin from the Royal Meteorological Society noted, from the absence of cameras from most meteorological observatories' instrumental arsenals.[45]

Volunteer observers were asked to contribute satisfactory photographic records in numbers that were not tenable before the interwar period, a point acknowledged by Cave in his summary of the survey's results.[46] Professionals could rely on amateur photographic contributions after the First World War, when photography was transformed by its surge in popularity, affordability, and convenience.[47] Hill's camera promised to further democratise photography, as its reception shows. A crucial benefit of Hill's camera was its economy. When Cave's British chapter of the International Survey of the Sky called for five photographs of the sky per appointed hour, Hill submitted only one, which was actually deemed sufficient, even preferable. This was a significant improvement on an expensive technology in a rapidly expanding meteorological research infrastructure, which remained reliant on the generosity of volunteers.[48] Descriptions of the camera consistently emphasise this crucial material benefit.[49]

Hill's camera was received by meteorologists intent on reversing Abercromby's earlier claim that stabilising cloud classifications required the free circulation of photographs from centre to periphery. Whereas late-nineteenth- and early-twentieth-century members of the Royal Meteorological Society distributed photographs to serve as visual standards, thus extending cloud literacy to a broad

44 'Study of Clouds', *Times*, 8 September 1923, p. 11.
45 See 'International Photographic Survey of the Sky', *Quarterly Journal of the Royal Meteorological Society*, 49 (1923), p. 136. The author writes that 'A photographic camera is not included in the normal equipment of an official meteorological station.'
46 Cave, 'The International Survey of the Sky'.
47 C. Ford, *The Story of Popular Photography* (London: Century, 1989), p. 10.
48 Cave 'The International Survey of the Sky'; and Gregory, 'Amateurs as Pioneers'.
49 See CUL MS Add.9267 [C], Hill to 'Amateur Photographer'; Cave, 'The International Survey of the Sky'; Hill, 'A Lens for Whole-Sky Photographs'; and Douglas, 'Review of *Les systèmes nuageux*'.

community, that community of amateurs was now asked to communicate photographic evidence to their local meteorological office. Doing so limited the need for cloud literacy to a small, elite, largely homogeneous group of experts. Where non-professional volunteers once produced tables recording the hourly occurrence of cloud types, they were now to take only photographs, thus centralising the practice of classification. Untrained observers wielding cameras could be even more assimilated into meteorological research without risk of error due to the camera's objective gaze. Thorny problems regarding proper classification no longer left the Royal Meteorological Society's door.

Hill, himself an amateur (albeit with scientific training), provided an innovative solution to a serious material problem. That solution seemed to promise the future success of the synoptic photographic survey, and in turn to restructure relations of power in meteorology. Recognition of this goal can be discerned in an essay by Sir Richard Gregory, President of the Royal Meteorological Society in 1929. Gregory praised amateurs for their contributions to meteorology, but emphasised the qualitative differences between amateur and professional knowledge: the former is non-expert, enthusiastic, and expressive of a 'love of the subject', while the latter is abstruse, mathematical, and analytic.[50] Professional meteorological meetings, he argued, must be conducted in a sufficiently accessible language to be understood by amateurs, lest their passion for the subject wane. In Gregory's tenure, Hill's camera promised to resolve debates about the accessibility and ease of cloud terminology – debates conducted by the likes of Ley and Whipple – by exploiting the camera's mechanical discipline to rein in amateurs' willful enthusiasms.

Conclusion

Meteorologists' vision of amateurs supplying thousands of sharp photographs to centralised surveys never came to pass. The International Survey of the Sky was a failure, short on photographic submissions and even shorter on usable images.[51] Hill's cloud camera met a similar fate: few meteorologists, much less amateurs, used it after 1935. Trust in photographs wavered, and ultimately proved unequal to the promise of newer technologies such as radar that even further excluded amateurs from the intellectual work of

50 Gregory, 'Amateurs as Pioneers', p. 104.
51 Cave, 'The International Survey of the Sky'.

meteorology. Since then, the cloud camera's function has become increasingly alien to our notions of meteorological practice.

The Whipple Museum has many such objects. While the use of many artefacts is obvious and intuitive to our present scientific sensibilities – objects like early microscopes, calculators, and globes – others speak to encounters between scientists, their instruments, and nature that appear bizarre and inexplicable. Why would early inventors of the fish-eye lens all identify clouds as its proper object of depiction? Why did meteorologists care so much about the taxonomy of clouds, those most formless and ephemeral of things, in the first place? Although they may speak to historical dead-ends, our most anachronistic, whimsical, and weird objects attest to an important fact: the choreography of encounters between scientists, their tools, and nature is momentary, mutable, and justified by reference to changing contexts. By examining Hill's cloud camera and seeking to explain its origins, its successes, and the conditions under which it achieved brief fame, connections between clouds, photographs, synoptic maps, meteorologists, and amateur volunteers are brought into focus.

13 ❧ Chicken Heads and Punnett Squares: Reginald Punnett and the Role of Visualisations in Early Genetics Research at Cambridge, 1900–1930

MATTHEW GREEN

One might not expect to find eleven immaculately painted plaster chicken heads (Figure 13.1) in a museum of the history of science such as the Whipple. The heads are cast from individual birds: they each share with their originals the same lifelike heft, the same scarlet comb and wattle with the same stippled reptilian feel, the same plumage colouring – even a few of the same feathers, transferred during the moulding process. A slice from the right side of each head has been removed to form a flat surface, with the back edges bevelled and painted in black to bring the head's profile into relief when displayed on a table. The heads were made in the early 1930s and have been attributed to Reginald Punnett, Alfred Balfour Professor of Genetics at the University of Cambridge from 1912 to 1940. His experimental notebooks, held by the Cambridge University Library, reveal that during each of these twenty-nine years he conducted detailed breeding experiments with chickens – his commitment to poultry was such that in 1923 he admitted that 'the hen has seldom been out of my thoughts.'[1] Punnett's chicken-breeding experiments are bound up with the invention for which he is known today, a form of visualisation quite different from the Whipple's chicken heads: the Punnett square, a tabular array still used in genetics to represent the outcome of a cross between two organisms.

While interest in three-dimensional scientific models has grown among historians over the past two decades, scholars also acknowledge that such models remain an 'understudied historical resource'.[2] In investigating the various roles that models and visualisations can play in science, Punnett's case is a particularly fruitful one, because

---

1 R. C. Punnett, 'Preface' to *Heredity in Poultry* (London: Macmillan, 1923).
2 J. Nall and L. Taub, 'Three-Dimensional Models', in B. Lightman (ed.), *A Companion to the History of Science* (Chicester: Wiley Blackwell, 2016), p. 572; and S. de Chadarevian and N. Hopwood (eds.), *Models: The Third Dimension of Science* (Stanford: Stanford University Press, 2004).

Figure 13.1 Eleven painted plaster chicken heads, attributed to Reginald Punnett, early 1930s. Image © Whipple Museum (Wh.6547).

his scarcely studied career spans a period in which Mendelian genetics in Britain was an emerging science that operated at the geographical and academic fringes of the University. Funding and academic reputation were at stake in debates over the practical benefits of this recently rediscovered way of understanding heredity; new research techniques were being developed to prove and extend Mendel's laws; and new students had to be attracted to the discipline and trained in its theory and praxis. In this chapter, I am interested in how both Punnett's square and his chicken-head models, qua visualisations, played different but related roles in all three of these areas during this crucial period in the development of genetics in Britain. Historians of three-dimensional models have often under-stood their work as addressing a conspicuous void in broader studies of representational media in science, with the result that there is a dearth of scholarship that treats three-dimensional models together with other visual media. I follow Nick Hopwood in endeavouring to show that models and their uses in science are most clearly illumin-ated when their relations to and differences from other forms of visual media, including flat material such as the Punnett square, are made clear.[3]

I begin by describing Punnett's partnership with the geneticist William Bateson, and with the help of their famous comb experi-ment I explain the function of the Punnett square, which was

3 N. Hopwood, *Embryos in Wax: Models from the Ziegler Studio* (Cambridge: Whipple Museum for the History of Science, 2002).

developed around the same time. Scholarship on the Punnett square has tended to focus on its genesis and its use in research; I build on these accounts by exploring how the square, along with the chicken-head models, were used in Punnett's teaching.[4] While the practice of constructing a Punnett square imparted knowledge of Mendel's laws and the theoretical basis of genetics to students, the chicken heads served as in-class teaching aids, their lifelike detail helping students gain the phenotypic literacy so essential to the practice of breeding experiments in early genetics. I describe how these functions, along with the use of the Punnett square as a conceptual tool in research, dovetailed with the efforts of Bateson, Punnett, and others, in fundraising campaigns, to pitch genetics as a science with important practical yields. I then explore the differing afterlives of the square and the heads, explaining these differences in reproduction and dissemination not only as a function of their materiality but also in terms of their uses in the theory and practice of a rapidly changing science. This leads me to conclude with a brief philosophical discussion of how my account of Punnett's chicken-head models aligns with and informs practice-centric accounts of the structure of knowledge in genetics.

The Comb Experiment and the Punnett Square

On Christmas Day, 1903, Punnett received a letter from his older colleague William Bateson, with an exciting request. Bateson, a fellow of St John's College, had already been studying variation and heredity for over a decade when he was made aware of Gregor Mendel's hybridisation experiments in 1900;[5] he quickly became an ardent defender of Mendelism, and sought to demonstrate and extend Mendel's laws by conducting carefully controlled breeding experiments with chickens and other organisms, largely at his home

4 On the genesis of the Punnett square, see A. W. F. Edwards, 'Punnett's Square', *Studies in History and Philosophy of Biological and Biomedical Sciences*, 43 (2012), pp. 219–24; and A. W. F. Edwards, 'Punnett's Square: A Postscript', *Studies in History and Philosophy of Biological and Biomedical Sciences*, 57 (2016), pp. 69–70. For use in research, see W. C. Wimsatt, 'The Analytic Geometry of Genetics: Part I: The Structure, Function, and Early Evolution of Punnett Squares', *Archives of the History of Exact Sciences*, 66 (2012), pp. 359–96.
5 This date has been disputed: see R. Olby, 'William Bateson's Introduction of Mendelism to England: A Reassessment', *British Journal for the History of Science*, 20.4 (1987), pp. 399–420.

in Grantchester.[6] By 1903, Bateson needed help: his wife and assistant Beatrice would be 'incapacitated' (i.e., pregnant) for the next season, and the breeding experiments – the menial and technical tasks involved in raising hundreds of chicks and carefully recording their characteristics – were 'not a one-man job'.[7] The sole Fellow at a Cambridge college engaged in genetics, Bateson's academic position was lonely and precarious. He had to cobble together an income and financial support for his research from different studentships, bequests, and individual donors, and would not secure a professorship until 1908, and then only in biology.[8] Luckily for him, Punnett's response to his request came less than a week later, and was enthusiastic: 'There is nothing I should like better.'[9]

Thus began a six-year partnership that introduced Punnett to chicken breeding and cemented his interest in Mendelism. One of Mendel's key insights was that *factors* (roughly equivalent to today's genes) were separate from but responsible for certain observable traits; by following patterns in the inheritance of observable traits, Mendel was able to define several laws that governed the inheritance and expression of factors. Early geneticists such as Bateson and Punnett experimentally confirmed many of Mendel's findings, but also encountered traits with inheritance patterns that weren't easily explained by his laws. One of their most famous experiments from this period was a demonstration of epistasis, or interaction between different genes, in four comb types in chickens: the rose, the pea, the

6 A. G. Cock and D. R. Forsdyke, *Treasure Your Exceptions: The Science and Life of William Bateson* (New York: Springer, 2008); and M. Richmond, 'The "Domestication" of Heredity: The Familial Organization of Geneticists at Cambridge University, 1895–1910', *Journal of the History of Biology*, 39 (2006), pp. 565–605.

7 B. Bateson, *William Bateson, F.R.S., Naturalist: His Essays and Addresses, Together with a Short Account of His Life* (Cambridge: Cambridge University Press, 1928), p. 87; Cambridge University Library Manuscripts & University Archives, William Bateson: Scientific Correspondence and Papers (hereafter Bateson Papers), 25 December 1903 (Add.8634/H.31). Until then, the breeding experiments *had* been a one-man job, since Bateson's collaborators and assistants were almost all women: his wife, Beatrice Bateson, and various female scientists associated with Newnham College, including Edith Saunders, Hilda Blanche Killby, and Muriel Wheldale (all of whom worked primarily with plants); see Richmond, 'The "Domestication" of Heredity'; and M. Richmond, 'Women in the Early History of Genetics: William Bateson and the Newnham College Mendelians, 1900–1910', *Isis*, 92.1 (2001), pp. 55–90.

8 Bateson, *William Bateson*, pp. 317–33.

9 Bateson Papers, 30 December 1903 (Add.8634/H.31).

walnut, and the single.[10] These observable characteristics, now collectively referred to as an organism's phenotype, were known to Punnett as *unit-characters*, and like Mendel he termed *factors* the 'somethings' which corresponded to these unit-characters and were contained in and inherited through parental gametes.[11] When Punnett and Bateson crossed a rose-comb chicken and a pea-comb chicken, they might have expected that the first generation produced (F_1) would be comprised of roses and peas in a ratio of 3 : 1, suggesting that the more common form of comb was dominant over the other. This would have aligned with Mendel's laws of independent assortment and of dominance. Instead, F_1 consisted entirely of chickens with a third form of comb, the walnut. Furthermore, a walnut × walnut cross produced a second generation (F_2) in which chickens with four types of comb – the walnut, the rose, the pea, and the single – appeared in a ratio of 9 : 3 : 3 : 1. Mendel had observed this same ratio with two independent pairs of dominant–recessive unit-characters, such as seed colour and seed shape in pea plants, but never with the same unit-character. This suggested that comb type, as a unit-character, was determined by two interacting factors – an example of epistasis.

The design of the experiment is not complicated, but it can be difficult to grasp the inheritance pattern with only a verbal description. When Punnett explained this experiment in the 1911 edition of his short but popular book *Mendelism*, he illustrated the inheritance pattern in F_2 using a method he had invented in 1906, now called the Punnett square (Table 13.1).[12] I summarise his description here. Both parents of F_2 are walnut-combed, with the factors *RrPp*: the capital letters represent dominant factors, and the lower-case recessive. Each gamete from each parent contains only one factor from each dominant–recessive pair, so the egg cell and sperm cell that form the zygote will each randomly contain *RP*, *Rp*, *rP*, or *rp*. Chicks end up with varying combinations of these factors. Any chick with at least one dominant *R* factor and one dominant *P* factor will go on to exhibit a walnut comb, like its parents; any chick with *R* as its only dominant will develop a rose comb, and likewise with *P* for the pea comb; and the rarest chick with no dominant factors will have a

10 W. Bateson, R. C. Punnett, and E. R. Saunders, 'Experimental Studies in the Physiology of Heredity', *Reports of the Evolution Committee of the Royal Society*, 2 (1905), pp. 1–131 and 3 (1905), pp. 1–53; and F. B. Hutt, *Genetics of the Fowl* (New York: McGraw-Hill, 1949).

11 R. C. Punnett, *Mendelism* (London: Macmillan, 1911), p. 31.

12 Punnett, *Mendelism*, p. 38.

TABLE 13.1 An example of a Punnett square

RP
RP
   Walnut
	RP
	Rp
	   Walnut
		RP
		rP
		   Walnut
			RP
			rp
			   Walnut

Rp
RP
   Walnut
	Rp
	Rp
	   Rose
		Rp
		rP
		   Walnut
			Rp
			rp
			   Rose

rP
RP
   Walnut
	rP
	Rp
	   Walnut
		rP
		rP
		   Pea
			rP
			rp
			   Pea

rp
RP
   Walnut
	rp
	Rp
	   Rose
		rp
		rP
		   Pea
			rp
			rp
			   Single

single comb. There are sixteen possible combinations of factors, which Punnett represented in a 4 × 4 square tabular array. To construct the square, one begins by filling the cells in the top row with *RP*, the second row with *Rp*; the third with *rP*; and the fourth with *rp*. These represent the factors that each zygote receives from the egg cell. Then one follows the same pattern in the columns, representing the factors from the sperm cell. Each square in the completed table represents a chick, and the four letters that appear in each square represent that chick's combination of factors, from which one can infer its comb type. The genius of the Punnett square is that it reveals all of the possible combinations of factors and resulting unit-characters for any given cross; it demonstrates how one arrives at the ratio of 9 : 3 : 3 : 1 for those unit-characters from the combination of factors; and it does both of these elegantly and economically.

The biochemist Dorothy Needham attended Punnett's undergraduate course in genetics in 1917–18, when she was a student at Girton College. Her lecture notes reveal that Punnett used the square to illustrate basic breeding experiments and illuminate Mendel's laws: there are three Punnett squares in her notes for the first lecture of the course, including a square identical to the one in Table 13.1, used to illustrate epistasis.[13] He gave these lectures, which were intended for students reading Zoology for Part I of the Natural Sciences Tripos, on Tuesdays and Thursdays for the duration of his professorship.

---

13 Girton College Archive & Special Collections, Personal Papers of Dorothy Needham, Undergraduate Notebooks, volume 23, 3/2/23.

Needham was a meticulous note-taker and may have recorded more information than the average student, but it is likely that Punnett encouraged all students in his classes to construct Punnett squares themselves. In his book *Heredity in Poultry*, a 'handy guide' to Mendelian inheritance in chickens, he not only includes several Punnett squares, but actually takes the reader through every step of creating one, ordering her to 'draw' the lines, 'write' the letters to fill in each cell, and 'examine' the finished product to draw conclusions from the information thus represented.[14] Punnett understood that the student of genetics needed to learn by making and doing: by constructing the square herself repeatedly, she would be led through the logic of genetics – indeed, through a demonstration of Mendel's laws.

Since each parent contributes two of four possible 'letters' to each row or column, and there are four rows or columns, each with an equal number of squares, it is evident that one of each pair of factors is inherited from each parent, and that the likelihood of getting either is random, resulting in equal distribution. This is Mendel's first law, the law of segregation. Once the whole square has been filled with letters, the reader then fills in the observable traits according to the capital letters in each square: in doing so she demonstrates Mendel's third law, the law of dominance, which states that dominant alleles (capital letters) override the expression of recessive alleles (lower-case letters). Finally, the reader tallies up the occurrences of each trait to arrive at the ratio of 9 : 3 : 3 : 1, which is a demonstration of Mendel's second law, the law of independent assortment.[15] Amazingly, with a single visualisation, Punnett is able to guide a student through the demonstration of all three of Mendel's laws of heredity. It is the teacher's hope that the completion of a Punnett square entails a kind of 'a-ha!' moment of pattern recognition and understanding; the didactic power of the square lies in its ability to allow the reader to see and understand *for herself* how the laws of genetics are borne out in particular experiments.

14 Punnett, *Heredity in Poultry*, preface, and pp. 14–15, 24, 28, 30, 33.
15 This law states that different factors are inherited independently of one another. The Punnett square in Table 13.1 does show this, but it would be most clearly illustrated with a dihybrid cross square, which Punnett actually used more frequently in his books: see Punnett, *Heredity in Poultry*, p. 14.

'How to See' a Chicken Head

In 1910 Bateson departed Cambridge for the John Innes Horticul-
tural Institute, but he remained a mentor and correspondent of
Punnett, who stayed at Cambridge and continued working with
chickens for the rest of his life, as Alfred Balfour Professor of
Genetics. In the late 1920s, with the financial support of the National
Poultry Institute, Punnett and his assistant Michael Pease developed
the Cambar: the first autosexing poultry breed whose chicks could be
sexed at birth by their plumage, allowing egg producers to immediately rid themselves of cockerels and reduce costs.[16]

The Whipple's accession catalogue states that Punnett's plaster
chicken heads may have been a part or a product of his research
surrounding the Cambar, but this seems unlikely for several reasons.
Punnett and Pease acknowledged the creation of the Cambar in a
paper published in 1930, the year their partnership ended and two
years before the earliest date inscribed on the chicken heads; at least
one of the chicken heads clearly exhibits a rose comb, which is not
a trait associated with any of the breeds involved in Punnett and
Pease's research into autosexing; and finally, it is unclear what such
models would have added to Punnett's research, since he and Bateson had already developed a consistent system of notation for
breeding experiments, in use and virtually unchanged by Punnett
since 1903.[17]

If Punnett had wanted to record the chickens' traits for his
research, he would only have had to note them down; if he
had wanted a representation of those traits, he could have taken
photographs, which he had done in his research for the Cambar.[18]
Instead, he went to the trouble of creating painted, textured three-
dimensional models, suggesting that the chicken heads were meant
not only to be viewed but also to be interacted with: to be touched, to
be compared with each other, and perhaps to be brought into the
classroom and used as a teaching aid. The use of models and physical
specimens as teaching aids had a precedent in Punnett's own life:

16 Punnett's relationship with the poultry-breeding community and industry is
    fascinating, but falls outside the scope of this chapter: see J. Marie, 'For Science,
    Love and Money: The Social Worlds of Poultry and Rabbit Breeding in Britain,
    1900–1940', *Social Studies of Science*, 38.6 (2008), pp. 919–36.
17 R. C. Punnett and M. S. Pease, 'Genetic Studies in Poultry: VIII. On a Case of
    Sex-Linkage within a Breed', *Journal of Genetics*, 22 (1930), p. 397; and R. C.
    Punnett, 'Genetic Studies in Poultry: XI. The Legbar', *Journal of Genetics*, 41
    (1940), pp. 1–9.
18 Punnett and Pease, 'Genetic Studies in Poultry: VIII', Plate XVII.

he was at an early, formative period in his career when he spent three years as a demonstrator at the University of St Andrews, where specimens from the departmental museum were fundamental to the curriculum. Professors used the specimens in lectures to illustrate key points, and students were also questioned on the specimens for their examination.[19] Professors at Scottish universities were known for making use of models in their lectures because, Margaret Maria Olszewski argues, their salary depended on lecture attendance, and the engaging models often succeeded in attracting students.[20] While Punnett's professorship guaranteed a salary, his genetics course was not covered by the Zoology composition fee, and cost £1 1s per term to attend.[21] It is not clear whether Punnett received the fee himself, but in any case to students the charge constituted an obstacle, which it was in Punnett's interest to overcome by making his lectures engaging and hands-on.

The chicken heads are an ideal size for handling, and the level of detail both in the painting and in texture (Figure 13.1) suggests that they were meant to be observed from up close. No other medium, short of real live chickens – highly impractical in the lecture room – would be able to convey this sensory information with such vividness and specificity. I have described the heads as sharing many of the same characteristics, and even some of the same matter, as their originals. In her 2004 study of natural history displays in early-twentieth-century German museums, Lynn Nyhart argues that such models, with their lifelike detail, were intended to teach the lay public 'how to see – how to look thoughtfully at objects ... [and] to take in their meaning'.[22] Similarly, Hopwood observes that the use of Ziegler's wax embryo models in the classroom was meant to teach students how to see microscopically.[23] But did Punnett's students really need to be taught how to see a chicken?

In order to answer this, we need to understand the skills involved in the practice of genetics in this period. Bateson asserted that there

---

19 F. A. E. Crew, 'Reginald Crundall Punnett, 1875–1967', *Biographical Memoirs of Fellows of the Royal Society*, 13 (1967), p. 312.

20 M. M. Olszewski, 'Auzoux's Botanical Teaching Models and Medical Education at the Universities of Glasgow and Aberdeen', *Studies in History and Philosophy of Biological and Biomedical Sciences*, 42 (2011), 285–96.

21 *Cambridge University Reporter* (Cambridge: Cambridge University Press, 1933–4), p. 807.

22 L. Nyhart, 'Science, Art, and Authenticity in Natural History Displays', in S. de Chadarevian and N. Hopwood (eds.), *Models: The Third Dimension of Science* (Stanford: Stanford University Press, 2004), pp. 307–35, on p. 315.

23 Hopwood, *Embryos in Wax*, p. 33.

was no other way to learn the laws of heredity and variation than by 'the direct examination of the phenomena ... [which] can only be provided by actual experiments in breeding'.[24] For Bateson and for Punnett, research in Mendelian genetics took place not in the laboratory, but outdoors: Punnett kept poultry pens and a shed with incubators on the University Farm, two miles northeast of the city centre but quite close to Whittingehame Lodge, where Punnett lived and kept chickens in the adjoining rooms and yard; he also carried out sweet-pea experiments at the Botanic Gardens.[25]

Punnett's experimental notebooks provide a glimpse into how these experiments proceeded. The essential goal was to track patterns in the inheritance of particular traits, and to do this one needed very large sample sizes. On average, Punnett bred about 500 chicks per year – though sometimes as many as 1,000 – and each had its own page in that year's notebooks. At the top of this page Punnett noted the chick's lay date and hatch date, a code for its parentage, and an identifying number that corresponded to a brass label clipped around the chick's leg. A list of dated observations followed as the chicken developed, with the death date concluding the entry. However, many of the chick's relevant traits could be recorded straight after hatching: the comb type, plumage colour, number of toes, presence of feathers on the legs, and so on.[26] With so many chicks to assess and so many traits to record, Punnett and Bateson initially developed a series of abbreviations that Punnett continued to use in the notebooks for the rest of his life: for example, 'lt., nts., r.c., n.e., f. l.' meant 'light down, no coloured ticks seen, rose comb, no extra toe, feathering on leg'.[27] Speed was valuable, but so was precision: traits could be ambiguous and require further description, and it was also important to note similarities between birds of different lineages as they matured.

For the geneticist, then, seeing a chicken did have to be taught. One needed a practised eye for detail and the ability to isolate and

24 B. Bateson, 'Heredity and Variation in Modern Lights', in A. C. Seward (ed.), *Darwin and Modern Science: Essays in Commemoration of the Centenary of the Birth of Charles Darwin and of the Fiftieth Anniversary of the Publication of the Origin of Species* (Cambridge: Cambridge University Press, 1909), p. 92.
25 D. L. Opitz, 'Cultivating Genetics in the Country: Whittingehame Lodge, Cambridge', in D. N. Livingstone and C. W. J. Withers (eds.), *Geographies of Nineteenth-Century Science* (Chicago: University of Chicago Press, 2011), pp. 73–98; and Crew, 'Reginald Crundall Punnett', p. 312.
26 Cambridge University Library Manuscripts & University Archives, Bateson–Punnett Notebooks (MS Add.10161); see, for example, 1931 notebook, p. 57.
27 R. C. Punnett, 'Early Days of Genetics', *Heredity*, 4.1 (1950), p. 6.

identify traits, which explains why having lifelike models was so important for Punnett. Drawn in by the heads' detail and novelty, the student would notice the subtle differences in specific characteristics: how one comb type might be distinguished from another, how to identify ambiguous traits within a type, how traits such as wattle size and plumage might differ between otherwise similar males and females (the sex of most models is noted on the rear), and so on. For Punnett, the careful observation and comparison of specific traits was the very foundation of studies in genetics, and the chicken-head models would have been indispensable in teaching students 'how to see' in the mode that was necessary for early genetics research at Cambridge.

Genetics: A New and Changing Discipline

While the Punnett square taught Mendelian theory through the practice of constructing a combinatorial diagram, the chicken heads imparted the visual skills necessary for the practice of early experimental genetics at Cambridge. In tandem, the two visualisations made pursuing genetics accessible and appealing to students. Attracting and training new talent would have been of great importance to Punnett, since genetics in this period was far from an established discipline in Britain. During Punnett and Bateson's collaborative period, leading journals such as *Nature* had refused to publish their research, which led them to jointly establish the *Journal of Genetics* in 1911.[28] They also helped form the Genetical Society of Great Britain in 1919, but academic enthusiasm for the discipline remained limited. By 1924, the society comprised mostly private individuals and plant and animal breeders; fewer than a quarter of the 108 members were affiliated with universities, and this was not to increase significantly until after the Second World War.[29]

Difficulties were also encountered on the path to the endowment of Punnett's 1912 Alfred Balfour Professorship, and published fundraising pleas from supporters of the protracted campaign for a genetics chair offer insight into the self-understanding and self-fashioning of an emerging discipline. Bateson, along with Punnett, Adam Sedgwick, Arthur Balfour, and other well-connected friends of

28 Punnett, 'Early Days of Genetics'.
29 W. Leeming, 'Ideas about Heredity, Genetics and "Medical Genetics" in Britain, 1900–1982', *Studies in History and Philosophy of Biological and Biomedical Sciences*, 36.3 (2005), pp. 538–58.

his, petitioned individual donors, the Evolution Committee of the Royal Society, the University of Cambridge Council of the Senate, and the Cambridge community for almost ten years, insisting that research into Mendelian genetics had a 'sure prospect of future success'.[30] Generally, this group adopted a strategy that focused on the practical benefits that a deeper understanding of Mendelian heredity might yield. Their 'Plea for Cambridge', published in the *Quarterly Review* in 1906, takes such an approach:

> The extreme importance of these studies [in genetics], which, if they prove a key to heredity, will place in man's hands an instrument as powerful as Watt's application of steam, is shown by the fact that Mr [Rowland] Biffen has already discovered that susceptibility to rust in wheat is Mendelian, and is thus a property which may be eliminated by breeding.[31]

According to this projection, the success of genetics research could be measured in part by the number of different isolated traits that could be shown to fall under Mendel's laws of inheritance, so that they could be bred out (or, in the case of the Cambar, bred in) for practical purposes. Punnett took a similar tack in *Mendelism*, where he argued that the Mendelian laws had been 'found to hold good' for everything from coat colour to the waltzing habit of Japanese mice, and that it would be reasonable to expect that, over time, more traits would be 'brought into line in the light of fuller knowledge'.[32]

Attracting and retaining new students helped recruit the manpower needed to conduct the labour-intensive breeding experiments that would uncover Mendelian patterns in the heredity of more and more traits. But Punnett's visualisations also served the research programme of early genetics in a more direct sense. As William Wimsatt observes, the Punnett square was not only a didactic aid, but in fact constituted a conceptual tool that could be used to make inferences from observed phenotypic patterns to possible genetic explanations of them.[33] The square's way of integrating and ordering information permitted the extension of Mendel's laws to new

30 Petition circulated by Arthur Balfour, quoted in Opitz, 'Cultivating Genetics in the Country', p. 86.
31 Cambridge University Association, 'A Plea for Cambridge', *Quarterly Review*, 204 (1906), pp. 521–2.
32 Punnett, *Mendelism*, pp. 29–30. For more on waltzing mice, see W. H. Gates, 'The Japanese Waltzing Mouse, Its Origin and Genetics', *Proceedings of the National Academy of Sciences of the United States of America*, 11.10 (1925), pp. 651–3.
33 Wimsatt, 'The Analytic Geometry of Genetics', p. 373.

hereditary patterns, by working backwards from observed character-istics to the genetic factors at play. To the extent that geneticists, aided by the Punnett square, could bring an ever-wider scope of observable traits within the regular order of hereditary laws, they justified the existence of their discipline in the university and pro-moted the endowment of their science.[34]

Having discussed the various functions of the square and the chicken heads in the particular context of early genetics at Cam-bridge, further and broader insight into the nature of models in science may be gleaned by elucidating the afterlives of these visual-isations. The Punnett square was rapidly disseminated in articles, letters, books, and textbooks; it became so useful and ubiquitous that no biologist educated after 1920 could have avoided encountering it.[35] Meanwhile, Punnett's chicken heads seem to have fallen into obscurity, with (apparently) no one after Punnett taking much notice of them at all until the Whipple Museum acquired them in 2013. Why might this be so?

At least some of the credit for the spread of the Punnett square should be given to Punnett himself, who included eight examples in the 1911 edition of *Mendelism* (which went through seven editions and several translations) and about as many in the less popular *Heredity in Poultry*.[36] Reviewers noted that *Mendelism* was 'richly illustrated with figures and coloured plates'.[37] Plates were attractive and clearly an asset, but they were also expensive to reproduce, while simple tables such as the Punnett square could be easily and cheaply typeset, which contributed to their propagation.[38] The square's sim-plicity also meant that it was somewhat flexible, and subsequent copiers of Punnett's square made important additions that became canonical: Herbert Walter added gamete types in the margins in 1913, and Edmund Sinnott and Leslie Clarence Dunn added visual

---

34 In this capacity, the Punnett square could be understood as what Ursula Klein has termed a 'paper tool': a visible and manoeuvrable 'tool' that, while not physically interacting with the object of study like a laboratory tool, still permits the manipulation and comparison of relevant representations of the research object. While potentially fruitful, this comparison has to do with the use of Punnett squares in research, which is not my primary focus in this chapter. See U. Klein, *Experiments, Models, Paper Tools: Cultures of Organic Chemistry in the Nineteenth Century* (Stanford: Stanford University Press, 2003), pp. 245–7.

35 Wimsatt, 'The Analytic Geometry of Genetics', p. 393.

36 Punnett, *Mendelism*; and Punnett, *Heredity in Poultry*.

37 L. Doncaster, 'Review: *Mendelism*, Third Edition', *Eugenics Review*, 4.2 (1912) p. 206; and G. H. Shull, 'Review of *Mendelism* by R. C. Punnett', *Botanical Gazette*, 52.3 (1911), pp. 235–6.

38 Wimsatt, 'The Analytic Geometry of Genetics', p. 363.

representations of characters in the individual cells in 1925.[39] Wimsatt notes that the square's open-ended structure permitted 'enormous adaptive radiation into a variety of new contexts where [the square has] played a role in conceptualizing and solving a number of diverse problems'.[40] Punnett first used the square to represent a dihybrid pea-plant cross, but the square was never bound to the particular content for which it was initially conceived: it was not only simple, but also easily adaptable.

In contrast, the chicken-head models are by nature irreproducible in a strict sense, because they are plaster casts of individual, short-lived birds. Plaster hardens over time and becomes fragile, which would have restricted the models' movement. For the uses I have described, the choice of particular bird is unimportant so long as variety exists amongst the models, so in theory one could have made similar models from different individuals without any loss in utility. However, this is certainly much more difficult, time-consuming, and expensive than typesetting and printing a black-and-white table.

This cannot be the only reason why the heads were never reproduced – after all, three-dimensional models such as Ziegler's wax embryos and Auzoux's papier-mâché anatomical and botanical models achieved a fairly wide circulation, even though they were expensive and difficult to make compared with books or other flat media.[41] Rather, the chicken-head models were rooted in the particular practice of genetics in the service of which they were created and used – that is, a science based on the tracking of directly observable phenomena in 'backyard' breeding experiments. The practice of genetics changed rapidly over the course of the following decades.[42] T. H. Morgan's experiments with fruit flies precipitated that organism's dominance in genetics experimentation, and shifted the locus of research away from the chicken pen and botanical garden into the laboratory. Teaching geneticists 'how to see' no longer meant telling a rose comb from a walnut but, for example,

39 Wimsatt, 'The Analytic Geometry of Genetics', pp. 388–9; p. 371.
40 Wimsatt, 'The Analytic Geometry of Genetics', p. 393.
41 Hopwood, *Embryos in Wax*; and Olszewski, 'Auzoux's Botanical Teaching Models'.
42 Some of these developments were already under way when Punnett made the heads in 1932–4, but Cambridge was somewhat isolated and not an important centre for genetics research at the time: see M. Ashburner, 'History of the Department', University of Cambridge Department of Genetics website, www.gen.cam.ac.uk/department/department-history (accessed 11 December 2018).

learning to see through a microscope.[43] As the practical knowledge involved in doing genetics changed, Punnett's highly specific chicken-head models quickly became obsolete, while his square lived on, kept alive thanks to its simplicity and theoretical adaptability, as well as the continued relevance of Mendelian laws to the study of genetics.

Still, in the context of the particular kind of genetics research being carried out in Cambridge in the early twentieth century, the practical knowledge imparted by the use of the chicken heads in teaching was fundamental. This insight both aligns with and illuminates philosophical accounts of early genetics in the United States by scholars such as Kenneth Waters, who have similarly attempted to supplant theory-centric accounts by emphasising the complementary role of practice.[44] According to Waters, 'philosophers typically assume that scientific knowledge is ultimately structured by explanatory reasoning and that research programs in well-established sciences are organized around efforts to fill out a central theory and extend its explanatory range.'[45] This account is by no means incorrect: indeed, the extension of the explanatory range of a theory is precisely the kind of process at play in the use of the Punnett square in research as a conceptual tool to extend Mendelian laws, as described above.

However, the chicken heads testify to the deficiency of such an account. The heads and the practical know-how they impart are not intended to explain anything about heredity – rather, they make possible the transmission of certain *investigative strategies*, namely the observation of chickens in breeding experiments.[46] Waters's larger point is that these investigative strategies, underpinned by practical knowledge, are central to research programmes, but are often overlooked in philosophical accounts of the structuring of scientific knowledge. We have seen that the square and the heads, particularly their use in teaching, made possible the dissemination of

43 R. Kohler, *Lords of the Fly: Drosophilia Genetics and the Experimental Life* (Chicago: University of Chicago Press, 1994).

44 C. K. Waters, 'A Pluralist Interpretation of Gene-Centered Biology', in S. H. Kellert, H. E. Longino, and C. K. Waters (eds.), *Scientific Pluralism*, Minnesota Studies in Philosophy of Science 19 (Minneapolis: University of Minnesota Press, 2004), pp. 190–214.

45 Waters, 'A Pluralist Interpretation of Gene-Centered Biology', p. 783.

46 C. K. Waters, 'What Was Classical Genetics?', *Studies in the History and Philosophy of Science*, 35 (2006), pp. 783–809.

experimental expertise and theoretical understanding, processes that were crucial to research programmes in early genetics.

Not only does my analysis of Punnett's chicken heads support Waters's argument that practical know-how was as important as theoretical explanation in early genetics, but it also helps explain why the former is frequently overlooked: it can be difficult to recover. In the case of early genetics at Cambridge, practical knowledge was most clearly embodied by fragile plaster models whose reproduction would have been difficult and time-consuming, if it were even possible. Luck, institutional resources, and curatorial diligence were all fundamental in allowing the uncovering of this knowledge and the reconstruction of a highly specific research method – and such a combination of resources is seldom guaranteed.

Punnett's visualisations – the square and the heads – played important but different roles in the classroom: while the square helped students understand Mendelian laws, the heads trained them to isolate, identify, and differentiate particular traits. To the extent that the Punnett square served as a conceptual tool used to infer the underlying factors at play in inheritance patterns, it also contributed to the narrative, advanced by Bateson, Punnett, and others, that genetics could yield practical benefits by enabling greater control over particular traits in domesticated animals and plants. I have shown that visualisations played a crucial role in establishing genetics as an academic discipline at Cambridge by disseminating theoretical and practical knowledge through educational channels and by helping to justify the endowment of a professorial Chair in 1912. I have also described how the differing afterlives of the square and the heads are the result of important material and scientific conditions, including the differing rates of change in the theory and practice of genetics in the first half of the twentieth century. If the roles of visualisations in science are both as fundamental and as varied as I have described, then further study of such visualisations and of the nuanced differences in their use and reproduction has the potential to illuminate the didactic strategies, self-fashioning, and research practices of additional scientific disciplines.

14 ❧ Stacks, 'Pacs', and User Hacks: A Handheld History of Personal Computing[*]

MICHAEL F. MCGOVERN

Introduction

In 1988, a churlish columnist for *The Daily Telegraph* by the name of Boris Johnson remarked upon the Whipple Museum's recent acquisition of Cambridge architect Francis Hookham's extensive handheld calculator collection. Ironically applauding the museum's curatorial foresight, the author encouraged it to 'branch out from mere science and become a major tourist attraction for its peerless collection of obsolete gadgets of every kind'.[1] This equation of calculators with worn socks and kitchen appliances pithily suggests how rapidly perceptions of earlier computing technology changed as 'personal' desktop computers became commonplace.[2] Conventional wisdom locates the origins of the personal computer (PC) in the Jobs family garage *circa* 1976 – the more erudite in the January 1975 issue of *Popular Electronics* announcing the Altair 8800 – but the first device actually marketed as such was a different 'PC' altogether: the HP-65, a programmable calculator launched in 1974.[3] In this

* For guidance through an earlier version of the paper and the opportunity to publish here, I thank Joshua Nall, Richard Staley, and Liba Taub. For sharing their personal histories and providing generous input, I thank Richard J. Nelson, Ralph Bernstein, and Garth Wilson. Conference participants in Trondheim, Norway, and Dearborn, Michigan, offered valuable feedback and encouragement. Finally, Erika L. Milam, Emily Thompson, and David Dunning helped me reckon with an earlier version of my academic self with conversation and acute editorial input.

1 The name is no coincidence: this was an early gig for the future Mayor of London and Conservative Foreign Secretary, Boris Johnson: 'Enter the Age of the Instant Antique', *Daily Telegraph*, 26 October 1988.

2 As David Edgerton argues, innovation-centred thinking conceals the more fundamental world of technologies-in-use: D. Edgerton, 'Innovation, Technology, or History: What Is the Historiography of Technology about?', *Technology and Culture*, 51.3 (2010), pp. 680–97.

3 P. E. Ceruzzi, *A History of Modern Computing*, 2nd edn (Cambridge: MIT Press, 2003), p. 213. John Markoff has argued that the term can be traced back into the 1950s: J. Markoff, 'How the Computer Became Personal', *New York Times*, 19 August 2001, www.nytimes.com/2001/08/19/business/how-the-computer-became-personal.html.

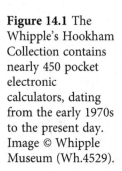

Figure 14.1 The Whipple's Hookham Collection contains nearly 450 pocket electronic calculators, dating from the early 1970s to the present day. Image © Whipple Museum (Wh.4529).

chapter, I work from the Hookham collection (Figure 14.1) to investigate the material culture and moral economy of handheld programming, a granular approach to one of the more intractable questions in the history of modern technology: how did computing become personal?

Influential accounts of the history of computing in the United States have shown how computers were marked from the outset by their development within the Cold War military–industrial complex, but came to be embraced by the counterculture movement that coalesced in protest against it.[4] This appreciation for social politics has given texture to a narrative arc that privileges invention, market strategy, and pioneering industries that took on the overhead of

---

4 Paul Edwards's account of computers becoming thinkable only within a kind of 'closed-world' discourse of Cold War defence systems has been elaborated upon by Janet Abbate, who shows how the defence research network, ARPANET, formed the basis for the Internet, and Rebecca Slayton, who shows how ideas about expertise in missile defence were a moving target. See P. N. Edwards, *The Closed World: Computers and the Politics of Discourse in Cold War America* (Cambridge: MIT Press, 1996); J. Abbate, *Inventing the Internet* (Cambridge: MIT Press, 1999); and R. Slayton, *Arguments That Count: Physics, Computing, and Missile Defense, 1949–2012* (Cambridge: MIT Press, 2013). While Fred Turner provides the best contextualised account of a counterculture information ethos evolving in tandem with defence research culture, he draws on earlier work including Steven Levy's earlier journalistic account of hacker culture. See F. Turner, *From Counterculture to Cyberculture: Stewart Brand, the Whole Earth Network, and the Rise of Digital Utopianism* (Chicago: University of Chicago Press, 2006); S. Levy, *Hackers: Heroes of the Computer Revolution* (London: Penguin, 1994); T. Bardini, *Bootstrapping: Douglas Engelbart, Coevolution, and the Origins of Personal Computing* (Stanford: Stanford University Press, 2000); and Markoff, 'How the Computer Became Personal'.

expanded information processing, a historiographical trail blazed by historians of technology working from business history.[5] Historians of science, on the other hand, have largely sought answers to the question set by Jon Agar in his influential article 'What Difference Did Computers Make?' by highlighting the continuities between pre-digital data practices and the forms of organisation required to successfully apply computers to scientific problems.[6] In the midst of software *and* hardware crises in the early 1970s, programmable calculators were readily calibrated to existing computational practices, but what's more is that they followed their users home.[7]

The appeal of early PCs was largely limited to a subset of electronics hobbyists: hardware enthusiasts who typically assembled the machines themselves. In contrast, the thousands of engineers, scientists, and professionals who got their first taste of programming through calculators made these commodities *work* in novel ways. Thus, following their uptake allows us to consider the merits of two increasingly powerful approaches within the history of computing. The first goes beyond the sociological conceptualisation of 'users' towards a politically grounded social history.[8] This project is perhaps

5 Some noteworthy examples of this literature include Ceruzzi, *A History of Modern Computing*; M. Campbell-Kelly, *From Airline Reservations to Sonic the Hedgehog: A History of the Software Industry* (Cambridge: MIT Press, 2004); and J. Yates, *Structuring the Information Age: Life Insurance and Technology in the Twentieth Century* (Baltimore: Johns Hopkins University Press, 2005).
6 J. Agar, 'What Difference Did Computers Make?', *Social Studies of Science*, 36.6 (2006), pp. 869–907. Atsushi Akera approaches the institutional development of computing expertise as an 'ecology of knowledge', while other historians of science – many associated with a Max Planck Institute research group, 'Historicizing Big Data' – stress how scientific problems drove the application of new technologies. See A. Akera, *Calculating a Natural World: Scientists, Engineers, and Computers during the Rise of U.S. Cold War Research* (Cambridge: MIT Press, 2006); E. Aronova, C. von Oertzen, and D. Sepkoski (eds.), *Data Histories*, Osiris 2nd Series 32 (Chicago: University of Chicago Press, 2017).
7 For a comparative perspective on Hewlett Packard versus Texas Instruments calculators, and a parallel development in Soviet computing, see D. Ristanović and J. Protić, 'Once upon a Pocket: Programmable Calculators from the Late 1970s and Early 1980s and the Social Networks around Them', *IEEE Annals of the History of Computing*, 34.3 (2012), pp. 55–66; and Ksenia Tatarchenko, 'How Programmable Calculators and a Sci-Fi Story Brought Soviet Teens into the Digital Age', *IEEE Spectrum: Technology, Engineering, and Science News*, 27 September 2018, https://spectrum.ieee.org/tech-history/silicon-revolution/how-programmable-calculators-and-a-scifi-story-brought-soviet-teens-into-the-digital-age.
8 The move to re-centre the user in the sociology of technology came in response to criticisms that the Social Construction of Technology approach took only the engineer's perspective. See N. Oudshoorn and T. Pinch, 'Introduction: How Users and Non-Users Matter', in Nelly Oudshoorn and Trevor Pinch (eds.),

best characterised by the work of Joy Lisi Rankin, who argues forcefully that the educational contexts of early computing experiments did much to teach the machines themselves, and reconstructs how a masculine culture of computing *citizens* crystallised out of more capacious attempts at inclusion – only to be later diverted towards consumption.[9] The second approach engages with historical epistemology and material culture to emphasise the discontinuities brought about by technical constraints and opportunities.[10] Stephanie Dick's work on mathematical proofs exemplifies the payoffs of this approach: the affordances which arise from the interaction between hardware and software shape disciplines and machines downstream.[11] Here, I seek a middle ground between these approaches, following the interplay between the affordances of devices and forms of social organisation that made them work in new ways.

This chapter takes up the development of HP's programmable calculators within the broader technological and organisational

How Users Matter: The Co-construction of Users and Technology (Cambridge: MIT Press, 2005), pp. 1–25; and R. S. Cowan, 'The Consumption Junction: A Proposal for Research Strategies in the Sociology of Technology', in Wiebe E. Bijker, Thomas P. Hughes, and Trevor Pinch (eds.), *The Social Construction of Technological Systems*, anniversary edn (Cambridge: MIT Press, 2012), pp. 261–80. Drawing on labour histories of white-collar work, an emerging historiography situates the office politics of data processing within broader developments in labour discrimination and the social movements that emerged in response. See T. Haigh, 'Inventing Information Systems: The Systems Men and the Computer, 1950–1968', *The Business History Review*, 75.1 (2001), pp. 15–61; R. Slayton, 'Revolution and Resistance: Rethinking Power in Computing History', *IEEE Annals of the History of Computing*, 30.1 (2008), pp. 96–7; N. Ensmenger, *The Computer Boys Take Over: Computers, Programmers, and the Politics of Technical Expertise* (Cambridge: MIT Press, 2010); J. Abbate, *Recoding Gender: Women's Changing Participation in Computing* (Cambridge: MIT Press, 2012); and M. Hicks, *Programmed Inequality: How Britain Discarded Women Technologists and Lost Its Edge in Computing* (Cambridge: MIT Press, 2017).

9 J. L. Rankin, *A People's History of Computing in the United States* (Cambridge: Harvard University Press, 2018).

10 Examples of this approach within the history of science include P. Galison, *Image and Logic: A Material Culture of Microphysics* (Chicago: University of Chicago Press, 1997); and H.-J. Rheinberger, *An Epistemology of the Concrete: Twentieth-Century Histories of Life* (Durham: Duke University Press, 2010).

11 S. Dick, 'AfterMath: The Work of Proof in the Age of Human–Machine Collaboration', *Isis*, 102.3 (2011), pp. 494–505; and S. Dick, 'Of Models and Machines: Implementing Bounded Rationality', *Isis*, 106.3 (2015), pp. 623–34. Hallam Stevens argues along similar lines that focusing on data organisation obscures the process by which data are transformed by algorithms and digital work practices. See H. Stevens, 'A Feeling for the Algorithm: Working Knowledge and Big Data in Biology', *Osiris*, 32.1 (2017), pp. 151–74.

Figure 14.2
Included with
Francis Hookham's
donation to the
Museum was a wide
range of calculator
ephemera,
advertisements, and
instruction manuals.
Image © Whipple
Museum (Wh.4529).

context of computing in the 1960s and 1970s. As mainframe systems and their attendant bureaucratic order were on the decline, handheld devices were viable alternatives that permitted a kind of personalisation without impinging upon institutional information-processing norms. One might even argue that handheld calculators prefigured the mass market for desktop personal computers and the software to run them. My aim here is a more modest exercise: to glimpse an aspect of a broader social and technical transformation through a focal and materially rich case, allowing me to foreground the shifting forms of agency at stake in the personalisation process. Ephemera and devices within the diverse Hookham collection (Figure 14.2) evidence a kind of mutualistic exchange between user and manufacturer that eventually soured. I analyse this material to show how the moral, monetary, and material economies underlying this user network offer insight into the changing cultures of computing during this crucial period.

Challenging a Computer: the HP-35 'Electronic Slide Rule'

Tom Osborne was an unemployed engineer in 1964 when he set his hand at building a calculator in his apartment. In his recollection, he turned out a machine with 'more computing power per unit volume than had ever existed on this planet' – power that must have been so unfathomable that it failed to land him a job.[12] Only after countless pitches did an HP employee recommend his assistance on a project that became the HP-9100A: a programmable desktop computer that utilised Reverse Polish Notation (RPN), a logical syntax that facilitated the storage of variables to increase coding efficiency.[13] Its 64-bit read-only memory (ROM) gave it the memory-retrieval and processing capabilities equivalent to most computers of the day, or so its 1968 marketing material declared by calling it equal parts 'personal computer' and 'electronic genie'.[14] Initial market research, however, suggested that the device would sell better as a *calculator*, and HP reconsidered its approach.[15] In this section, I unpack how HP constructed a market for scientific handheld calculators during a time of rapid industry upheaval.

Mainframes in the 1960s were conceived and sold as *systems*. The apotheosis of this model was IBM's System/360, launched in 1964: a line of mutually compatible computers designed to preserve software compatibility as business and research programs grew.[16] IBM's modus operandi was to lease hardware, while offering software and support free of charge. This bundle of red-and-blue tape cabinets and consoles consolidated IBM's dominance over its competitors – labelled the 'Seven Dwarfs' – for a time, but the hegemony of bundled software was on the wane. One such dwarf, the Digital Equipment Corporation, sold 'minicomputers' to clients at a fraction of IBM's lease fees and encouraged independent modification rather than providing full support. They simply gave away manuals to support their flagship model, the PDP-8. This strategy appeared to work: over 50,000 such machines were installed, beginning in 1965. While not in direct competition with large mainframes, a slew of independent rental companies provided support for these less expensive computers, and pressure from an antitrust suit eventually

12 T. Osborne, 'Tom Osborne's Story in His Own Words', *HP9825.COM* (blog), 11 November 2004, www.hp9825.com/html/osborne_s_story.html.
13 For a more detailed discussion of RPN, see below.
14 Hewlett-Packard, '[HP-9100 Advertisement]', *Science*, 162, no. 3849 (October 1968), p. 6.
15 Bill Hewlett thought, 'If we had called it a computer, it would have been rejected by our customer's [*sic*] computer gurus because it didn't look like an IBM.' Hewlett-Packard, 'History of the 9100A Desktop Calculator, 1968' (2012), www.hp.com/hpinfo/abouthp/histnfacts/museum/personalsystems/0021/0021history.html.
16 Ceruzzi, 'The Go-Go Years and the System/360, 1961–1975', in *A History of Modern Computing*, Chapter 5.

led IBM to unbundle its software from its hardware by the end of the decade.

The 1970s saw the development of microprocessor technology: thousands of integrated circuits printed onto a single chip.[17] Declining costs of equipment lowered the bar of entry for new computer firms, and concerns about market saturation led to industry-wide unrest.[18] Manufacturers of computer parts, like Texas Instruments (TI), turned towards consumer electronics to bolster their bottom line, secure brand recognition, and stay competitive in chip development.[19] Industry insiders were acutely aware that chip technology was not autonomous, and represented an educational problem first and foremost. As computer manufacturers turned from selling primarily to businesses towards targeting individuals, the new users needed to be made aware of problems before they could be sold on solutions.[20] This brief portrait of computing *circa* 1970 suggests an unsettled industrial matrix in which the prime movers were well aware that the idea of *where* computers belonged and with *whom* was in flux.

While one would be remiss to neglect how the centrifugal social politics of the 1960s provided a milieu for re-imagining the computer as a consumer good, the connections between Osborne's device and the counterculture are surprisingly direct.[21] Just as much of the New Left came together to protest against the military–industrial complex, some groups began to accommodate the ethos of technological reconversion.[22] The year 1968 marked the first publication of

---

17 On the origins of the semiconductor industry as an outgrowth of the wartime defence firms, see C. Lécuyer, *Making Silicon Valley: Innovation and the Growth of High Tech, 1930–1970* (Cambridge: MIT Press, 2006).

18 C. Beardsley, 'Forecast for '72', *IEEE Spectrum*, 9.3 (1972), pp. 4–8.

19 D. Mennie, 'Designers' Tools: The Big Roundup of Small Calculators: Hand-Held Calculators Are Getting Smaller and Smarter; They Make the Slide Rule a Museum Piece', *IEEE Spectrum*, 11.4 (1974), pp. 34–41; and Ceruzzi, *A History of Modern Computing*, 214.

20 Osborne reflected in 1976, 'I know it will be an order of magnitude easier to design and manufacture any futuristic "personal computer" than it will be to teach people how to use it.' See T. Osborne, 'Personal Thoughts on Personal Computing', *Computer*, 9.12 (1976), p. 23; and D. C. Brock (ed.), *Understanding Moore's Law: Four Decades of Innovation* (Philadelphia: Chemical Heritage Foundation, 2006).

21 On how the consumer politics of the 1960s departed from post-war mass consumption, see L. Cohen, *A Consumers' Republic: The Politics of Mass Consumption in Postwar America* (New York: Alfred A. Knopf, 2003).

22 Matthew Wisnioski has shown that engineers were themselves heavily involved in pushing for socially responsible technology, even though such efforts ended up reifying technology as an uncontrollable force existing outside of politics:

Stewart Brand's *Whole Earth Catalog*, a compendium of 'tools' for what historian Fred Turner has described as the 'new communalist' movement. Nestled between advertisements for beads, yarn, and buckskin, the HP-9100A was described as 'a superb inquiry machine' and fitted the bill on the basis of its futuristic appeal rather than its $4,900 price tag.[23] Brand's advocacy of back-to-basics living and 'techno-utopianism' derived from the collaborative work practices of Cold War technologists led many to retroactively label him a pioneer of personal computing. When bootstrapped onto coeval imaginaries of power redistribution and autonomy, devices like the 9100A could be seen as more than mere number crunchers.

Nonetheless, a machine that took up half a desk was hardly suitable to a yurt. In 1972, HP engineers redesigned the 9100A to fit into a chassis the size of a shirt pocket, in the fashion of a transistor radio. The HP-35 was announced as 'the world's first pocket calculator that challenges a computer', and promised '[slide]-rule portability and computer-like power for just $395', a price initially deemed too high by corporate consultants.[24] This hardly deterred eager customers, and 50,000 units were sold in the first year as the device won accolades for its accuracy and likeness in operation to computers. HP's 'challenge' lay in the 35's portability, leveraging the popularity of Japanese four-function calculators to unsettle IBM and the Seven Dwarfs, and it helped establish calculators as the leading edge of consumer electronics using advanced chips. Put in context, the 35's success depended on and bolstered in turn the legitimacy of a notion that computer power could and should be uncoupled from the mainframe.

Ultimately, the fulfilment of HP's aggressive marketing claims required that the HP-35 fit into existing norms of calculation, routinised practices of reckoning that preceded the first digital computers by millennia. In fact, the device was marketed as an 'electronic slide rule' to suggest continuity with the ubiquitous instrument for logarithmic calculations.[25] RPN was the linchpin of this strategy.

M. H. Wisnioski, *Engineers for Change: Competing Visions of Technology in 1960s America* (Cambridge: MIT Press, 2012).

23 Turner, *From Counterculture to Cyberculture*, p. 138.
24 Hewlett-Packard, '[HP-35 Advertisement]', *IEEE Spectrum*, 9.8 (1972), inside cover; and C. H. House and R. L. Price, *The HP Phenomenon: Innovation and Business Transformation* (Stanford: Stanford Business Books, 2009), p. 165.
25 T. M. Whitney, F. Rodé, and C. C. Tung, 'The "Pocket Powerful": An Electronic Calculator Challenges the Slide Rule', *Hewlett-Packard Journal*, June 1972, Francis Hookham Archive, Whipple Museum (Wh.4529).

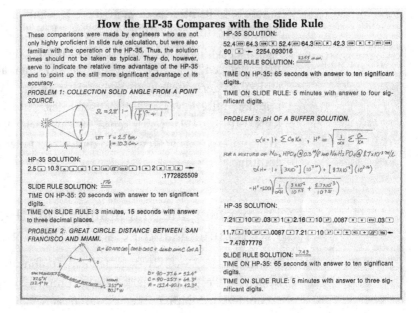

Figure 14.3 Racing the HP-35 against a slide rule. From *Hewlett-Packard Journal*, June 1972, p. 7, Hookham Collection. Image © Whipple Museum (Wh.4529).

Earlier scientific calculators were deemed too inaccurate for precision uses, as memory limitations required early rounding that affected the final result. RPN was a postfix notation that stored operands before their operators, assigning values to a 'stack' in one of the calculator's memory registers, where they could be called up when needed through successive operations.[26] While its speed was almost taken for granted as a benefit, HP foregrounded the calculator's *accuracy* vis-à-vis a slide rule in promotional materials (Figure 14.3).[27] This emphasis was reflected in the calculator's very design. Electronic calculation was not a matter of idly plugging in numbers and pressing the '=' button, because the HP-35 had no such

26 RPN was a notation system based on the development of prefix mathematical logic by the Polish logician Jan Łukasiewicz that placed the operator at the end of a string of inputs: Hewlett-Packard, 'Father of RPN', *The Hewlett-Packard Personal Calculator Digest*, 1980, Francis Hookham Archive, Whipple Museum (Wh.4529); and S. Davis and C. Eagleton, 'Touching Numbers', *Anthropological Theory*, 10 (2010), pp. 192–7, on p. 194. Davis and Eagleton give the example of (3 + 4) × 2 in infix notation, which would be written as *3 [enter] 4 [enter] 2 [enter] + x* in RPN. Another aspect of RPN was its reputation in finance, a reputation garnered by the HP-12C, which became ubiquitous beginning in the 1980s; see K. Peterson, 'Wall Street's Cult Calculator Turns 30', *Wall Street Journal*, 4 May 2011, Technology Section.

27 Whitney, Rodé, and Tung, 'The "Pocket Powerful"', p. 7. Donald MacKenzie explores the various ways accuracy is claimed and reassembled in computational practices, see D. MacKenzie, 'Nuclear Missile Testing and the Social Construction of Accuracy', in Mario Biagioli (ed.), *The Science Studies Reader* (New York: Routledge, 1999), pp. 343–57.

button – in its place was an arrow '→' signifying that the 'program' was ready to run. RPN can and should be understood as a kind of programming *language,* and more generally as a *technique* that replaced slide-rule movements with mental manoeuvres that enthralled generalists with technical training.[28] While TI produced popular, affordable, and functional scientific calculators without RPN, HP users derived a sense of superiority from the likeness of their more expensive device to a digital computer, a trend that would continue.[29]

Dismantling the Mainframe: The HP-65, Programming, and Personal Autonomy

When the HP-35 wildly exceeded expectations, the company did not hesitate to design and market its next model explicitly as a 'personal computer', exhibiting a keen awareness of how users might exploit its major modification: full programmability.[30] The HP-65 was brought to market in 1974 and featured interchangeable magnetic cards as storage media for factory and user programs. Once loaded, a card could be slid into place above five assignable keys to remind a user of the functions and variables unique to the program.[31] Early press accounts of the device corroborated this likeness to a computer, but highlighted obvious differences in memory capabilities rather than the relative ease of personalisation.[32] HP countered this perceived shortcoming in its marketing material by showing the calculator in a variety of settings: in the laboratory, office, living room, and even next to an imposing mainframe (Figure 14.4).[33] By playing up the contrast between the centre–peripheral model of computing and the flexibility of handheld programmability, HP elevated a vision of autonomy predicated upon the deconstruction of the existing computing culture, a process that, as we have seen, was already in motion.

28 On the social history of programming languages, see Ensmenger, *The Computer Boys Take Over.*

29 Ristanović and Protić, 'Once upon a Pocket', p. 57.

30 C. C. Tung, 'The "Personal Computer": A Fully Programmable Pocket Calculator', *Hewlett-Packard Journal,* May 1974.

31 Mennie, 'Designers' Tools'.

32 R. Woolnough, 'They All Add Up to a Handy Way of Doing Sums', *The Guardian,* 27 August 1974; and V. K. McElheny, 'Hewlett-Packard Markets Pocket Calculator Doing Computer's Job', *New York Times,* 17 January 1974.

33 Hewlett-Packard, '[HP-65 Leaflet]', 1974, Francis Hookham Archive, Whipple Museum (Wh.4529).

Figure 14.4
A 1974 HP-65
advertisement,
showing the
calculator in use in a
variety of settings.
Hookham
Collection. Image
© Whipple Museum
(Wh.4529).

The *New York Times* initially cast doubt on the market for the HP-65, citing the prevalence of commercial and academic time-sharing systems.[34] While large mainframes had hitherto been a technological necessity, solutions to give more users access to a single machine provided a compelling principle of social organisation. Time-sharing was largely a creation of Cold War defence research, a model in which users at different terminals – even miles away – could access computer time on a mainframe, using batch processing to run programs so no precious computer time would be wasted.[35] In contrast, HP emphasised the importance of 'specialised' applications for different professions and branded their devices as '[gifts] for a lifetime', intimating a thread of computational continuity between job changes in a fragile economy.[36] Personalised electronics were premised upon ownership, a

34 McElheny, 'Hewlett-Packard Markets Pocket Calculator Doing Computer's Job'.
35 Time-sharing provided a key technological support for what Paul Edwards has termed 'closed-world' discourse, while Janet Abbate shows how the early ARPANET attempted to adopt a 'resource-sharing' framework on similar terms that was abandoned in favour of local files and programs, though the ability to communicate broadly over the network with applications like email stuck around: Edwards, *The Closed World*; and Abbate, *Inventing the Internet*.
36 Hewlett-Packard, '[Application Pac Leaflet]', 1974, Francis Hookham Archive, Whipple Museum (Wh.4529); and Hewlett-Packard, '[Gift for a Lifetime Ad]', *IEEE Spectrum*, 11.12 (1974), p. 21.

departure from the modality of centralised control under which early
computing had taken shape.

Such autonomy came at a price: $795 to be exact, effectively
double the price of the HP-35. Many potential customers were
driven to TI's SR-52, their answer to the HP-65 launched the
following year at half the price with more memory, though devotees
felt the 65 was 'clearly better in many ways ... HP's keyboards *felt*
like quality, and were reliable, unlike TI's.'[37] Ralph Bernstein, an
IBM employee working on government research contracts, used the
calculator to demonstrate results while travelling and to develop
algorithms at his desk, minimising expensive mainframe usage,
ironically, at the world's largest mainframe producer. He felt that
the calculator engendered 'some kind of possessive instinct and
pride of ownership', especially insofar as it allowed one to circum-
vent the sovereignty of system operators.[38] Programmable calculators functioned in a kind of technological ecosystem alongside
mainframes, and could be more flexibly incorporated into work
practices, providing their users with a sense of independence. American astronauts used the HP-65 to correct their flight course – it was
more powerful than the on-board computer – during the 1975
Apollo–Soyuz Test Project, itself a symbol of the new geopolitics of
détente: a shift from closed-world policy parallel to the transition
from mainframe to microcomputer.[39]

The promise of autonomy, however, did not work alone to lure
new calculator users. What was most remarkable about the HP-65
was the sheer amount of support material and social infrastructure
built into its marketing apparatus. Within the Hookham collection,
one can find boxes upon boxes of cards: many from specialised 'pacs'
advertised with a forward-looking iconography – uncannily reminiscent of the 'apps' that run on smart phones – and many more
programmed by Hookham himself (Figure 14.5).[40] While it is easy to

37 Some of the following material and perspective are derived from chat-room and
    email conversations with HP enthusiasts conducted in December 2013. See also
    'New Product Applications', *IEEE Spectrum*, 12.11 (1975), pp. 57–60; and
    G. Wilson, 'Re: Project on Programmable Calculator User Networks', *HP41.
    Org*, 24 December 2013, http://forum.hp41.org/viewtopic.php?f=11&t=246&
    sid=4d7bd9025fb2b0966c4473d32d43b720#p683. On the personal dimensions
    of the calculator, see Davis and Eagleton, 'Touching Numbers'.
38 Ralph Bernstein, personal communication, 15 December 2013.
39 Hewlett-Packard, 'HP-65 in Space with Apollo–Soyuz', *Scientific American*,
    September 1975, Francis Hookham Archive, Whipple Museum (Wh.4529);
    and Ceruzzi, *A History of Modern Computing*, p. 189.
40 Hewlett-Packard, '[Application Pac Leaflet]'.

Figure 14.5 An HP-65 with its quick reference guide and magnetic program cards. Image © Whipple Museum (Wh.4529.227A).

think of calculators as workaday tools, to many users they courted intrigue, even obsession. A hobbyist who shelled out for an HP device might soon become preoccupied with unlocking its 'secrets'. Promotional letters for the HP-35 were headed by a well-known quote from polymath Gottfried Wilhelm Leibniz: 'it is unworthy of excellent men to lose hours like slaves in the labor of calculation.' As users came to devote considerable time, work, and social networking to augmenting their programming skills, the phrase 'labor of calculation' took on an ironic new meaning. Richard J. Nelson, who headed the most prominent HP calculator user group, observed that scientific calculators promised that practising engineers would be able to consolidate their libraries:

> I remember sitting in my living room with all of the books of math tables that I had collected. The stack was over two feet high . . . WOW, I could replace all of these gigantic heavy books with this handheld marvel![41]

While it is tempting to say that these unwieldy, if not quite towering, stacks of tabular compendia were replaced by stacks of assignable calculator memory, materials in the Hookham collection (Figure 14.3) show that these media were in fact only *displaced* by heaps of newsletters, program sheets, and books devoted to programming.[42] The programmable calculator as such owes its success to the reinvestment of infrastructure and literary culture it was intended to replace.

---

41 R. J. Nelson, 'Remembering the HP-35A', hhuc.us (2007), http://hhuc.us/2007/Remembering%20The%20HP35A.pdf.

42 For a parallel claim on the importance of writing practices in computing history, see S. Dick, 'Machines Who Write', *IEEE Annals of the History of Computing*, 35.2 (2013), pp. 85–8.

Programming Infrastructure: HP's User Library, Publications, and Authorial Norms

The rise of microcomputers and their desktop successors posed a problem for the division of labour that had made electronic data processing successful: without teams of coders, operators, programmers, and system administrators, how could an individual make a computer useful?[43] The keynote speaker at the 1974 IEEE Computer Society Computer Elements Technical Committee held up the HP-65 as an example of how the power of newly advanced integrated circuits and microchips could be adapted to the needs and learning abilities of new users flying solo.[44] Alongside their 'Application Pacs', HP launched the 65 with two major social innovations: a library of user-submitted programs and a newsletter for users to voice concerns and share developments. To attract new customers, HP continued to expand its in-house product support, marketing specific applications geared towards ease of use and producing a voluminous product literature, both of which the PC industry lacked into the 1980s.[45] HP's support, however, largely stopped there. By 1975, over 25,000 HP-65s were being used by engineers, lawyers, financiers, and other professionals who had learnt most of the programming basics on their own.[46] HP helped engender an *infrastructure* in which it regulated moral, monetary, and material economies of program exchange, turning ordinary professionals into programmers by enrolling them in a sociotechnical system that redefined the calculator.[47]

Included with the HP-65 was a one-year membership of the HP-65 User's Club, which provided access to their user's library,

43 On the definition of computational labour roles, see A. Akera, 'Voluntarism and the Fruits of Collaboration: The IBM User Group, Share', *Technology and Culture*, 42.4 (2001), pp. 710–36; and Ensmenger, *The Computer Boys Take Over*.

44 M. H. Eklund, 'Technology in the Real World', *Computer*, 8.5 (1975), pp. 56–57. On developments in integrated circuits, see Ceruzzi, *A History of Modern Computing*, p. 195.

45 J. Raskin and T. M. Whitney, 'Tutorial Series 4 Perspectives on Personal Computing', *Computer*, 14.1 (1981), pp. 62–73.

46 Ceruzzi, *A History of Modern Computing*, p. 215.

47 According to information scholars Susan Leigh Star and Karen Ruhleder, infrastructure *emerges* when 'a complex constellation of locally-tailored applications and repositories, combined with pockets of local knowledge and expertise ... interweave themselves with elements of the formal infrastructure to create a unique and evolving hybrid'. See S. L. Star and K. Ruhleder, 'Steps toward an Ecology of Infrastructure: Design and Access for Large Information Spaces', *Information Systems Research*, 7.1 (1996), p. 132.

program catalogs, and user newsletter, *HP-65 Key Notes*, all of which were nodes in a material economy of program exchange. The club subsequently cost $15 per year and programs $3 each to offset shipping and maintenance of the library in Corvallis, Oregon, with currency-adjusted prices for a parallel European library in Geneva.[48] HP encouraged users to submit personal programs to the user library, providing official templates that were often shared between users. Initially, the only remuneration that HP provided for an accepted program was a free program from the library in return; this was later elaborated into a points system through which one could purchase other HP products.[49] The magnetic cards themselves were the subject of much discussion in *Key Notes*: the introduction of cardholders met with much fanfare and users recommended different strategies for writing on and making the most of the cards (Figure 14.5).[50] Throughout the life of the periodical, HP followed the 65 with further fully programmable devices: the 67 and 97 upgraded the memory, with a built-in thermal printer added to the latter model to compete with TI, while the 41C brought the programmable calculator up to pace with advances in personal computing, adding four expandable ports and alphanumeric capabilities. The newsletter provided continuity between successive models.

This system of exchange was the cornerstone of a community visible in the pages of *Key Notes*, an essential complement to the emerging moral economy of HP programmers.[51] While finding programmers and regulating their activity had proved a persistent problem for business management since the 1960s, the dispersed, personal nature of calculator programming required a solution that would honour intellectual autonomy without turning the manufacturer into a publishing house.[52] HP adopted a model of authorship in which individuals published program abstracts in the journal, providing the

---

48 Hewlett-Packard, 'Users' Library Corner', *HP-65 Key Notes*, 1.2 (1974), p. 2; and Hewlett-Packard, 'Hewlett-Packard Order Form', *HP Key Notes*, 3.1 (1979), p. 8.
49 Hewlett-Packard, 'HP-65 Users Library Europe Program Submittal Form', 1974, Francis Hookham Archive, Whipple Museum (Wh.4529); and Ristanović and Protić, 'Once upon a Pocket', p. 60.
50 Hewlett-Packard, 'Accessories Update', *HP Key Notes*, 1.1 (1977), p. 1; and Hewlett-Packard, 'How Small Can You Write?', *HP Key Notes*, 4.2 (1980), p. 6.
51 For a discussion of the notion of 'moral economy', which has different resonances in the history of science than in E. P. Thompson's original formulation, see L. Daston, 'The Moral Economy of Science', *Osiris*, 2nd Series, 10 (1995), pp. 2–24.
52 Ensmenger, 'The Black Art of Programming' and 'Chess Players, Music Lovers, and Mathematicians', in *The Computer Boys Take Over*, Chapters 2 and 3, respectively.

social fulfilment sought by hobbyists, and gave these authors exclusive editing rights to pre-empt intellectual property disputes.[53] The library maintained a high rejection rate to ensure program quality and required extensive documentation with examples. Nonetheless, *Key Notes* also contained a column devoted to corrections, allowing both authors and HP officers to address bugs on a regular basis.

Finally, the periodical helped to put the 'personal' into personal computing by making the community visible to itself. The newsletter described a thirteen-year-old from Texas named Nickey who used the HP-65 to plot solar eclipses, 'celebrity' calculator enthusiasts working in the White House, and printed pictures of homemade HP rugs and T-shirts.[54] HP used the publication to cast calculators as companions and points of entry to new professional territory:

> Do you like challenging calculator games? Or – are you contemplating starting a photographic darkroom? Going into a small business? Learning more about Forestry? You'll find programs for all of those – and more.[55]

These user publications represent a mixed genre of advertising and technical support. Cultural historians of marketing have argued that advertisements should be read as expressions of norms, forms of consent building rather than bald-faced statements about social realities.[56] With this in mind, we can see HP's exhortations towards learning and regulated information exchange as expressing a kind of idealised curious consumer, cleansed of the transgressive hacker ethos that was taking hold as the dominant computational way of life.[57]

53 R. J. Nelson, personal communication, 16 January 2014; and Hewlett-Packard, 'When Contributing Programs', *HP-65 Key Notes*, 1.2 (1974), p. 3. On the complication of property in open-source programming, see S. Weber, *The Success of Open Source* (Cambridge: Harvard University Press, 2004), pp. 16–17.
54 Hewlett-Packard, 'An Amazing Young Man!', *HP-65 Key Notes*, 1.5 (1980), p. 1; P. W. Weiss, 'I Owe It All to My HP', *HP Key Notes*, 4.3 (1980), p. 7; and S. Seeherman, 'Some True Believers!', *HP Key Notes*, 1.2 (1977), p. 6.
55 Hewlett-Packard, 'Here Come the Solutions!', *HP Key Notes*, 1.3 (1977), p. 1. This was an advertisement for the forty volumes of *Users' Library Solutions* sold by HP, containing programs devoted to a specific application.
56 For influential examples, see R. Marchand, *Advertising the American Dream: Making Way for Modernity, 1920–1940* (Berkeley: University of California Press, 1985); and T. J. Jackson Lears, *Fables of Abundance: A Cultural History of Advertising in America* (New York: Basic Books, 1994).
57 Levy, *Hackers*; Adrian Johns, 'From Phreaking to Fudding', in *Piracy: The Intellectual Property Wars from Gutenberg to Gates* (Chicago: University of Chicago Press, 2009), Chapter 16.

'The World's Largest (and Poorest) Personal Computing Club'

HP's marketing strategies have driven much of the action in this narrative. By late 1977 there were nearly 5,000 programs in HP's user library, a testament to its runaway success. However, *Key Notes* and the user libraries continued only until 1983 when the editor retired, stating that the company had not yet found a replacement: 'I cannot foresee the future of KEY NOTES. I can tell you only that HP knows the value of staying in touch with you. I am sure an alternative to this newsletter will be found.'[58] While HP initially channelled the activity of users into a commercial resource, infrastructure requires invest-ment, and a feature of the developing digital world was the ability for alternative systems to bootstrap onto successful ones. In this last section I want to make good on my promise to consider agency in a more expansive sense by highlighting the afterlife of the official user network through a grassroots one that emerged in parallel. As we will see, this group attempted to work in tandem with the manufac-turer to leverage its interest in developing a product glitch as a community resource, resulting in fallout that led to its dissolution.[59]

When *Key Notes* was dissolved, its subscribers were referred to another group known as PPC, which stood for nothing in particular but could mean Personal Programmer Club.[60] PPC had been founded by Richard J. Nelson as the HP-65 Users Group not long after the calculator was launched in 1974.[61] Nelson was a seasoned hobbyist. Having been actively involved in amateur radio in the

---

58 H. C. Horn, 'My Last "Editorial"', *HP Key Notes*, 8.2 (1983), p. 1.

59 It should be stated that this group *did* continue in various guises. The UK Handheld and Portable Computer Club (HPCC) published an edited volume celebrating twenty years of activity in 2002, and remains active in 2019. See W. A. C. Mier-Jędrzejowicz and F. Wales (eds.), *RCL 20: People, Dreams & HP Calculators* (London: W. Mier-Jędrzejowicz, 2002); and 'HPCC: Handheld and Portable Computer Club', http://hpcc.org/ (accessed 13 March 2019). The HP Handheld Community has been meeting since at least 1999 and held its most recent conference, 'Celebrating 50 Years of HP Programmables' in September 2018, with roughly fifty attendees: 'HHC 2018: HP Handheld Conference, 29–30 September 2018, San Jose, California', http://hhuc.us/2018/index.htm. (accessed 26 October 2018).

60 R. J. Nelson, 'PPC Journal', *PPC Journal*, 5.1 (1978), pp. 1–2. The expandability of the abbreviation mirrored the flexibility of the community: 'The Personal Programmers Club does Prolific and Productive Computing with Hewlett-Packard Personal Programmable Calculators.'

61 R. J. Nelson, 'Starting a Calculator Club', in Mier-Jędrzejowicz and Wales, *RCL 20*. The group began with over 600 members in nine countries and had 3,100 members by 1981: Hewlett-Packard, 'HP-65 Users Club', *HP*, 2.1 (1975), p. 5; and R. J. Nelson, 'Member Letter', *PPC Journal*, 8.1 (1981), p. 16.

United States and the Philippines in the 1960s before the calculator consumed his interests, he got three issues of his own *65 Notes* newsletter out before HP launched *Key Notes*.[62] As a grassroots organisation, PPC employed print as the primary means for distributing programs – mailed or faxed in using HP's program sheets, then reproduced as facsimiles in the newsletter. Its monthly organ, *PPC Journal*, often exceeded fifty pages. PPC facilitated face-to-face interaction between members through meetings of various group chapters across the United States and Europe and larger annual conferences. The standards of program publication were also different: PPC maintained an open 'Share-A-Program' listing, while Nelson acted as the central arbiter for programs published in the newsletter. Here, unlike through HP, users were able to submit modifications of prior programs due to the programs' status as communal property.[63] Programming calculators was not always serious work: one published program, called 'DRINKS', was intended for users to monitor their drinking habits.[64] Though unrepresentative of the maths- and engineering-based programs regularly published in the newsletter, 'personal' programs such as these are artefacts of PPC's predominantly masculine hobbyist culture, wherein the boundary between 'personal' and 'computer' was eroded along particular lines of identity.

Nelson was also known to push boundaries, and got into trouble for publishing user information and estimates of the company's manufacturing output in the newsletter.[65] However, HP recognised the group's influence, and its Corvallis Division published a regular column. Nelson's philosophy for PPC explicitly addressed the inability of hardware manufacturers to support their products, maintaining that involvement with them must be limited to supporting the equitable exchange of information. He spurned the notion of for-profit developments from community resources, claiming that 'few, if any, commercial software, application, or even accessory concerns have been financially successful'.[66] Nonetheless, PPC would seek

---

62 There is a documented affinity between amateur radio enthusiasts and early personal computer users. M. Campbell-Kelly and W. Aspray, *Computer: A History of the Information Machine* (Boulder: Westview Press, 2004), p. 207. On the culture of amateur radio, see K. Haring, *Ham Radio's Technical Culture* (Cambridge: MIT Press, 2007).

63 Nelson, personal communication.

64 B. Lomasky, 'Drinks [Program]', *65 Notes*, 3.2 (1976), p. 19.

65 Nelson, 'Starting a Calculator Club'.

66 Nelson, 'PPC Journal'.

incorporation as a business to further develop technologies central to the activity of its members.

The watershed moment for PPC came in 1979 when Nelson and John Kennedy discovered a hack, and published a complete table of special display characters on the HP-41C that were inaccessible to standard user code.[67] More available characters meant more sophisticated programs, and this exposed machine code spurred the elaboration of a practice called 'synthetic programming', in which users could make more of existing memory registers by reassigning basic instructions elsewhere.[68] Realising the extent of interest, HP published an extensive description of the 41C system architecture in the following *PPC Journal*. Numerous programs utilising synthetic functions were published in the journal and even spawned a book. Nelson saw an opportunity to support the new practice with a manufactured component. He proposed that PPC build its own ROM unit – which could be inserted into one of the 41C's four ports – that would give programmers easier access to standard subroutine calls through the synthetic programming method. This would be a literal piece of collective memory for a group previously maintained by paper and conversation, a new technology to expand a shared practice.

The ROM project took nearly two years and strained the PPC's human resources, but HP agreed to produce a run of 5,000 for the group to purchase and distribute. It is not clear whether the company was uninterested in spinning off the project for conventional users, or whether PPC wished to keep the technology to itself. In order to handle financial matters more officially, PPC filed for non-profit incorporation in January of 1982.[69] The obligatory institution of a voting board compromised Nelson's autocratic position within the organisation, and the board fired him for refusing to make changes they demanded.[70] Nelson's sudden absence unsettled members of the group, and their unrest was fed by a rumour that he had been forced out of the PPC clubhouse by a security guard.[71] PPC continued until 1987, with its core comprised of device-specific user groups charging steep membership fees, some equivalent to the cost of previous devices. In attempting to expand its technological

---

67 J. Kennedy, 'HP-41C Combined Hex Table', *PPC Journal*, 6.5 (1979), pp. 22–5.
68 Ristanović and Protić, 'Once upon a Pocket', p. 61.
69 As a condition, PPC also had to support devices from other manufacturers – see Ristanović and Protić, 'Once upon a Pocket', p. 62.
70 Nelson, personal communication.
71 D. E. White, 'Member Letter', *PPC Journal*, 11.7 (1984), pp. 1–5.

capacity, the group's hobbyist ethos was supplanted by institutional formality. If the poor could, in fact, inherit the Earth in the early days of personal computing, seismic shifts in corporate interest and power would soon render the landscape unrecognisable.

Conclusion

Calculator user groups supported an infrastructure through which their device of interest became redefined as part of a way of life. This was a fragile consensus built upon shifting ground, but the economies of programmable calculators offer a focal window into the development of personal computing from a hobbyist practice to a traditionally consumer-driven one. As the story goes, Steve Wozniak sold his HP-65 after quitting his job with the company to fund Apple Computer.[72] This raises an important question that as yet remains unanswered: what, if any, were the convergences between calculator programmers and the PC users that followed them? PPC launched its own separate PC publication in 1979 to provide similar product support for these newer machines, though it lasted only two years.[73] Paul Ceruzzi claims that the major difference between calculator and PC enthusiasts was a bifurcated set of interests in personal computing: PC people cared about the *personal* whereas calculator people cared about the *computer*.[74] However, if we disabuse ourselves of the absolute distinction between programming as a practice and application use, itself a product of the commercial software revolution yet to come, this interpretation does not hold.[75]

There are remarkably forward-looking features of the culture of programmable calculators. Synthetic programming arose from the communalistic norms of PPC in a way that suggests an analogy with the more contemporary politics of open-source software. Through his ethnography of Free Software, Chris Kelty has developed the analytic of recursive publics: technological communities 'capable of speaking to existing forms of power through the production of

72 Levy, *Hackers*, p. 253.
73 R. J. Nelson, 'Member Letter', *Computer Journal of PPC*, 1.1 (1982).
74 Ceruzzi, *A History of Modern Computing*, p. 216.
75 Projecting his own vision, Nelson drew up a table of different computers in the first *Computer Journal of PPC* to contrast desktop 'Personal' computers with personal programmable calculators, touting the advantages of the latter. See Nelson, 'Member Letter'.

actually existing alternatives'.[76] PPC tried to leverage community knowledge to substantially augment the capabilities of their device of choice, but this gambit led to its dissolution. The hex codes that users unearthed provided a community niche, albeit one enmeshed in a recalcitrant matrix of production. Calculator users thought globally and acted locally, as the saying went. Their devices were a platform germane to techno-utopian futures, from the *Whole Earth Catalog* to HP's own marketing materials. An issue of *Hewlett-Packard Calculator Digest* in 1979 ran a cover story on a sci-fi dramatisation of its calculators as the personal assistants of the future, uncannily reminiscent of the iPhone and Siri despite retaining the button-based design of a calculator.[77] People dreamed about their present and future with calculators, though we have yet to fully understand the meaning of such dreams. Lest we cede curatorial authority to the Boris Johnsons of the world, collections of these materials in museums like the Whipple can help us to do so.

76 C. M. Kelty, *Two Bits: The Cultural Significance of Free Software* (Durham: Duke University Press, 2008), p. 3.
77 G. Dickson, 'Thank You, Beep . . .!', *The Hewlett-Packard Personal Calculator Digest*, 5 (1979), pp. 2–3.

Appendix: Student Research Conducted on the Whipple Museum's Collections since 1995

1995

'Nineteenth-century wave motion machines', MPhil essay (Wh.2007; Wh.3747; Wh.4517; Wh.4558).

1996

'How late is late medieval? Some preliminary notes on an armillary sphere in the Whipple collection', MPhil essay (Wh.0336).

'Educating the astronomer: the use and collection of gothic teaching instruments', MPhil dissertation (Wh.0336).

1997

'Making waves: a history of the wave machine', MPhil essay (Wh.2007; Wh.3747; Wh.4517).

'Representing Euclid in the eighteenth century', MPhil essay (Wh.0368).

'Circles of heaven: an unusual silver celestial planisphere in the Whipple Museum', Pt II dissertation (Wh.1762).

1998

'The 19th-century papier-mâché models of human and comparative anatomy by Louis Thomas Jérôme Auzoux', MPhil essay (various Auzoux models in the Whipple collection).

'Mathematical models and the visual expression of theory', MPhil essay (various mathematical models in the Whipple collection).

'Aspects of a Korean astronomical screen of the mid-eighteenth century from the Royal Palace of the Yi Dynasty (Choson Kingdom, 1392 to 1910)', Pt II dissertation (Wh.0935).

'Navicula de Venetiis: construction and characteristics', Pt II dissertation (Wh.0731; Wh.5902).

1999

'Geography as a game: the case of a puzzle globe', MPhil essay (Wh.4608).

'A study of a nineteenth century jigsaw globe from the Whipple collection', Pt II dissertation (Wh.4608).

2000

'The decline and fall of the astrolabe', MPhil essay (various astrolabes in the Whipple collection).

'On the use of the globe: the Earl of Castlemaine's English Globe and Restoration mathematics', MPhil essay (Wh.1466).

'The Victorian scientific instrument-maker's trade catalogue: marketing & patronage', MPhil essay (various trade and sales literature in the Whipple collection).

'A case study of two armillary spheres made by Richard Glynne, 1715 and 1725', MPhil essay (Wh.0784; Wh.0785).

'Representing time and motion: Sekiya, the Gilbreths, and chrono-photographic art', MPhil essay (Wh.3461).

'The Duddell oscillograph: making waves visible', MPhil essay (Wh.4328; Wh.3331).

'An early Italian globe? A critical study of a terrestrial globe in the Whipple Museum', MPhil essay (Wh.0365).

'An eighteenth century Japanese celestial globe', MPhil essay (Wh.5617).

'Study of an astrological astrolabe in the Whipple collection', Pt II dissertation (Wh.4552).

'Bearing the heavens: astronomers, instruments, and the communication of astronomy in early-modern Europe', PhD thesis (Wh.0336)

2001

'Phrenology goes bust!: the material culture of a nineteenth century popular science', MPhil essay (Wh.2744; Wh.4618).

'Sounding the depths: trials and tribulations in the development of sounding machines', MPhil essay (Wh.2970).

2002

'A casket of useful knowledge: study of a case of geological specimens in the Whipple Museum', MPhil essay (Wh.3395).

'Thoughts relating to a study of the British Drug Houses' capillator (1924)', MPhil essay (Wh.5244).

'The many "odd things which a microscopist delights to own": aspects of nineteenth-century popular microscopy', MPhil essay (Wh.1844).

'Some preliminary notes on a manuscript in the Whipple collection', MPhil essay (Wh.5358).

'Writing the history of astronomy: Flamsteed and Sherburne', Pt II dissertation (E268).

2003

'Reforming mathematics: late nineteenth century mathematical models', MPhil dissertation (various mathematical models in the Whipple collection).

'Nineteenth century vacuum techniques and applications', MSc dissertation (various air pumps in the Whipple collection).

'Embodying the abstract: mathematical models in Cambridge', MPhil essay (various mathematical models in the Whipple collection).

'The mind of the frontispiece: myth, meaning and motivation in Sherburne's *Manilius*', Pt II dissertation (E268).

'Sherburne's library and its relation to his history of astronomers', Pt II dissertation (E268).

'A study of the Lusuerg instruments in the Whipple Museum', Pt II dissertation (Wh.0323; Wh.0865; Wh.1609; Wh.1612).

2004

'Using globes and celestial planispheres in Restoration England', PhD thesis (Wh.1466; Wh.1762).

'The social life of observatories and their scientific instruments from the late seventeenth to the early nineteenth centuries', MA dissertation (various astronomical instruments in the Whipple collection).

'"Every Boy & Girl a Scientist": construments and the domestication of scientific instruments in interwar Britain', MPhil essay (Wh.4565).

'Instrument-making families', Pt II dissertation (various instruments in the Whipple collection).

'Why make fakes?', Pt II dissertation (Wh.0226; Wh.0306; Wh.0365; Wh.0563; Wh.1148; Wh.1149; Wh.1639).

'Instruments in context: telling the time in England, 1350–1500', PhD thesis (Wh.0731; Wh.1264; Wh.5902).

2005

'Shagreen, science and status: a study of the materials used to make early telescopes', Pt II dissertation (Wh.2662; Wh.0251).

2006

'Papier-mâché flowers, fruits and seeds: the botanical models of Louis Thomas Jerôme Auzoux', MPhil essay (various botanical models by Auzoux in the Whipple collection).

'Cultures of science, magic and masculinity in twentieth-century toy chemistry sets', MPhil essay (various chemistry sets in the Whipple collection).

'The use of instruments in propagating Newtonianism: the Musser Copernican Planetarium', MPhil essay (Wh.5812).

'Are orreries "Newtonian"? A consideration of the material, textual and pictorial evidence', Pt II dissertation (Wh.1275).

'Touching numbers: the pocket electronic calculator in advertising', MPhil essay (Francis Hookham Collection of Hand Held Electronic Calculators).

2007

'The images that accompany *The Sphere of Marcus Manillius* by Edward Sherburne', Pt II dissertation (E268).

'Mathematics in motion: understanding Olivier's movable hyperbolic paraboloid model', MPhil essay (Wh.5795).

2008

'Mogg's Celestial Sphere (1812): a catalogue of conversation', MPhil essay (Wh.5620).

'Remodeling the electrocardiograph at the Cambridge Scientific Instrument Company', MPhil essay (various ECG machines and related ephemera by the Cambridge Scientific Instrument Company).

'Practical mathematics in one foot: the story of a slide rule from the 18th century', MPhil essay (Wh.1451).

'Finding out about an X-ray crystallography camera', MPhil essay (Wh.3469).

'Models of the eye in the Whipple Museum', Pt II dissertation (Wh.2037; Wh.5825; Wh.6068; Wh.6194; Wh.6202).

'An exploration of a seventeenth-century terrestrial globe held in the Whipple Museum collection', Pt II dissertation (Wh.2691).

2009

'Teacher, toy or calculator? Reflections of mathematics, education and society in a 20th-century American object', MPhil essay (Wh.5821).

'Botany of the air: experiments, airships and agriculture in 1930', MPhil essay (Wh.5826).

'Materializing Edinburgh in Egypt: the five inch Great Pyramid standard of Charles Piazzi Smyth', MPhil essay (Wh.1155).

'Materials for a history of science in Cambridge: meanings of collections and the 1944 scientific instrument exhibition at the University of Cambridge', MPhil dissertation (history of the Whipple Museum and its collection).

'Long in the tooth: a study of a papier-mache set of horses' teeth', MPhil essay (Wh.6135).

'The Edinburgh Stereoscopic Atlas of Anatomy (1905–6): anatomical representation between two and three dimensions', MPhil essay (Wh.6247).

'Designer nature: the papier-mâché botanical teaching models of Dr. Auzoux in nineteenth-century France, Great Britain, and America', PhD thesis (various botanical models by Auzoux in the Whipple collection).

2010

'Telescopic tracings: astronomy, art, and the Varley family', MPhil essay (Wh.0069).

'Cube roots: A. H. Frost's model of a magic cube in the Whipple Museum', MPhil essay (Wh.1251).

'Robin Hill's cloud camera: an analysis of the development of the fisheye lens at Cambridge University', MPhil essay (Wh.1635; Wh.2170; Wh.4416; Wh5732).

'Modelling nature: the case of the Whipple Museum's pomological models', MPhil essay (Wh.6267).

2011

'"The art in science and the science in art": glass models of flowers and fungi', MPhil essay (Wh.5826).

2012

'Something in the world: looking into a Spanish globe', MPhil essay (Wh.5892).

'Negretti & Zambra's scientific instruments: a new dimension to the Victorian culture of travel', Pt III essay (various instruments and sales literature by Negretti & Zambra in the Whipple collection).

'Bumps across borders: towards a transnational historiography of phrenology c. 1838', MPhil essay (various objects and ephemera relating to phrenology in the Whipple collection).

'Astrolabes in context: a reappraisal of medieval astronomical instruments', MPhil dissertation (Wh.1264).

'Sounding in silence: the mechanics of discipline in the early nineteenth-century Royal Navy', MPhil essay (Wh.2970).

'Philips' Popular Manikin: the culture of flap anatomies around 1900', MPhil essay (Wh.5852).

2013

'Playing with the eyes: a comparative history of two rare stereoscopic instruments', MPhil essay (Wh.2902; Wh.2117).

'Alexander Crum Brown's knitted mathematical models', MPhil essay (Wh.4469; Wh.4470).

'Models as mathematics: intellectual functions of physical models in nineteenth-century mathematical practice', MPhil essay (Wh.5175).

'The Edinburgh stereoscopic atlas of anatomy and anatomy teaching in medical schools, c. 1905–1930', MPhil essay (Wh.6247).

'Discipline and pedagogy: molecular model kits and the doing of synthetic organic chemistry', MPhil essay (Wh.5815).

'Kurt Ziesing's "Tectonic Globe of the Earth": a case of tectonics without plates', MPhil essay (Wh.6383).

'An early nineteenth-century "museum microscope" and cultures of collecting', Pt III dissertation (Wh.0200).

'News from Mars: transatlantic mass media and the practice of new astronomy', PhD thesis (Wh.1264; Wh.6067; Wh.6211; Wh.6238; Wh.6604; Wh.6605).

2014

'Stacks, pacs, and system hacks: handheld calculators as an alternative history of personal computing', MPhil essay (Francis Hookham Collection of Hand Held Electronic Calculators).

'Robert S. Whipple, as collector, donor and historian (1871–1953)', Pt III dissertation (history of the Whipple Museum and its collection).

'Models of authority: the place of geological models in the visual language of geology', Pt III dissertation (Wh.1581; Wh.6529).

'"Educated at the shrine of nature": Eliza Brightwen's Bible Album and the study of natural theology', Pt III essay (Wh.6517).

'Men, mines, and machines: Robert Were Fox, the dip circle and the Cornish system', Pt III dissertation (Wh.6538).

2015

'The teaching diagrams of John Stevens Henslow: botany in 19th-century Cambridge', MPhil essay (sixty-six botanical teaching diagrams by Henslow in the Whipple collection).

'Logic, labour, and organisation: establishing EDSAC in Cambridge', MPhil essay (Wh.5901).

'Colour values: Joseph Lovibond's Tintometer and the scientific meanings of vision', MPhil essay (Wh.4521; T338).

'The harmonograph and its locations in late Victorian Britain: public display and laboratory settings', Pt III essay (Wh.2033; Wh.6243).

2016

'Cantabrigian collaborative commercialisation: collaborations between Cambridge University scientists and scientific instrument manufacturers, *circa* 1890–1960', MPhil dissertation (various scientific instruments in the Whipple collection).

'Hill's cloud camera in the Whipple Museum: meteorological communication, cloud classification', MPhil essay (Wh.1635; Wh.2170; Wh.4416; Wh5732).

'The philosophical foundation of Maxwell's induction model', MPhil essay (Wh.2455).

'Using historical microscopes to understand microscopic anatomical observations by Marcello Malpighi and Nehemiah Grew', MPhil essay (Wh.0211).

'The Morden, Berry and Lea terrestrial globe (*c.* 1683)', MPhil essay (Wh.2691).

'"The whole matter of interference microscopes is … becoming rather complex": Sir Andrew Huxley and the design and dissemination of his custom interference microscope', MPhil essay (E519; Wh.6574).

'Signpost to a forgotten science: Stokes's Capital Mnemonical Globe (1870s)', MPhil essay (Wh.6600).

'Charles Elcock and the Postal Microscopical Society: a nineteenth-century scientific community', MPhil essay (Wh.6601).

'Things of science: science kits and educating young scientists 1940–1980', Pt II dissertation (Wh.6615).

2017

'Unpacking boxes in British maritime history (18th–19th century)', MPhil essay (various instrument boxes and packing crates in the Whipple collection).

'A Korean astronomical screen at the Whipple Museum: parsing a composite cosmography', MPhil dissertation (Wh.0935).

'False measures: seventeenth-century English gauging instruments', MPhil essay (Wh.6239).

'The moonshot Briton and the media: Francis Bacon's hydrogen–oxygen fuel cell', MPhil essay (Wh.6081).

2018

'Wollaston hypsometers vs. mountain barometers in nineteenth-century surveying', MPhil essay (Wh.2883; Wh.2890).

'(Un)folding proteins: Courtaulds chemical models, British industrial fibre development, and the search for the α-helix', MPhil essay (Wh.5815).

'Collecting habits and valuable antique scientific instruments: what can annotated sales catalogues tell us?', Pt III essay (history of the Whipple Museum and its collection).

'Characterising collections: on the preservation of old scientific apparatus at the Cavendish Laboratory and the Whipple Museum, Cambridge', Pt III dissertation (history of the Whipple Museum and its collection).

'Chicken heads & Punnett squares: Reginald Punnett and the role of
   visualization in early genetics research, Cambridge, 1900–1930',
   MPhil essay (Wh.6547).
'Classifying a calculator of cranial categorisation: investigating
   knowledge claims embedded in Professor Arthur Thomson's tri-
   gonometer', MPhil essay (Wh.6638).

Index

Printed in the United States
By Bookmasters